L. German S. Zemskov (Eds.)

New Fluorinating Agents in Organic Synthesis

With Contributions by
L. A. Alekseeva, V. V. Bardin, L. S. Boguslavskaya, A. I. Burmakov,
N. N. Chuvatkin, G. G. Furin, B. V. Kunshenko, L. N. Markovskii,
F. M. Mukhametshin, V. E. Pashinnik, L. M. Yagupolskii,
Yu. L. Yagupolskii

Springer-Verlag Berlin Heidelberg New York
London Paris Tokyo Hong Kong

Dr. Lev Solomonovich German

Institute of Organoelement Compounds,
Vavilov St. 28, 117813 Moscow, USSR

Professor Dr. Stanislav Valerianovich Zemskov

Institute of Inorganic Chemistry, Pr. Lavrent'eva 3,
630090 Novosibirsk, USSR

ISBN-13:978-3-642-74765-6   e-ISBN-13:978-3-642-74763-2
DOI: 10.1007/978-3-642-74763-2

Library of Congress Cataloging-in-Publication Data
New flourinating agents in organic synthesis / with contributions by
L. A. Alekseeva ... [et. al.] ; L. German, S. Zemskov, eds.
Includes index.
ISBN-13:978-3-642-74765-6(U.S.)
1. Chemistry, Organic--Synthesis. 2. Flouridation.
I. Alekseeva, L. A. II. German, L. (Lev), 1931-. III. Zemskov, S. (Stanislav), 1934-.
QD262.N46   1989   547.2--dc20   89-10086 CIP

© Springer-Verlag Berlin Heidelberg 1989
Softcover reprint of the hardcover 1st edition 1989

Typesetting: Macmillan, Indien

2151/3020-543210

In Memoriam

Professor S. V. Zemskov, one of the editors of the Russian edition of *New Fluorinating Agents in Organic Synthesis,* recently died. He was known as an outstanding researcher, a specialist in inorganic fluorine chemistry.

S. V. Zemskov was born in 1934. In 1960 he began work at the Institute of Inorganic Chemistry in Novosibirsk. Later he founded the laboratory of noble metal halide compounds. The scientific heritage of Professor Zemskov includes more than 150 publications and 30 patents, but the best memorial to his activities is the scientific school, his followers and this book.

Preface

Fluoroorganic compounds find extensive use in many fields of science, medicine, industry, and agriculture. However fluorination of organic compounds with elemental fluorine sometimes presents a serious problem. Along with classical fluorinating agents (HF, alkali metal fluorides, antimony fluorides, etc.), efficient agents for synthesis of fluoroorganic compounds are xenon fluorides, compounds with the O–F bond, N–F bond, and the higher fluorides of group V and VI elements. For most of these compounds, the preparative procedures have been developed only recently. Information on these synthetically interesting agents is scattered in many periodicals and patents. This book is intended to help researchers select the most suitable fluorinating agents and reaction conditions. For that purpose the book contains the necessary data on the new fluorinating agents illustrated by syntheses of some fluoroorganic compounds.

Recently new fields of application of sulfur tetrafluoride and fluorosulfuranes in fluoroorganic synthesis have been found. Therefore we considered it necessary to include reviews of applications of these fluorides. Due to space limitations, the mechanistic analysis of fluorination is incomplete; some interesting works dealing with this problem remained beyond the scope of this volume.

We hope that this book will stimulate further search for efficient fluorinating agents and applications of those already available.

July, 1989

L. German
S. Zemskov

Contents

1 Xenon Difluoride

Vadim Viktorovich Bardin[1] and Yuri L'vovich Yagupolskii[2]

[1] Institute of Organic Chemistry, Pr. Lavrent'eva 9, 630090 Novosibirsk, USSR
[2] Institute of Organic Chemistry, Murmanskaya St. 5, 252094 Kiev, USSR

Contents

1.1 Introduction

Three binar xenon fluorides are known (XeF_2, XeF_4 and XeF_6), among which the difluoride is the most studied. Relatively easy availability of XeF_2 allows it to be used to obtain organofluorine compounds in mild conditions. The peculiarity of xenon difluoride is its ability to carry out the "electrophilic" fluorination of organic compounds, i.e. addition of fluorine atoms proceeds as if an attacking species were "F". The reactivity of XeF_2 increases sharply in the presence of Lewis acids. At the same time under certain conditions (gas phase or irradiation) xenon difluoride may be a radical fluorinating agent.

This chapter considers the use of xenon difluoride in the synthesis of fluorine-containing organic and organoelement compounds of different types and demonstrates the synthetic utility of XeF_2. The mechanisms of these reactions, though also of interest, are not discussed here because of space limitations.

1.2   Synthesis and Some Properties of Xenon Difluoride. Interaction with Lewis Acids and Bases

The literature reports several methods for the preparation of xenon difluoride using various fluorinating agents [1,2]. Among them, the most popular is the photochemical method which involves the sunlight or the UV irradiation of a mixture of xenon and fluorine (1:1, mol) in a pyrex or quartz reactor [3]. This gives 99% pure xenon difluoride in amounts of several grams per day. To obtain larger amounts of XeF_2 there is a more convenient method based on the catalytic fluorination of xenon with chlorine trifluoride [4]. Besides, xenon difluoride formed in the process contains practically no higher xenon fluoride admixtures. As ClF_3 is commercially available and easier to handle than fluorine (see, e.g. [5]), Sect. 1.10 includes both preparations of XeF_2 [3,6]. Under normal conditions (20 °C, 1 atm) xenon difluoride is a colourless crystalline substance subliming upon heating (subl. p. 114 °C [2]). Purified XeF_2 melts at 141 °C (under pressure) [7]. The enthalpy of formation of xenon difluoride in the gas phase is -122 ± 3 kJ mol^{-1}, the heat of sublimation 51.5 ± 0.8 kJ mol^{-1}, the mean value of Xe–F bond thermochemical energy 140.6 kJ mol^{-1} [7]. The IR, Raman, ^{19}F and ^{129}Xe NMR, UV, X-ray and photoelectron spectral data for XeF_2 are reported in [1,2].

At 20 to 25 °C xenon difluoride is well-soluble in HF, BrF_3, IF_5, $NOF \cdot 3HF$, SO_2, SO_2FCl, CH_3CN, $(CF_3CO)_2O$; not so soluble in WF_6, MoF_6, CCl_4, diethyl ether, dimethyl sulfoxide; and slightly soluble in liquid ammonia, chlorine, bromine, trichlorofluoromethane, hydrocarbons. The solubilities of XeF_2 in some of these substances are given below, g in 1000 g of solvent (°C) [1,2,8].

HF	1329(12)	$NOF \cdot 3HF$	2755(17)	CH_3CN	167(0)	$SiCl_4$	1.9(22)
	1680(30)		5670(80)		309(21)		
IF_5	1538(20)	BrF_5	1893(25)	CCl_4	5(22)		

The acetonitrile solution of XeF_2 is stable at -10 °C, decomposes slowly at 20 °C, and quickly—when boiled. Xenon difluoride is moderately soluble in water (25 g in 1000 g of water), forming a molecular solution which decomposes with evolution of xenon (half-life period 7 h). In an alkaline medium, XeF_2 decomposes very quickly [2].

The mechanism of catalytic action of strong aprotic Lewis acids (SbF_5, NbF_5 etc.) involves ionization of the xenon difluoride molecule, leading to formation of fluoroxenonium salts $XeF^+MF_n^-$ or $Xe_2F_3^+MF_n^-$. These salts are strong oxidants: the electron affinity of XeF^+ is 10.8 ± 0.4 eV [2]. Less strong acceptors of the fluoride ion (HF, BF_3, BF_3OEt_2) polarise the XeF_2 molecule; an intermediate here is rather a complex of type F–$\overset{\delta+}{Xe}$ . . . F . . . $\overset{\delta-}{BF_3}$ than the ion-structured salt.

Protic oxygen-containing acids (CF_3COOH, HSO_3F, HNO_3) react with xenon difluoride, forming unstable ethers FXeOY or $Xe(OY)_2$. Due to polarisation of bond $\overset{\delta+}{FXe}$–$\overset{\delta-}{OY}$, the xenon atom has a partial positive charge; that is why strong protic acids are used as fluorination catalysts. At the same time, ethers $Xe(OY)_2$ are apt to decompose by the radical pathway, which leads to by-products [2,9].

Xenon difluoride practically does not react with fluoride ion donors at room temperature. Studies of the system XeF_2–CsF by the differential thermal analysis method have shown that only at the temperature of about 400 °C does slow disproportionation occur with the formation of $Xe + Cs_2XeF_8$ [10]. For this reason some cases of the catalytic effect of fluoride ion donors in the reactions of XeF_2 with organic substrates should be explained by interaction of F^- with these substrates rather than with XeF_2.

1.3 Fluorination at the Tetrahedric Carbon Atom

At room temperature xenon difluoride is insoluble in hexane and does not react with it [11]. Heating the XeF_2–hexane mixture to 80–90 °C in a sealed glass tube leads to its inflammation. The reaction products are HF, Xe, and tarry substances. Upon heating a mixture of C_6H_{14} and XeF_2 (5:1, mol) at 105 °C in a Teflon jacket inserted in a stainless steel reactor, the product is a mixture of 1-, 2- and 3-fluorohexanes (28:42:30) (total yield 15–20%).

$$C_6H_{14} + XeF_2 \xrightarrow[105\,°C,\ 2.5\,h]{} 1\text{-}FC_6H_{13} + 2\text{-}FC_6H_{13} + 3\text{-}FC_6H_{13}$$

In these conditions cyclohexane is converted to fluoro-cyclohexane in a 18% yield.

Heating of adamantane with xenon difluoride (105 °C, 70 min) leads to formation of a mixture of mono-, di-, and trifluoroadamantanes in a 18–20% yield. In this case, fluorine is almost exclusively substituted for the hydrogen atoms bound with the tertiary carbon, the main reaction product being 1-fluoroadamantane. In carbon disulfide, xenon difluoride fluorinates adamantane at temperatures as low as -15 to 0 °C, giving 1-fluoroadamantane in yields of up to 35% [12].

Fluorination of tris(fluorosulfonyl)methane with xenon difluoride in CF_2Cl_2 leading to $CF(SO_2F)_3$ has been described, but no details of synthesis and yield of the product were given [13].

Substitution of hydrogen by fluorine is observed upon treatment of alkyl(aryl)methyl sulfides with xenon difluoride in acetonitrile [14–17]. Methyl-phenyl sulfide is transformed at room temperature to fluoromethylphenyl sulfide in a good yield and further to difluoromethylphenyl sulfide [14].

$$C_6H_5SCH_3 + XeF_2 \xrightarrow[25\,°C]{HF/CH_2Cl_2} C_6H_5SCH_2F \xrightarrow[25\,°C]{XeF_2/HF,\ CH_2Cl_2}$$

$$67\%$$

$$C_6H_5SCHF_2$$

$$58\%$$

Such fluorination will only take place if at the α-carbon atom there is at least one hydrogen atom. For example, phenylisopropyl sulfide reacts with XeF_2 (CH_3CN, −10 °C, 0.5 h) giving phenyl(α-fluoroisopropyl) sulfide in a 90% yield, whereas fluorination of *tert*-butylphenyl sulfide and diphenyl sulfide produces the respective difluorosulfuranes in good yields [15]. It should be noted that phenyl(α-fluoroalkyl) sulfides (alkyl ≠ CH_3) are unstable and with increase of temperature there increases the probability of HF elimination. This is vividly illustrated by the results of fluorination of phenylisopropyl sulfide by XeF_2 at different temperatures [15].

$$C_6H_5SCHMe_2 + XeF_2 \xrightarrow[-10\,^\circ C,\ 30\ min]{MeCN} C_6H_5SCFMe_2\ (90\%)$$

$$C_6H_5SCHMe_2 + XeF_2 \xrightarrow[-5\,^\circ C,\ 10\ min]{MeCN} C_6H_5SCFMe_2\ (60\%) + C_6H_5SC(Me){=}CH_2\ (40\%)$$

$$C_6H_5SCHMe_2 + XeF_2 \xrightarrow[0\,^\circ C,\ 2\ min]{MeCN} C_6H_5SC(Me){=}CH_2\ (90\%)$$

Apparently for the same reason it was impossible to isolate the products of fluorination of *cis*-2,6-diphenyltetrahydro-1-thio-4-pyrone and thiochroman-4-one. Instead of these compounds the reaction gives unsaturated products with good yields [14].

65%

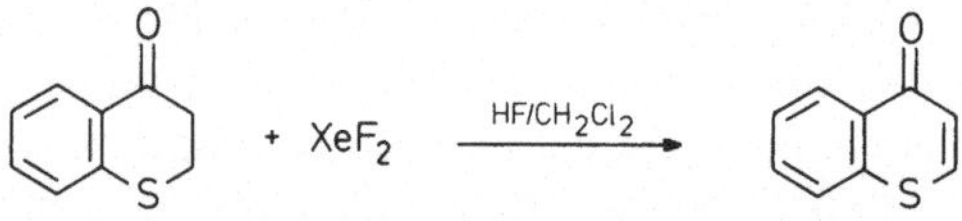

Yanzen and his co-workers [17] used xenon difluoride to obtain fluorine-containing derivatives of methionine and methionylglycine. It appeared that substitution of hydrogen by fluorine proceeds regiospecifically, and it is exclusively methyl at the sulfur atom that is fluorinated.

$$CH_3SCH_2CH_2CH(NHCOR^2)COR^1 + XeF_2 \xrightarrow[-20\ to\ 20\,^\circ C,\ -Xe,\ -HF]{MeCN} CH_2FSCH_2CH_2CH(NHCOR^2)COR^1\ (70\text{–}90\%)$$

$R^1 = OMe$, $R^2 = CF_3$; $R^1 = 4\text{-}NO_2C_6H_4O\text{-}$, $R^2 = CF_3$; $R^1 = 4\text{-}NO_2C_6H_4O\text{-}$, $R^2 = t\text{-Bu}$; $R^1 = EtOCOCH_2NH$, $R^2 = PhCH_2O$

There are no products of fluorination of the α-methylene group, nor of deep fluorination of the methyl group.

By contrast with diorganyl sulfides, thiols are not fluorinated by xenon difluoride, but quantitatively oxidised to disulfides. Neither the structure of organic radical at the sulfur atom (R = Me, i-Pr, Ph) nor the ratio of thiol to XeF_2 affect the reaction course [15]. It is interesting that aliphatic alcohols (C_1–C_4) are oxidised by xenon difluoride to the respective aldehydes [18].

With carbon tetrachloride, xenon difluoride markedly reacts at 180 °C [8]. Introduction of catalytic amounts of antimony, niobium, tantalum fluorides and even SiO_2 leads to substitution of chlorine by fluorine at 20 to 30 °C with formation of chlorofluoromethanes.

$$CCl_4 + XeF_2 \xrightarrow[\text{20 to 30 °C}]{\text{Ct}} CFCl_3 + CF_2Cl_2 + CF_3Cl + CF_4$$

Ct	Yields %			
SbF_3	56	39	12	3
TaF_5	68	28	4	1
SiO_2	85	14	—	—

1.4 Fluorination of Alkenes and Alkynes

The first time the use of xenon difluoride for the synthesis of difluoroalkanes from alkenes was suggested was in Ref. [19]. This report immediately attracted the attention of researchers, as before that time there had been no method of direct fluorination of unsaturated hydrocarbons. Xenon difluoride reacts at room temperature with ethylene to form 1,2-difluoroethane, 1,1-difluoroethane, and 1,1,2-trifluoroethane [20].

$$CH_2=CH_2 + XeF_2 \xrightarrow[\text{25° C, 4 days}]{} CH_2F–CH_2F + CHF_2–CH_3 + CHF_2–CH_2F$$

$$ 45\% \qquad\qquad 35\% \qquad\quad 20\%$$

Fluorination of propylene by XeF_2 leads to formation of 1,2-difluoropropane (12%), 1,1-difluoropropane (46%), 2-fluoropropane (24%), and 1,1-difluoroethane (9%). The authors of [20] explain the complex composition of the reaction mixtures, and, particularly, the presence of geminal difluorides, by the radical character of the fluorination reaction. However it cannot be excluded that in the course of the reaction, isomerisation of vic-difluorides by elimination-addition of HF may occur. This is indicated by the formation of the product of addition of HF to propylene in the reaction of the latter with XeF_2.

Xenon difluoride fluorinates hexachlorocyclopentadiene relatively easily [21]. The reaction proceeds at room temperature in $CFCl_3$ without a catalyst and results in hexachlorodifluorocyclopentenes with a 93% yield. In a similar way XeF_2 reacts with 3-fluoropentachlorocyclopentadiene. The residual carbon–carbon double bond containing two vinyl chlorines is not fluorinated even in the presence of HF (60 h, 25 °C).

Fluorination of alkenes by xenon difluoride is markedly facilitated if the reactions are carried out in the presence of Lewis acids. Warming of a mixture of 1-decene, XeF_2, BF_3OEt_2 and CH_2Cl_2 from -78 to 20 °C with subsequent stirring for 20 h leads to the formation of 1,1-difluorodecane (26%) and a mixture of 1-fluorodecene isomers (48%). However under the same conditions 2-fluoro-2,2-dinitroethyl vinyl ether is only converted to 1',2'-difluoroethyl 2-fluoro-2,2-dinitroethyl ether [22].

$$(NO_2)_2CFCH_2OCH=CH_2 + XeF_2 \xrightarrow[20\,°C,\ 20\,h]{BF_3OEt_2/CH_2Cl_2}$$

$$(NO_2)_2CFCH_2OCHFCH_2F$$
$$79\%$$

Investigation of the reaction of xenon difluoride with aliphatic 1,3-dienes in the presence of BF_3OEt_2 has shown that under mild conditions this reaction proceeds with kinetic control, to form mainly the products of 1,2-addition of fluorine atoms [23].

$$CH_2=CH–CH=CH_2 + XeF_2 \xrightarrow[-10\,°C]{BF_3OEt_2/CH_2Cl_2} CH_2FCHFCH=CH_2$$
$$87\%$$

$$+CH_2FCH=CHCH_2F$$
$$13\%$$

$$\underset{\substack{|\\ H_3C}}{CH_2=C}\underset{\substack{|\\ CH_3}}{–C}=CH_2 + XeF_2 \xrightarrow[25\,°C]{BF_3OEt_2/CH_2Cl_2} \underset{\substack{|\\ H_3C}}{CH_2F–\ CFC}\underset{\substack{|\\ CH_3}}{=CH_2}$$
$$100\%$$

The reaction of xenon difluoride with phenyl-substituted alkenes has been studied in detail. Treatment of 1,1-diphenylethylene with XeF_2 at room temperature in the presence of HF quickly leads to formation of 1,2-difluoro-1,1-diphenylethane and a small amount of deoxybenzoine. Xenon difluoride reacts likewise with 1,1-diphenyl-2-fluoroethylene and 1,1-diphenyl-1-propene [24].

$$Ph_2C=CHR + XeF_2 \xrightarrow[25\,°C,\ 20\,min]{HF/CH_2Cl_2} Ph_2CF–CHFR + PhCOCHRPh$$
$$65\text{–}95\%$$

R = H, F, Me

If CF_3COOH is used as a catalyst, then instead of deoxybenzoine the reaction gives 1-trifluoroacetoxy-1,1-diphenyl-2-fluoroethane.

Cis- and trans-stilbenes are fluorinated in the same conditions with the formation of 1,2-difluoro-1,2-diphenylethane diastereomers [25]. In this case, fluorination of trans-stilbene proceeds stereoselectively, which indicates the preferential anti-addition of fluorine to alkene. Fluorination of *cis*-stilbene proceeds non-stereoselectively.

$$PhCH{=}CHPh + XeF_2 \xrightarrow[\text{20 °C, 30 min}]{\text{HF/CH}_2\text{Cl}_2} PhCHF{-}CHFPh$$
$$90\%$$

cis	*erythro*:*threo* =	53:47
trans		62:38

As in the case of 1,1-diphenylethylene, fluorination of stilbenes by XeF_2 in the presence of CF_3COOH is accompanied by their fluorotrifluoroacetoxylation. In more detail the reaction of xenon difluoride with phenyl-substituted ethylenes in the presence of CF_3COOH has been analysed in [25–27].

Treatment of *trans*-1-phenylpropene and *trans*-stilbene with XeF_2 in CH_2Cl_2 catalysed by trifluoroacetic acid has been shown to lead to an about equimolar mixture of the products of fluorination and fluorotrifluoroacetoxylation of the starting alkenes. The same compounds are formed from the corresponding *cis*-isomers, but saturation of the double bond of *cis*-stilbene (2) (R =Ph) proceeds, by contrast with other alkenes, non-stereoselectively.

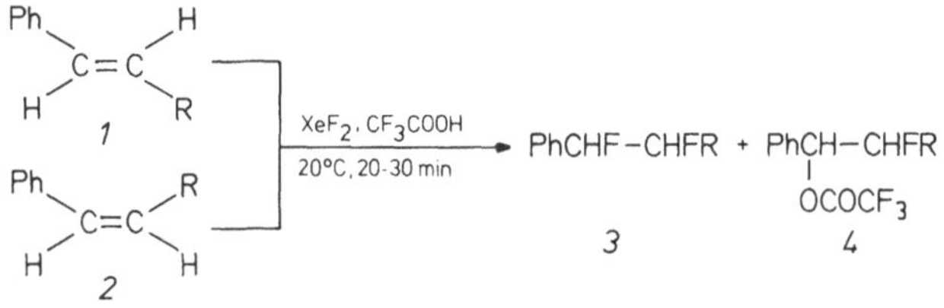

		erythro-3	*threo*-3	*erythro*-4	*threo*-4
1	R = Me	34	19	29	18
	R = Ph	30	18	35	17
2	R = Me	34	19	31	16
	R = Ph	26	24	26	24

It should be noted that fluorotrifluoroacetoxylation of compounds *1* and *2* (R = Me) proceeds regiospecifically.

In the reaction of xenon difluoride and CF_3COOH with styrene, apart from the products of fluorination and fluorotrifluoroacetoxylation of double bond, trifluoromethyl-containing compounds are formed [27].

$$PhCH{=}CH_2 + XeF_2 \xrightarrow[\text{25 °C, 1 h}]{\text{CF}_3\text{COOH}} PhCHF{-}CH_2F + PhCH{-}CH_2F$$
$$\qquad\qquad\qquad\qquad\qquad\qquad\qquad\qquad\qquad\qquad\qquad OCOCF_3$$
$$28\% \qquad\qquad 29\%$$

$$+ PhCHF{-}CH_2OCOCF_3 + PhCHF{-}CH_2CF_3 + PhCH{-}CH_2CF_3$$
$$\qquad\qquad\qquad\qquad\qquad\qquad\qquad\qquad\qquad\qquad\qquad OCOCF_3$$
$$8\% \qquad\qquad\qquad 26\% \qquad\qquad 9\%$$

Thus, even small changes in the structure of phenyl-substituted alkenes affect the product ratio in their reaction with XeF_2 and CF_3COOH. At the same time, a property common for all the reactions of xenon difluoride with alkenes in the presence of trifluoroacetic acid is the formation of trifluoroacetyl derivatives.

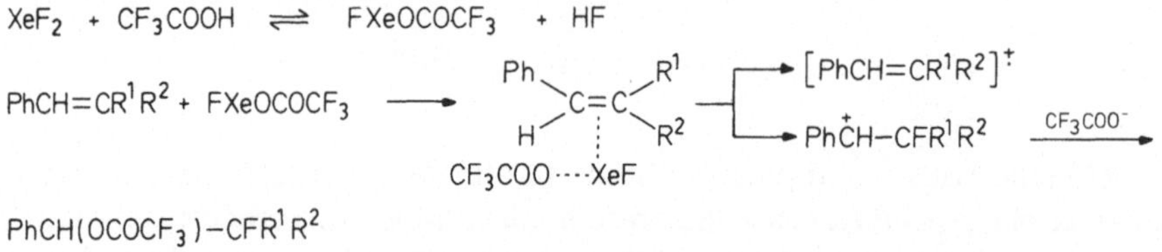

$$XeF_2 + CF_3COOH \rightleftharpoons FXeOCOCF_3 + HF$$

By contrast with hydrocarbon phenyl-substituted alkenes, the compounds containing halogen atoms at the double bond react slowly with xenon difluoride. Thus complete fluorination of 1,1-diphenyl-2-X-ethylene (X = F, Cl, Br) requires 25 h, with stirring at room temperature in the presence of HF. It is interesting that the products of 1,2-shift of the phenyl radical are not formed in this case [28] (cf. [24]).

$$Ph_2C=CHX + XeF_2 \xrightarrow[\text{25 °C, 25 min}]{\text{HF/CH}_2\text{Cl}_2} Ph_2CF-CHFX$$
$$50\%$$

$$X = F, Cl, Br$$

Introduction of substituents, such as Cl, Me, OMe, into position 3 or 4 of the aromatic ring of 1,1-diphenylethylene does not produce any marked effect upon the rate and stereochemistry of fluorination as compared with unsubstituted substrate [29].

Treatment of 1,1-diphenylethylene with xenon difluoride in the presence of bromine and HF leads to formation of 2-bromo-1-fluoro-1,1-diphenylethane instead of difluoroethane. Bromofluorination of 1,1-diphenylethylenes with halogen or methyl at position 2 proceeds in the same way. The reaction occurs quickly at room temperature and leads to the respective bromofluorides with high yields [30].

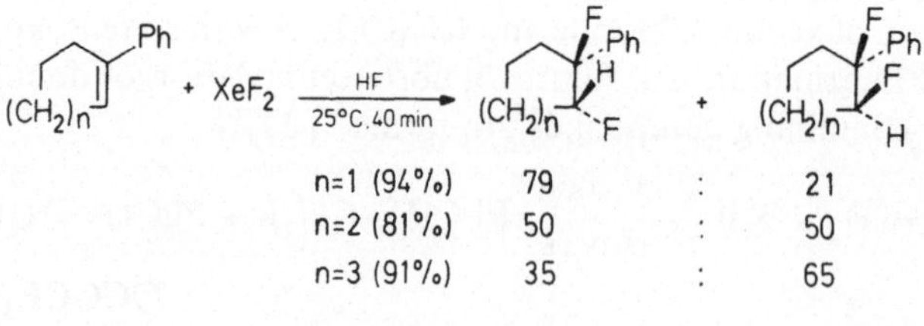

n=1 (94%)	79 :	21
n=2 (81%)	50 :	50
n=3 (91%)	35 :	65

According to [32], a change in the ratio of *cis*- to *trans*-difluorides from 1-phenylcyclopentene to 1-phenylcycloheptene is called for by steric reasons. Substitution of the phenyl group at the C=C bond by the 4-anisyl one does not affect the stereochemical outcome of fluorination of substituted cyclohexene, nor introduction of a bulky *tert*-butyl group into position 4 [33]. Treatment of

1-(4-anisyl)-4-*tert*-butylcyclohexene with xenon difluoride in the presence of chlorine or bromine source leads to the products of fluorohalogenation of the substrate [33].

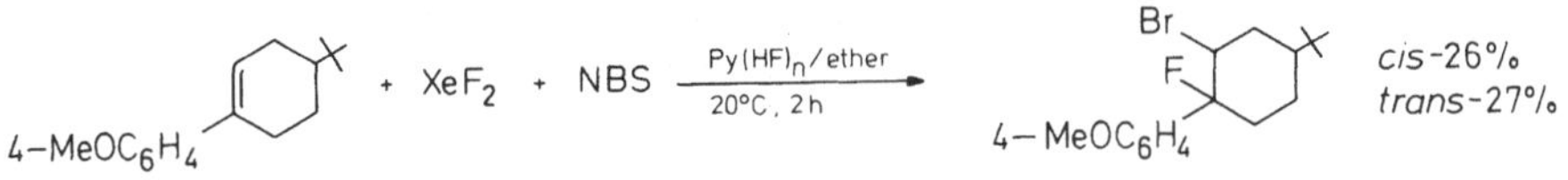

In [34,35], reactions of xenon difluoride with acenaphthylene, indene, 1,2-dihydronaphthalene and 1,4-dihydronaphthalene have been studied. The products of fluorination of indene are *cis*- and *trans*-1,2-difluoroindans. Addition of fluorine atoms to 1,2-dihydronaphthalene occurs in a similar way.

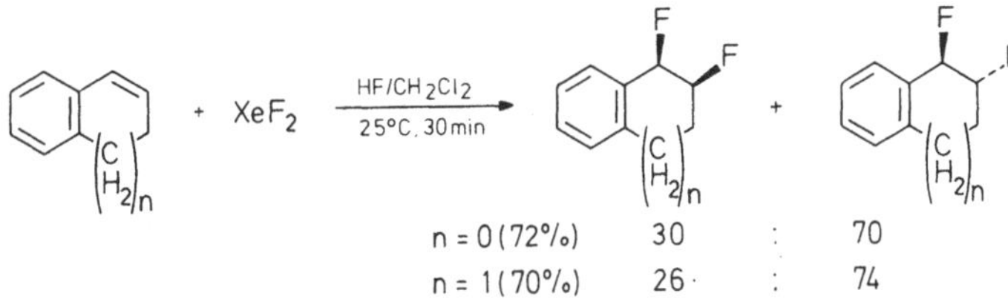

Treatment of acenaphthylene with XeF_2 under the same conditions as for indene and 1,2-dihydronaphthalene leads to resinification of the reaction mixture. To avoid this, the reaction should be carried out in a diluted solution for 5 min. Yield of 1,2-difluoroacenaphthene is 68% (*cis* to *trans* = 16 to 84%).

Interaction of xenon difluoride with 1,4-dihydronaphthalene led to the formation of a mixture of naphthalene, 1- and 2-fluoronaphthalene and 2-fluorotetralin instead of the expected 2,3-difluorotetralin. This must be due to the instability of 2,3-difluorotetralin under the reaction conditions [35].

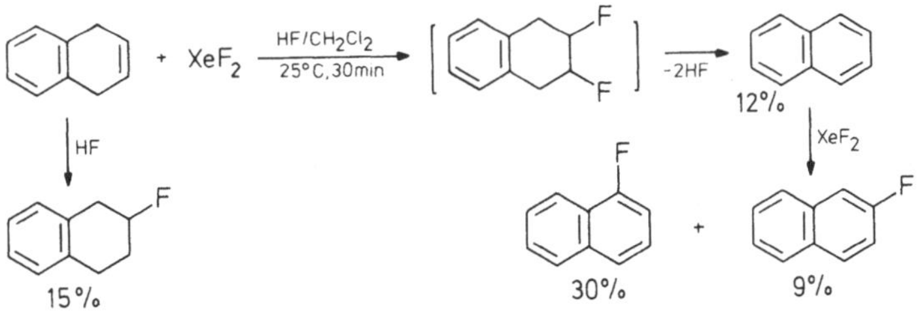

Treatment of alkene 5 with XeF_2 afforded the derivatives of 1,2-difluoro-1,2-dideoxyhexose [36].

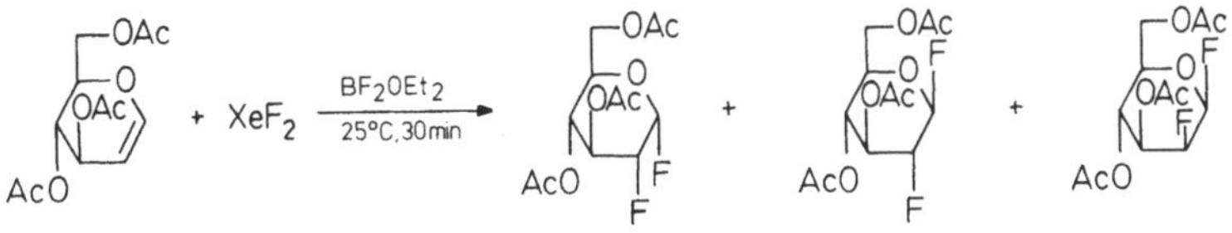

Fluorination of compounds of type *5* with a different arrangement of AcO groups proceeds likewise. The main product in this case is always a difluoride containing both fluorine atoms in the *trans, trans*-position to the acetoxy group at position 3. As a solvent for this synthesis the authors of [36] recommend a mixture of ether and benzene, as in ether under the action of BF_3OEt_2, isomerisation of target products takes place.

A rather complex mixture of products is formed in the reaction of xenon difluoride with norbornene in the presence of acid catalysts [37–41]. The HF-catalysed fluorination of norbornene at room temperature leads to a mixture containing at least 7 components, and their yields practically do not depend on the solvent (CCl_4, $CHCl_3$ or CH_2Cl_2) [37].

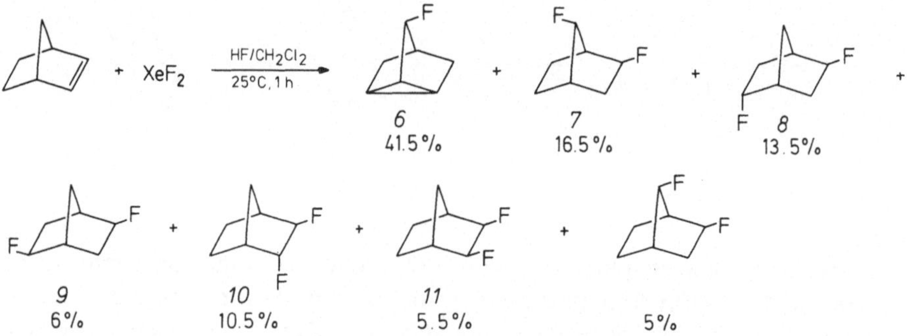

Under milder conditions (-78 to $26°C$, 22 h) and in the presence of boron trifluoride etherate, the reaction affords a mixture containing only 2 main products -2-*exo*-5-*exo*-difluoronorbornane (*9*) and 2-*endo*-5-*exo*-difluoronorbornane (*8*) (2:1) at a total yield of 51–76%. If the reaction is carried out at -46 to $-39°C$ for 75 min, the yields of compounds *8, 9* become almost equal (32 and 25% respectively), and the main product is 2-*exo*-7-*anti*-difluoronorbornane (*7*) (42%) [38]. The structure dependence of the products of fluorination of norbornene with XeF_2 on the solvent, temperature, reaction duration, nature of catalyst (HF, $Py(HF)_n$, BF_3, BF_3OEt_2, CF_3COOH, C_6F_5SH) and the routes of product isomerisation are analysed in more detail in [39,40]. If fluorination of norbornane with XeF_2 is carried out in $CFCl_3$ + MeCN without a catalyst and under UV irradiation, it gives, apart from the radical fluorination products *10* and *11*, a marked amount of compounds *12–14*. When the reaction is carried out in acetonitrile, the total yield of the latter increases [42].

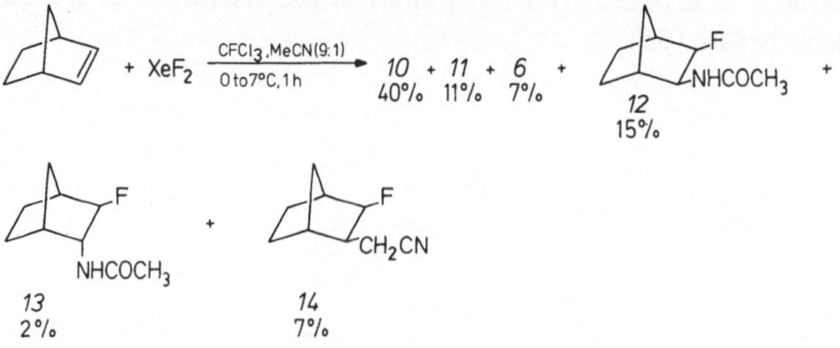

It should be noted that the ion fluorination products *8* and *9*, and the rearranged 2,5- and 2,7-difluoronorbornanes are not formed here.

Interaction of xenon difluoride with norbornadiene leads to 3-*endo*-5-*exo*-difluoronortricyclane, 3-*exo*-5-*exo*-difluoronortricyclane and 2-*exo*-7-*sin*-difluoro-5-norbornene [37,41,43]. Their relative yields change depending on the catalyst of fluorination.

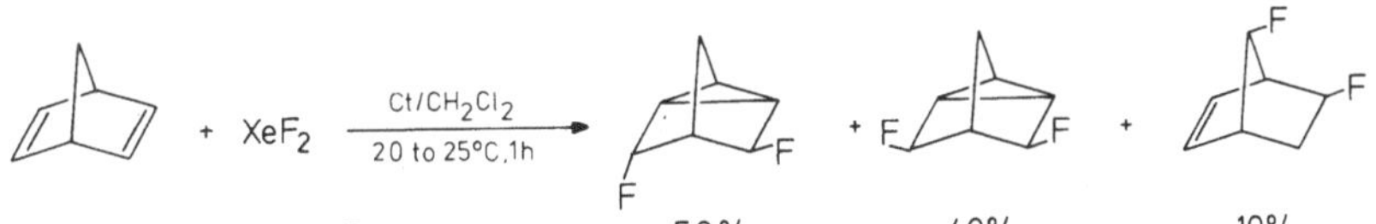

Ct = HF	50%	40%	10%
Ct = Py(HF)$_n$	27%	68%	5%
Ct = (P)—C$_5$H$_4$N(HF)$_n$	45%	44%	11%

Treatment of an alkene with xenon difluoride in the presence of methanol and acid catalyst (HF, BF$_3$) leads predominantly to methoxylation products [44]. Fluoroderivatives are formed in these conditions in low yields or not formed at all. Exclusions are, evidently, indene and dihydropyrane.

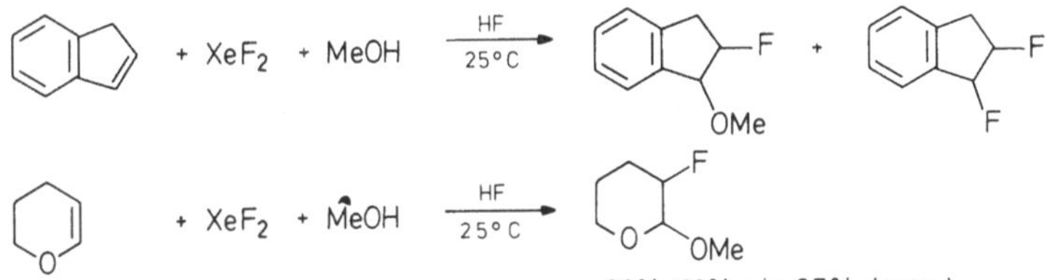

60% (33% *cis*, 67% *trans*)

At room temperature xenon difluoride fluorinates uracyl, giving 5-fluoro-uracyl, though the yield of the latter is small (10%) [45].

Literature also reports an unsuccessful attempt at fluorination of imidazo(1,2-b)pyridazine with xenon difluoride in CCl$_4$, CHCl$_3$ or CHBr$_3$ catalysed by HF or CF$_3$COOH [46]. The reaction product is 3-chloro-imidazo(1,2-b)pyridazine (or the corresponding brominated product); there were no fluorinated compounds in the reaction mixture. This reminds us once more that chlorinated methanes are not always inert in the reactions with xenon difluoride (see Sect. 1.3).

Xenon difluoride reacts with enols and their ethers. The products are α-fluorocarbonyl compounds. For example, treatment of 1-trimethylsilyloxy-cyclopentene or -hexene with an equimolar amount of XeF$_2$ leads to 2-fluorocyclopentanone and -hexanone respectively [47]. Acetates of the corresponding enols react in a similar way [48].

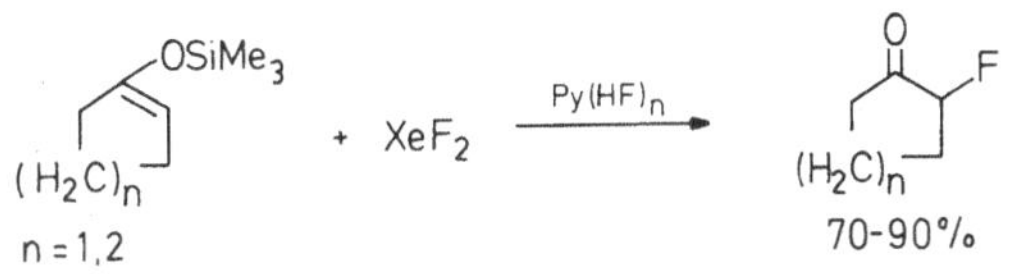

70-90%

The yield of 2-fluorocycloheptanone (n = 3) considerably decreases as a result of the side reaction of protodesilylation of enol ether which leads to cycloheptanone [47,48]. Fluorination of trimethylsilyl ethers of some steroid enols with XeF_2 proceeds without catalysts giving the least sterically hindered α-fluoroketones [49].

Enolised 1,3-diketones react with XeF_2 in the presence of protic acids, forming 2,2-difluorides. 2,4-Pentanedione reacts with xenon difluoride giving 3,3-difluoro-2,4-pentanedione (yield 70%). However fluorination of cyclic 1,3-diketones in the presence of HF is frequently complicated by side reactions and it is reasonable to use polymeric acids as a catalyst [48].

Fluorination of alkynes with xenon difluoride has been studied much less than fluorination of alkenes. In the absence of acid catalysts propyne has been noted [20] to react with XeF_2 more slowly than with propene, forming a complex mixture with 2,2-difluoropropane as the main component (yield 33%). Interaction of diphenylacetylene with xenon difluoride catalysed by an HF trace leads to 1,2-diphenyltetrafluoroethane [50]. This compound is also formed with a lack of XeF_2 (there also remains unreacted diphenylacetylene). Together with a relatively low reaction rate and the absence of difluorostilbenes in the reaction mixture this circumstance definitely shows that the reactivity of the triple carbon-carbon bond in the interaction with XeF_2 is much lower than that of the double bond.

$$PhC \equiv CR + 2XeF_2 \xrightarrow[25\,°C,\,6\,h]{HF/CH_2Cl_2} PhCF_2-CF_2R$$

$$R = Ph\,(50\%),\ Me\,(53\%),\ Pr\,(55\%)$$

In the presence of CF_3COOH, the interaction of diphenylacetylene with XeF_2 proceeds faster, but the yield of 1,2-diphenyltetrafluoroethane is only 4% [27]. The main reaction products are the trifluoroacetoxy- and trifluoromethyl-containing compounds.

1.5 Fluorination of Aromatic Compounds

Fluorination of benzene and some of its derivatives with xenon difluoride in gas phase is reported in [51]. Heating of benzene with an equimolar amount of XeF_2 in a nickel reactor gives fluorobenzene and a small amount of o- and p-difluorobenzenes. Nitrobenzene is converted at 120 °C to a mixture of o-, m- and p-fluoronitrobenzenes.

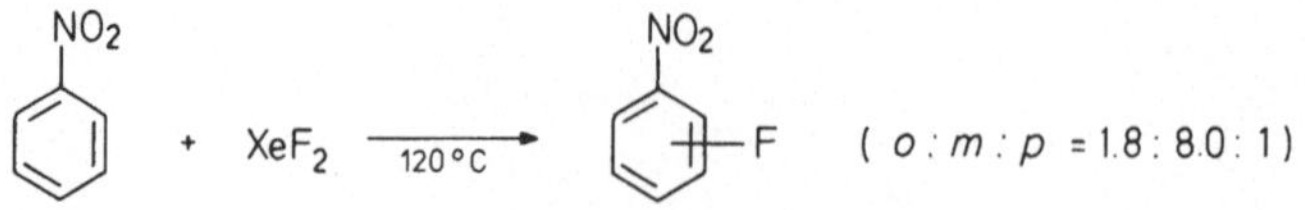

Fluorination of arenes in condensed phase has been studied more intensely. Benzene and its derivatives are fluorinated with xenon difluoride at room temperature and in the presence of HF to form predominantly monofluoro-arenes together with a small amount of difluorides. At the same time di- and polyphenyls are formed [52–56].

$$C_6H_5R + XeF_2 \xrightarrow[25\,°C]{HF/CCl_4} FC_6H_4R$$

$$R = H\,(68\%),\ F\,(47\%),\ Cl\,(66\%),\ Me\,(32\%),\ OMe\,(65\%),$$
$$CF_3\,(76\%),\ NO_2\,(81\%)$$

The isomer ratio corresponds with the orienting effect of the R substituents in the reactions with electrophilic agents. Thus, *ortho-*, *meta-* and *para-*chloro-fluorobenzenes are formed in yields 16.0, 3.2 and 46.3%, and fluorination of benzotrifluoride gives *meta-* and *para-*fluorobenzotrifluoride in yields 71.7 and 3.8% respectively [53].

Instead of HF, anhydrous hydrogen chloride may be used as a catalyst, but the yield of aryl fluorides is decreased by formation of aryl chlorides [55].

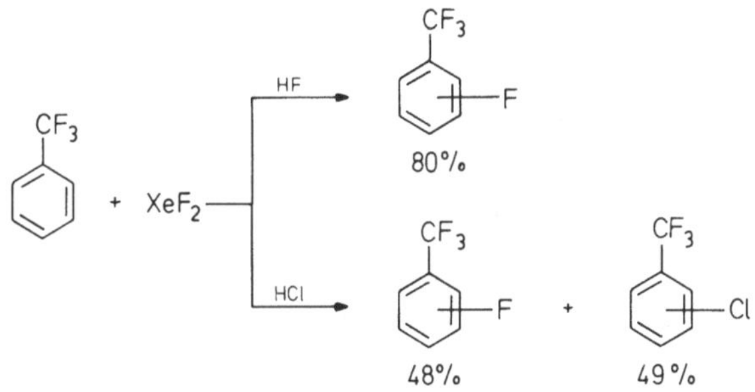

It should be noted that the catalytic activity of HCl is much lower than that of HF. For example, in HF ($-75\,°C$) benzotrifluoride immediately reacts with XeF$_2$ to form fluorobenzotrifluorides, whereas in HCl ($-75\,°C$) the compound remains unchanged for 2 days [55].

Interaction of polymethylbenzenes with xenon difluoride in the presence of HF leads to the products of fluorination of the aromatic ring, the methyl group remaining intact. Mesitylene reacts with two equivalents of XeF$_2$ to give 2,4-difluoromesitylene.

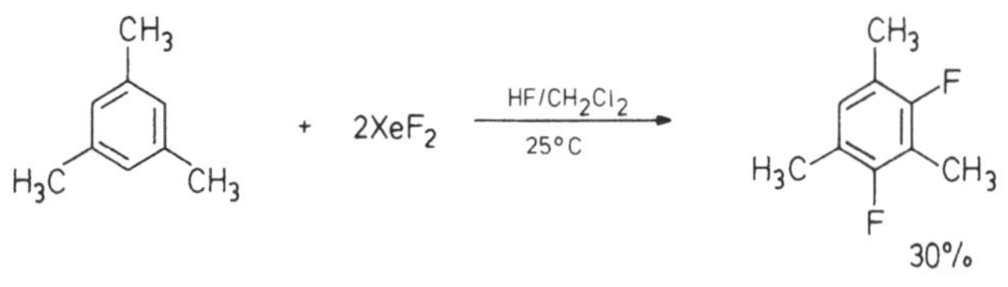

The HF-catalysed fluorination of 1,2,3-trimethylbenzene with XeF_2 (1:1, mol) gives a 1:2 mixture of 1-fluoro-3,4,5-trimethylbenzene and 1-fluoro-2,3,4-trimethylbenzene (yield 38%). Further fluorination of the mixture of these products leads to 1,2-difluoro-3,4,5-trimethylbenzene (yield 70%). While the reaction of 1,2,3-trimethylbenzene with XeF_2 (1:2, mol) gives a complex mixture of up to ten products, 1,2,4-trimethylbenzene is fluorinated by XeF_2 to 1-fluoro-2,4,5-trimethylbenzene.

Treatment of 1,2,4,5-tetramethylbenzene with an equimolar amount of XeF_2 (catalyst HF) leads to 1,4-difluorotetramethylbenzene. When the ratio of substrate to XeF_2 is increased to 1:2, the fluorination products are 1,4-difluorotetramethylbenzene and 1-fluoro-2,4,5-trimethylbenzene (3:2). Yields of these fluorine-containing arenes are small (approx. 7–9%) and together with these a considerable amount of tar is formed. At the same time, fluorination of polymethylbenzenes with xenon difluoride, catalysed by CF_3COOH, is complicated by the same processes of trifluoromethylation and trifluoroacetoxylation as the reactions of XeF_2 with olefins. In these reactions the trifluoromethyl group substitutes hydrogen atoms in the ring and the trifluoroacetoxy group – hydrogen atoms in the side chain [57].

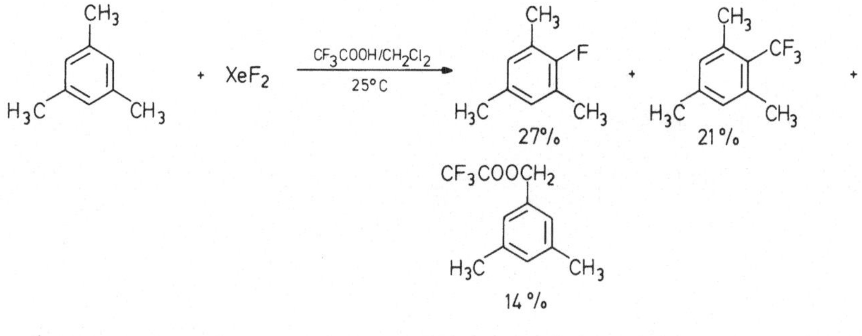

Xenon difluoride reacts with hexamethylbenzene in the presence of HF forming pentamethylbenzyl fluoride. Upon substitution of HF with trifluoroacetic acid there occurs trifluoroacetoxylation of hexamethylbenzene rather than fluorination, giving pentamethylbenzyl trifluoroacetate [58].

The reactions of xenon difluoride with benzocyclenes and related compounds have been studied [59–61]. Tetralin, 9,10-dihydroanthracene, and acenaphthene react with XeF_2 forming a mixture of monofluoroderivatives which contain fluorine in the aromatic ring. It should be noted that an isomer with the fluorine atom *ortho* to the methylene group is always formed in a small amount. A similar case is observed for the reaction of xenon difluoride with *o*-xylene. Indan [59] and fluorene [60] are almost completely fluorinated at the *meta*-position. Further fluorination of the resulting compounds leads to difluorides.

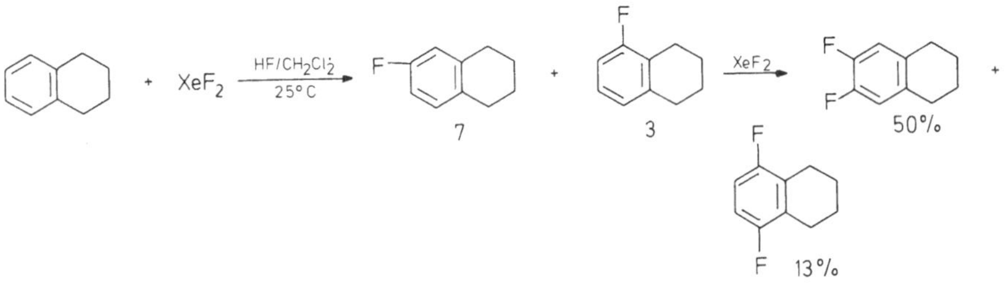

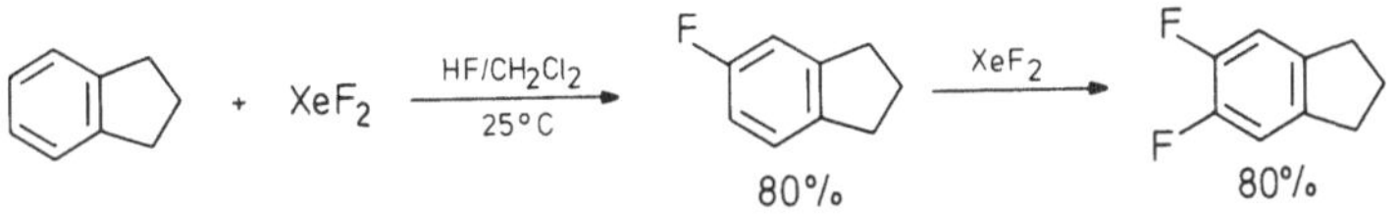

Fluorination of the aromatic ring of hydroxy- and alkoxyaromatic compounds by xenon difluoride proceeds easier than fluorination of arenes with such substituents as alkyl, F, Cl, CF_3, and NO_2. Thus conversion of anisole by XeF_2 to fluoroanisoles readily proceeds without catalyst, as well as fluorination of 1,2-dimethoxybenzene [62].

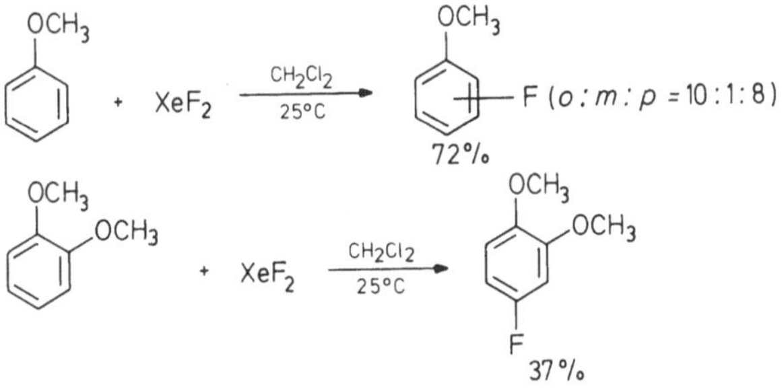

Phenol and pyrocatechol react with xenon difluoride in CH_2Cl_2 forming fluorophenols (47%) and 1,2-dihydroxy-4-fluorobenzene (38%) [62]. At the same time, upon treatment of hydroquinone and 4-tert-butyl-1,2-dihydroxybenzene with XeF_2 (CH_2Cl_2, 20 °C), instead of fluoroarenes one obtains benzoquinone and 4-tert-butyl-1,2-benzoquinone respectively [63].

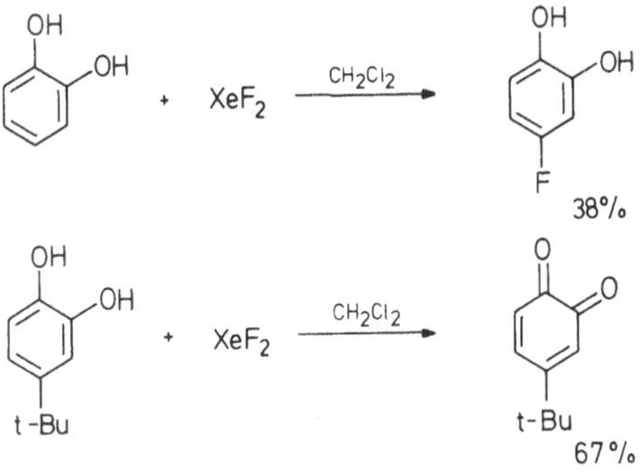

It is interesting that in water phenol is oxidised by xenon difluoride to p-benzoquinone, presumably with the intermediate formation of hydroquinone [64, 65].

Dutch scientists [66] reported that the product ratio in the reaction of XeF_2 with 2-bromo-4,5-dimethylphenol (15) (catalyst BF_3OEt_2) strongly depends on the temperature and duration of the process. The nature of solvent (CH_2Cl_2, $CHCl_3$ or 1,2-$C_2H_4Cl_2$) is of no importance. For example, treatment of the substituted phenol 15 with an equimolar amount of XeF_2 at -65 to $-25\,°C$ leads to 2-bromo-4-fluoro-4,5-dimethylcyclohexa-2,5-dienone. But if the CH_2Cl_2 solution of phenol 15 and XeF_2 is stirred for 6 h at $-10\,°C$, then the usual treatment affords 3-bromo-5,6-dimethyl-1,2-benzoquinone (2.6%), 2-bromo-6-fluoro-4,5-dimethylphenol (5%) and 2-bromo-6-fluoro-3,4-dimethyl-phenol (7%).

Fluorination of the aromatic ring of L-4-hydroxy-3-methoxyphenylalanine proceeds more smoothly and gives L-6-fluoro-3,4-dihydroxyphenylalanine (after protodemethylation with HBr) [67]. This was used later in the synthesis of L-6[^{18}F]dihydroxyphenylalanine [68].

Aromatic carboxylic acids react with xenon difluoride to form unstable ethers ArCOOXeF whose decomposition in excess benzene leads to aroyloxyl-ation of the latter [69–72]. Phthalic acid is dehydrated by XeF_2 (acetonitrile, $20\,°C$) to phthalic anhydride in a 26% yield, whereas diphenyl-2,2'-dicarboxylic acid is transformed in the same conditions mainly to 3,4-benzocoumarine (yield 30%) and the respective anhydride is formed in very small amounts [73]. Saccharin reacts with XeF_2 in excess benzene ($60\,°C$, 48 h) to form N-phenyl-saccharin (22%), and phthalimide is converted to N-phenyl-phthalimide [74]. It should be stressed that no fluorine-containing compounds are formed in all these reactions.

Aniline and benzylamine are fluorinated by xenon difluoride so that the N–H bonds remain intact [75]. The products of the reaction of aniline with XeF_2 are o-, m- and p-fluoroanilines (yields 37, 3 and 11% respectively). The ortho-orientation is still more pronounced in fluorination of benzylamine.

$$C_6H_5NH_2 + XeF_2 \xrightarrow[-196\ to\ -0\ °C]{MeCN} 2\text{-}FC_6H_4NH_2 + 3\text{-}FC_6H_4NH_2 + 4\text{-}FC_6H_4NH_2$$
$$\qquad\qquad\qquad\qquad\qquad\quad 37\% \qquad\qquad 3\% \qquad\qquad 11\%$$

$$C_6H_5CH_2NH_2 + XeF_2 \xrightarrow[-78\ to\ -20\,°C]{CH_2Cl_2} 2\text{-}FC_6H_4CH_2NH_2 + 4\text{-}FC_6H_4CH_2NH_2$$
$$\qquad\qquad\qquad\qquad\qquad\qquad 40\% \qquad\qquad\qquad 2\%$$

It should be noted that these amines, like phenols, react with XeF_2 without catalyst.

Interaction of xenon difluoride with benzenesulphonyl chloride and related compounds in acetonitrile is limited to substitution of chlorine by fluorine in the functional group ($25\,°C$, 4 to 10 h) [76]. In methylphenylchlorosilane, substitu-tion of chlorine by fluorine is faster (3 min), simultaneously fluorine is substituted for the hydrogen atom at silicon. A similar exchange reaction of XeF_2 occurs in the series of alkylchlorosilanes [77].

$$RSO_2Cl + XeF_2 \xrightarrow[25\,°C,\ 4-10\ h]{MeCN} RSO_2F$$
$$20-80\%$$

$$R = Me,\ Ph,\ C_6F_5,\ 4\text{-}CH_3C_6H_4$$

$$C_6H_5SiHClMe + XeF_2 \xrightarrow[25\,°C,\ 3\ min]{MeCN} C_6H_5SiF_2Me$$

The reactions of xenon difluoride with elementoaromatic compounds proceeding without oxidation of heteroatom in a side chain have been practically uninvestigated. Reutov and his co-authors [78,79] have shown that diarylmercury readily reacts with XeF_2 in mild conditions, forming aryl fluoride, arylmercuric fluoride, arene and diaryl. The reaction of xenon difluoride with dibenzylmercury and bis(phenylethynyl)mercury proceeds in a similar way.

$$R_2Hg + XeF_2 \xrightarrow[-45 \div 5\,°C]{CHCl_3} RF + RHgF + RH + R_2$$

$$R = 4\text{-}CH_3OC_6H_4,\ 4\text{-}Me_2NC_6H_4,\ 4\text{-}EtOCOC_6H_4,\ PhCH_2,\ PhC\equiv C$$

Because of complex composition of the reaction mixture and low yields, it is not practical to synthesize fluorides RF in this way.

Xenon difluoride is a convenient fluorinating agent for the synthesis of fluorine-containing polycyclic aromatic compounds many of which are difficult to obtain by conventional methods. But the reaction in the series of polycyclic compounds differs from fluorination of olefins and benzene derivatives. First, the reactivity of polycyclic arenes exceeds by far that of substituted benzenes and olefins, therefore in many cases the use of a catalyst is unnecessary. Second, in order to raise the yield of fluorinated arenes, the reagents should be mixed at a low temperature and the mixture slowly warmed to room temperature. Third, the reaction is carried out in very diluted solutions, otherwise the polymeric substance is formed in large amounts.

According to [80], naphthalene, anthracene and phenanthrene are fluorinated with xenon difluoride in the absence of a catalyst, giving a mixture of isomeric monofluoroarenes.

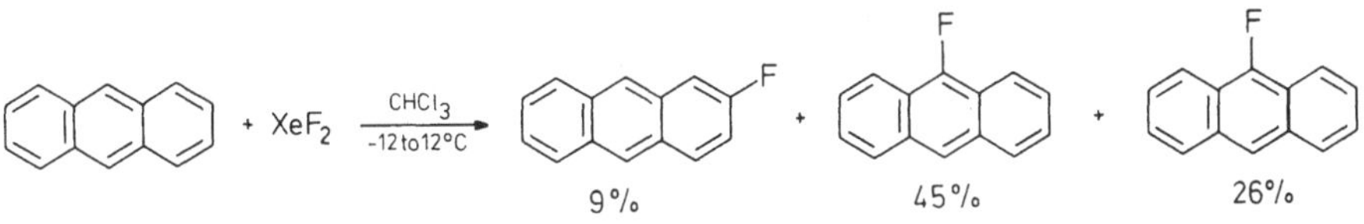

Later it has been found that fluorination of naphthalene gives in addition to 1- and 2-fluoronaphthalenes, 1,4-difluoronaphthalene (yields 46 and 15% respectively) [81]. The latter seems to be the product of fluorination of 1-fluoronaphthalene, but there is no direct evidence of it.

The reaction of xenon difluoride with phenanthrene was independently studied by three groups of researchers [80,82,83] who obtained similar results. Treatment of the diluted solution of phenanthrene ith XeF$_2$ in CH$_2$Cl$_2$ or CHCl$_3$ leads at first to 9-fluorophenanthrene (yield 33–60%). Raising the amount of XeF$_2$, addition of HF and prolonged keeping of the reaction mixture at room temperature raise the degree of fluorination, leading to 9,10-difluorophenanthrene, 9,9,10-trifluoro-9,10-dihydrophenanthrene and 9,9,10,10-tetrafluoro-9,10-dihydrophenanthrene. A special experiment has shown them to be the products of fluorination of 9-fluorophenanthrene [82].

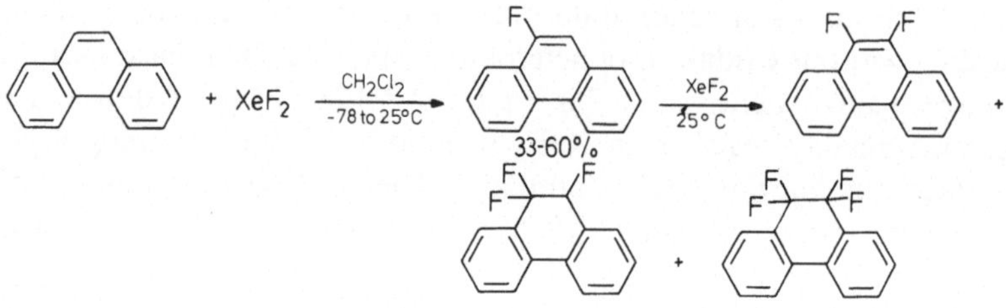

An attempt to find in the reaction mixture 9,10-difluoro-9,10-dihydro-phenanthrene—the most probable primary product of the reaction of phenanthrene with XeF$_2$—was unsuccessful.

At $-78\,°C$, XeF$_2$ fluorinates pyrene and benzo[a]pyrene without catalyst. The general yield of monofluoroarenes is 20 to 30%. At the same time the reaction gives a marked amount of di- and trimeric products, which are isolated by TLC [84,85].

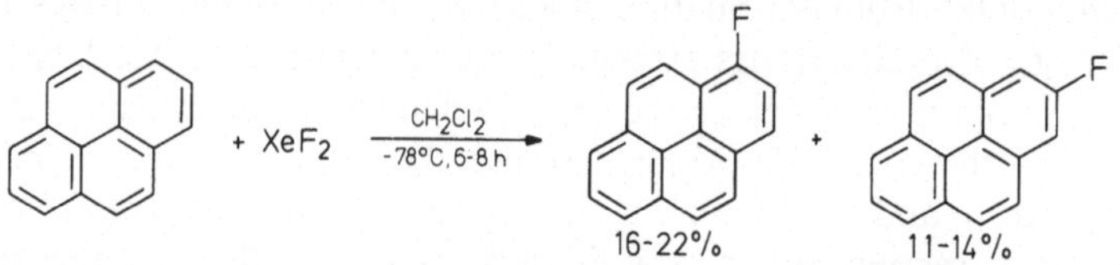

This method allowed to obtain in one step the hardly available 1-fluoropyrene and 6-fluorobenzo[a]pyrene.

Perylene is transformed by XeF$_2$ (25 °C, 30 h) to 1-fluoroperylene, 3-fluoroperylene and difluoroperylene of unknown structure, but conversion of perylene in this reaction is small (40%) [86].

Only one work reports the results of studies on the reaction of xenon difluoride with pyridine and quinoline derivatives [75]. By contrast with benzene derivatives, pyridine easily reacts with XeF$_2$ without a catalyst, forming 2-fluoropyridine, 3-fluoropyridine and 2,6-difluoropyridine.

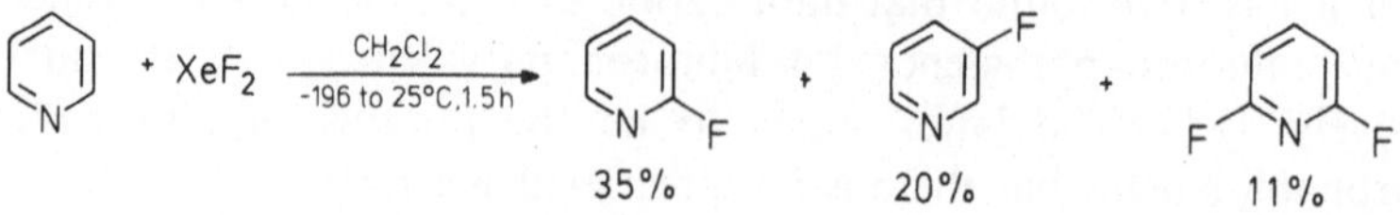

The ease of this reaction is the more surprising in that pyridine reacts with electrophilic agents in much more rigid conditions than benzene.

There are no data on the reaction of XeF$_2$ with quinoline, but 8-hydroxy-quinoline at 5 to 25 °C is readily converted to 5-fluoro-8-hydroxyquinoline (yield 35%) and a mixture of other compounds whose structure was not determined [75].

1.6   Fluorination of Polyfluoroaromatic Compounds

Polyfluoroaromatic compounds, like their hydrocarbon analogues, react with xenon difluoride in the presence of Lewis acids, but the fluorination proceeds by regiospecific 1,4-addition of fluorine even in the case of pentafluorobenzene and deuteropentafluorobenzene [87–91]. Position of the R substituent in the product of fluorination of substituted pentafluorobenzene C$_6$F$_5$R depends on its nature: with R = H, D, Cl, Br, C$_6$F$_5$ the reaction gives 1-R-heptafluoro-1,4-cyclohexadienes, with R = OAlk—a mixture of 1-R- and 3-R-heptafluoro-1,4-cyclohexadienes. The latter are unstable in the reaction conditions and are converted to hexafluoro-2,5-cyclodien-1-one.

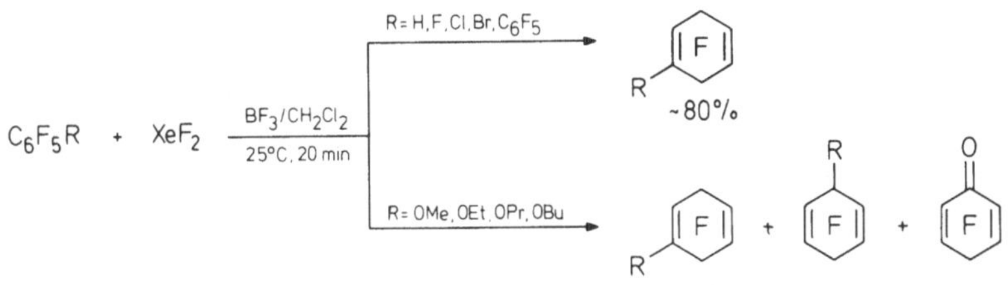

Fluorination of octafluoronaphthalene with xenon difluoride catalysed by boron trifluoride or tungsten hexafluoride leads to perfluoro-1,2- and -1,4-dihydronaphthalene in the ratio of 1 to 7 [90,92]. At the same time, the single product of fluorination of 2-methoxy- and 2-ethoxyheptafluoronaphthalene is 2-alkoxyperfluoro-1,4-dihydronaphthalene [92].

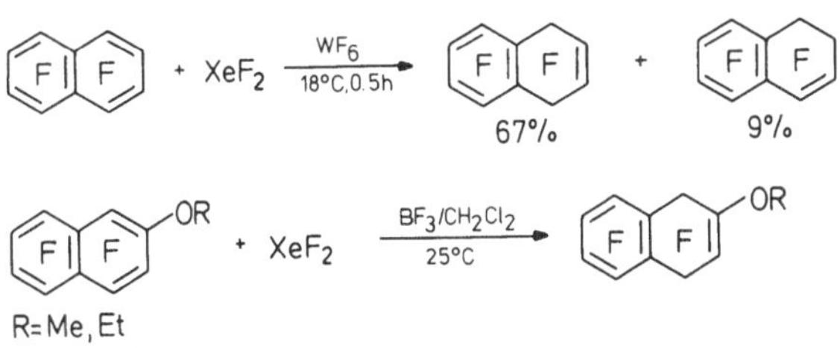

The above fluorination reactions of polyfluoroaromatic compounds with XeF$_2$ proceed at room temperature and are catalysed by HF, BF$_3$ or WF$_6$. However attempts to fluorinate pentafluorobenzonitrile and nitropentafluoro-benzene in the presence of BF$_3$ at 25 °C failed [89], though in the HF solution of xenon difluoride, nitropentafluorobenzene is transformed to 1-nitroheptafluoro-1,4-cyclohexadiene (yield 32%) [90]. Much stronger fluorinating agents are the fluoroxenonium salts, for example $XeF^+SbF_6^-$ [88, 90, 91]. Under the action of these fluorooxidants transformation of polyfluoroaromatic compounds to 1,4-cyclohexadiene derivatives proceeds even at -70 to $-80\,^{\circ}C$.

$$R=F, NO_2$$

Boron trifluoride and HF have no catalytic effect in these conditions. Hence, upon fluorination of polyfluoroaromatic compounds by XeF_2, the catalytic activity of Lewis acids decreases in the series $SbF_5 > HF > BF_3$ [90].

Literature reports no data on fluorination of polyfluoroarenes containing the hydrocarbon alkyl groups. There is a description of the transformation of pentafluorophenol by XeF_2 to compound $C_{12}F_{10}O_2$, to which the structure of $C_6F_5OOC_6F_5$ was initially assigned [93]. Later it was established that actually the reaction gives a mixture of isomeric perfluorophenoxycyclohexadienones, and the reaction carried out in HF yields in addition perfluoro-2,5-cyclo-hexadien-1-one [94].

Trimethylsilylperfluoroarenes react with xenon difluoride in the presence of potassium or caesium fluorides with the aryl–silicon bond cleavage. It is interesting that in the absence of a fluoride ion source the reaction fails to proceed [95].

$$Ar_FSiMe_3 + XeF_2 \xrightarrow[20\,^{\circ}C]{MeCN,\ CsF} Ar_FH + Ar_F{-}Ar_F + FSiMe_3 + Xe$$

$$R_F = C_6F_5,\ 4\text{-}CF_3C_6F_4,\ 4\text{-}C_5F_4N$$

A key intermediate here seems to be the perfluoroaryl radical. This is confirmed by the formation of Ar_FBr and $Ar_FC_6H_5$ when the reaction is conducted in the presence of $CHBr_3$ or benzene respectively, and by the observed chemically induced polarization of ^{19}F nuclei in C_6F_5H and $C_6F_5{-}C_6F_5$ [95].

1.7 Oxidative Fluorination of Organoelement Compounds

Since the synthesis of xenon difluoride, one of the most attractive prospects was its application for the synthesis of organoelement compounds where an element is in one of the highest stages of oxidation. In 1968, Bartlett [96] showed the possibility of oxidative fluorination of iodine to IF_5, and of sulfur dioxide—to SO_2F_2 by XeF_2 in the presence of HF or BF_3. Subsequently the Australian chemists [97] carried out an extensive investigation of xenon difluoride as a reagent for the synthesis of inorganic fluorides. As the starting compounds they used elements, their oxides, halides and even carbonyls. The authors of [97] used the 10^{-2} to 1 M HF solution of xenon difluoride, the reactions proceeded at room temperature for several minutes. Below are given the highest oxidation degrees obtained for different elements using XeF_2.

Group	II	III	IV	V	VI
Element	Hg (2+)	Tl (3+)	Sn (4+)	As (5+), Sb (5+), Nb (5+)	S (6+), Cr (4+), Mo (6+), W (6+)

Group	VII	VIII
Element	I (5+), Mn (3+), Re (6+)	Co (3+), Ni (2+), Ru (5+), Rh (4+), Os (6+)

Among the compounds which were synthesized, it is worthwhile to mention carbonyl fluorides of tungsten, molybdenum, rhenium and ruthenium as representing the class of compounds unavailable by other methods.

The oxidative fluorination of organoelement compounds by xenon difluoride was studied by investigating the reactions of phosphorus-, sulfur-, selenium-, tellurium-, antimony-, arsenic- and iodoorganic derivatives. The carbon-element bond remains intact in these reactions.

Transformation of phosphines by XeF_2 to difluorophosphoranes proceeds in mild conditions (low temperature, MeCN or CH_2Cl_2 as solvent) and with high yields. Fluorination of the phosphine containing the hydroxy group and synthesized from diphenylphosphine and hexafluoroacetone occurs exclusively at the phosphorus atom [98].

$$(C_6H_5)_2PC(OH)(CF_3)_2 + XeF_2 \xrightarrow[\substack{-63\,°C,\ 2\,h \\ -23\,°C,\ 1\,h}]{MeCN-CFCl_3} (C_6H_5)_2PF_2C(OH)(CF_3)_2$$

The P–H bond is stable towards XeF_2, therefore alkyl- or arylphosphines are readily transformed to difluorophosphoranes [99].

$$R_2PH + XeF_2 \xrightarrow[-15\,°C]{MeCN} R_2PF_2H$$

$$RPH_2 + XeF_2 \xrightarrow{MeCN} RPH_2F_2 \quad (R = Ph, NCCH_2CH_2)$$
$$>90\%$$

Chlorine-containing phenylphosphines react with XeF_2 forming fluorophosphoranes and chlorine; the intermediate formation of chlorofluorophosphoranes was not observed.

$$Ph_nPCl_{3-n} + XeF_2 \xrightarrow{-10\ to\ 0\,°C} Ph_nPF_{5-n}$$

In the case of bis(tert-butyl)chlorophosphine the reaction gives a mixture of the products of fluorination of methyl groups [99]. However the reaction with methyldiphenylphosphine or o-tolyldi(ethyl)phosphine leads exclusively to

difluorophosphoranes with the alkyl groups remaining intact [100]. Triphenyl-
and trimethyldifluorophosphorane may be obtained by fluorination of R_3P in a
90–100% yield [99,100].

$$R_3P + XeF_2 \rightarrow R_3PF_2 \quad (R = Me, Ph)$$

In compounds containing the bonds P–N, P–O, P–S, transformation of the
difluorophosphino group to the tetrafluorophosphoranyl one by XeF_2 is
complicated by a slight elimination of PF_5 and formation of the products
containing the P=X fragment, where X=N, O, S. Thus the reaction of
bis(difluorophosphino)methylamine with xenon difluoride proceeds with evolution of PF_5 and formation of difluorophosphino(tetrafluorophosphoranyl)methylamine, which further reacts with XeF_2 forming the iminophos-
phorane dimer and PF_5 [102].

$$MeN(PF_2)_2 + XeF_2 \xrightarrow[-PF_5]{XeF_2} MeN(PF_2)PF_4 \xrightarrow[-PF_5]{XeF_2} (MeNPF_3)_2$$

$$(PF_2)_2X + XeF_2 \rightarrow F_3PX + PF_5 \quad (X = O, S)$$

Investigation of the oxidative fluorination of sulfur-containing compounds
was started by Zupan [14] (see Sect. 3). The sulfides containing no α-hydrogen
atoms were shown to transform smoothly to difluorosulfuranes, whereas the
presence of methyl, ethyl, isopropyl or benzyl groups at the sulfur atom prevents
isolation of difluorosulfuranes. In this case the reaction gives α-fluorine-
containing sulfides – the product of further transformations of compounds with
the SF_2 group [15].

$$Ph_2S + XeF_2 \rightarrow Ph_2SF_2$$

$$PhSCMe_3 + XeF_2 \rightarrow PhSF_2CMe_3$$

$$RSCHR'_2 + XeF_2 \rightarrow RSCFR'_2 + HF$$

The fluorination is carried out in the acetonitrile solution at low temperatures. In
these conditions formation of sulfur (VI) is not observed. But the reaction of
xenon difluoride with diphenyl sulfide in the $MeCN$–$CFCl_3$ mixture in the
presence of the catalytic amount of HF with subsequent hydrolysis leads to
diphenyl sulfone [103,104]. Duration of the reaction is 15 h; after 1 h, mainly
diphenyl sulfoxide is isolated.

$$Ph_2S + XeF_2 \xrightarrow{15\ h} [Ph_2SF_4] \xrightarrow{H_2O} Ph_2SO_2$$

The literature contains a detailed report on the fluorination of dimethyl
sulfide by XeF_2 giving Me_2SF_2 [16]. Interaction of the reagents at 20 °C without
solvent leads to an explosion but upon dilution with $CFCl_3$ it yields $MeSCH_2F$
which reacts with excess dimethyl sulfide, giving $[Me_2SCH_2SMe]^+[F(HF)_n]^-$.
In the anhydrous hydrogen fluoride solution, the reaction product is
$[Me_2SF]^+[F(HF)_n]^-$; which was strictly proved by the ^{19}F NMR data and an

alternative synthesis of Me_2SF_2 from Me_2S and AgF_2.

$$Me_2S + XeF_2 \xrightarrow[-HF]{} MeSCH_2F \xrightarrow{Me_2S-HF} [Me_2SCH_2SMe]^+[F(HF)_n]^-$$

$$Me_2S + XeF_2 + HF \rightarrow [Me_2SF]^+[F(HF)_n]^-$$

Substitution of one methyl group by the trifluoromethyl one in dimethyl sulfide does not alter the character of the end product of fluorination, the product being CF_3SCH_2F. But bis(trifluoromethyl) sulfide is oxidised to difluorosulfurane [101].

$$CF_3SCH_3 + XeF_2 \rightarrow CF_3SCH_2F$$

$$CF_3SCF_3 + XeF_2 \rightarrow (CF_3)_2SF_2$$

Despite the fact that aryltrifluoromethyl sulfides do not add chlorine, these compounds were transformed with the help of XeF_2 to aryltrifluoromethylsulfur difluoride in the presence of the catalytic amount of HF [105].

$$4\text{-}XC_6H_4SCF_3 + XeF_2 \xrightarrow[40\ to\ 50\,°C]{} 4\text{-}XC_6H_4SF_2CF_3$$

$$X = H, Cl, NO_2$$

Thiols and thiophenols are oxidised by XeF_2 to disulfides [15] with evolution of HF, which allows to use C_6F_5SH as a catalyst in fluorination of organic compounds with XeF_2. But the disulfides can also react in rigid conditions with xenon difluoride. As shown by one of the authors, melting of decafluorodiphenyldisulfide with XeF_2 at 85 °C in a sealed tube for 3 h leads to the formation of a mixture of $C_6F_5SF_3$ and C_6F_5SOF, as determined by ^{19}F NMR.

$$2RSH + XeF_2 \rightarrow RS\text{-}SR + 2HF$$

$$C_6F_5SSC_6F_5 + XeF_2 \rightarrow 2C_6F_5SF_3 \xrightarrow{H_2O} C_6F_5SOF$$

Xenon difluoride is a very mild and highly selective fluorinating agent for compounds with the sulfur–nitrogen bond. Studies of the methyleneamidosulfenyl system $(CF_3)_2C=N-SX-$ have shown that at 20 °C, the oxidative fluorination proceeds with exclusive formation of the derivatives of S(IV), whereas upon addition of BF_3 as a catalyst, the products containing S(VI) are formed. Thus hexafluoroisopropylidenimino(trifluoromethyl) sulfide is quantitatively converted in mild conditions to sulfoximide [106].

$$(CF_3)_2C=N-SCF_3 + XeF_2 \rightarrow (CF_3)_2CF\text{-}N=SFCF_3$$

Hexafluoroisopropylidenimidosulfenyl isocyanate is transformed at room temperature to products 16 and 17 in high yields, corresponding to 1,3- and 1,5-difluorination of the substrate, the fluorination proceeding without catalyst

[107,108].

$$(CF_3)_2C=N-S-N=C=O + XeF_2 \xrightarrow[25\,°C]{} (CF_3)_2CFN=SF-NCO$$

16

$$+ (CF_3)_2CF-NSN-COF$$

17

In the presence of BF$_3$, the product of fluorination is N-(heptafluoroisopropyl)-N'-(carbonyl fluoride)-S,S-difluorothiodiimide [107]. The pseudotriene system $(CF_3)_2C=NSN=C(CF_3)_2$ is converted by XeF$_2$ to diimide, whereas in the presence of BF$_3$, further oxidation to the sulfur (VI) derivative occurs [106].

$$(CF_3)_2C=NSN=C(CF_3)_2 + XeF_2 \longrightarrow \begin{array}{l} \longrightarrow (CF_3)_2CF-NSN-CF(CF_3)_2 \\ \xrightarrow{BF_3} (CF_3)_2CF-N=SF_2=N-CF(CF_3)_2 \end{array}$$

N-Sulfinylperfluoroalkylamines are transformed by XeF$_2$ in the perfluorodecaline solution to N-(difluorosulfinyl)perfluoroalkylamines with about 50% yields [109].

$$R_FNSO + XeF_2 \xrightarrow[20\,°C]{} R_FNSOF_2 \quad (R_F = CF_3, C_2F_5)$$

Aryltrifluoromethyl selenides are fluorinated by XeF$_2$ under milder conditions than the corresponding sulfides, giving aryltrifluoromethyldifluoroselenanes in quantitative yields [105]. Difluoroselenanes are much more stable than difluorosulfuranes. Thus, dimethyldifluoroselenane is isolable at a high yield in the reaction of XeF$_2$ with dimethyl selenide [101]. In a similar process, the respective trifluoromethyl selenium derivatives are obtained.

$$4\text{-}XC_6H_4SeCF_3 + XeF_2 \xrightarrow[20\,°C]{} 4\text{-}XC_6H_4SeF_2CF_3 \quad (X=H, Br, CH_3, CF_3)$$

$$R\text{-}Se\text{-}R' + XeF_2 \rightarrow R\text{-}SeF_2\text{-}R' \quad (R=R'=CH_3, CF_3; R=CH_3, R'=CF_3)$$

85–95%

It is interesting that the interaction of decafluorodiphenyldiselenide with XeF$_2$ leads to C$_6$F$_5$SeF$_3$ which is readily hydrolysed by air moisture to C$_6$F$_5$SeO$_2$H.

The oxidative fluorination of 2,5-dihydrotellurophene with xenon difluoride at $-78\,°C$ leads to 2,5-dihydrotellurophene-1,1-difluoride [110]. The authors of [111] managed to synthesize aryltellurium (VI) fluorides with the help of xenon difluoride. The starting compounds are diaryl ditellurides or the Te(IV) derivatives.

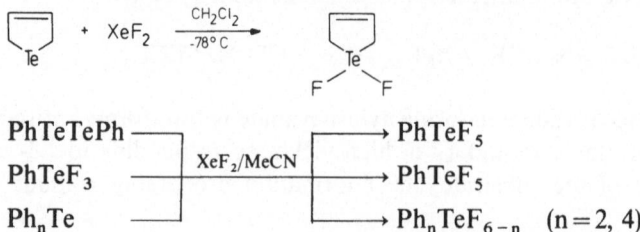

The perfluorinated dimethyl and diphenyl tellurides are easily transformed to the respective tellurium (IV) difluorides, but no Te(VI) derivatives are obtained in this reaction [112,113], which may be attributed to the electron-accepting nature of the R_F groups.

$$(R_F)_2Te + XeF_2 \rightarrow (R_F)_2TeF_2 \quad (R_F = CF_3, C_6F_5)$$

Xenon difluoride is a unique agent for the synthesis of organic derivatives of polyvalent iodine, especially of compounds containing hydrocarbon groups. Fluorination of methyl iodide by XeF_2 leads to CH_3IF_2. The reaction proceeds in excess of the starting iodide, and the product is rather stable in solution [99,114].

$$CH_3I + XeF_2 \xrightarrow[20\,°C]{} CH_3IF_2$$

The use of a catalyst, hydrogen fluoride, allows us to obtain (difluoro-iodo)methane is substantial amounts and to use it as the iodofluorinating agent for various types of unsaturated compounds—styrene, stilbene, 1-phenylpropene [115,116], other phenylalkenes [117] and -alkynes [118], and dihydron-aphthalenes [119].

Perfluoroalkyl iodides may be transformed by XeF_2 to (difluoro-iodo)perfluoroalkanes by merely mixing the reagents and keeping the mixture at 20 to 40 °C. Heating of (difluoroiodo)pentafluorobenzene with one more mole of XeF_2 leads to (tetrafluoroiodo)pentafluorobenzene at a quantitative yield [120].

$$R_FI + XeF_2 \xrightarrow[20\text{ to }40\,°C]{} R_FIF_2 \quad (R_F = C_3F_7, C_6F_5)$$
$$60\text{–}80\%$$

$$C_6F_5IF_2 + XeF_2 \xrightarrow[60\text{ to }65\,°C]{} C_6F_5IF_4$$
$$100\%$$

In a similar way, (difluoroiodo)benzene is synthesized from iodobenzene, but further fluorination of this product to the iodine (V) derivative proceeds with the introduction of fluorine into the aromatic ring [99,120]. Later it was found that the ring-substituted iodobenzenes may be transformed to the respective (difluoroiodo)benzenes in CH_2Cl_2 at 20 °C using HF as a catalyst [100,121,122].

$$XC_6H_4I + XeF_2 \xrightarrow[25\,°C]{HF} XC_6H_4IF_2$$

$$X = H,\ 3\text{-}CH_3O,\ 4\text{-}CH_3O,\ 3\text{-}Cl,\ 4\text{-}Cl,\ 3\text{-}NO_2,\ 2\text{-}CF_3,\ 3,5\text{-}Cl_2$$

Interesting fluorinating agents are (difluoroiodo)arenes on a polymer support [123,124]. Iodobenzene groups are formed on the polymer surface by iodination of polystyrene and transformed by XeF_2 to (difluoroiodo)arenes. Such polymers are used as fluorinating agents for the conversion of olefins (norbornene, phenylcyclopentene, -hexene) to the respective difluorides.

Nesmeyanov and his co-workers showed the possibility of the oxidation of aryl bromides by xenon difluoride at the bromine atom [125]. The (difluorobromo)arenes formed further react with aromatic substrates in the presence of BF_3 etherate, giving diarylbromonium salts whose structure was proved by an independent synthesis.

$$ArBr + XeF_2 \rightarrow [ArBrF_2] \xrightarrow{Ar'H,\ BF_3} [ArBrAr']^+\ BF_4^-$$

Xenon difluoride is a mild selective agent for the oxidative fluorination of the aromatic compounds of Sb(III). Thus diphenylfluorostibine and triphenyl-stibine are converted by XeF_2 to diphenyltrifluoroantimony and triphenyldi-fluoroantimony respectively [126].

$$Ph_2SbF + XeF_2 \xrightarrow[25\,^\circ C]{CH_2Cl_2} Ph_2SbF_3$$
$$98\%$$

$$Ph_3Sb + XeF_2 \rightarrow Ph_3SbF_2$$
$$95\%$$

The methyl derivatives of As(III) are also readily transformed to the corresponding difluorides [100,101].

$$CH_3AsR_2 + XeF_2 \rightarrow CH_3AsR_2F_2 \quad (R = CH_3,\ C_6H_5)$$

It is interesting that the trifluoroacetate group bonded with the antimony atom is substituted under the action of XeF_2 by the fluorine atom [126].

$$Ph_2SbOCOCF_3 + XeF_2 \xrightarrow[25\,^\circ C]{CH_2Cl_2} Ph_2SbF_3$$
$$75\%$$

Up to now there have been no extensive studies on the reaction of xenon difluoride with organoelement compounds containing the metal-carbon bond (except organomercury compounds, see Sect.1.5). The possibility of this bond being retained in oxidative fluorination would open up new prospects of synthesis of organometallic fluorides in the highest stages of oxidation. But the question of the metal-carbon bond stability in the presence of XeF_2 should be decided for each metal individually. Recently it has been found [127] that the methylene chloride solutions of organoaluminium compounds Et_3Al, i-Pr_3Al, i-Pr_2AlH, Et_2AlCl, and $EtAlCl_2$ treated by xenon difluoride intensively lumin-esce. The oxidation is believed by the authors to proceed with cleavage of the Al–C bond. Among the stable products of the reaction was Et_2AlF, as shown by the mass-spectral data. The detailed investigation of the reactions of alkylalum-inium compounds with xenon difluoride in toluene confirmed these results and led to a number of alkylaluminium fluorides R_2AlF (R = Et, i-Bu, EtO) at 80–97% yields [128].

It should be noted that the oxidative fluorination of organoelement com-pounds is a promising preparative method for the synthesis of compounds of new types, and work in this field, owing to the enhanced availability of xenon difluoride, will be intensively developed.

1.8   Other Fluorination Reactions

Some fluorination reactions of organic compounds with xenon difluoride may
not be assigned to any of the above types. These involve, for example, the
reactions of fluorodecarboxylation of alkanoic, arylalkanoic and aryloxyacetic
acids with XeF_2 [129,130]. Alkanoic acids react with xenon difluoride in
CH_2Cl_2 or $CHCl_3$ giving fluoroalkanes. The reaction may be performed at room
temperature and usually requires 8–16 h. Pure products are obtained simply by
washing the reaction mixture with sodium carbonate solution followed by
removal of the solvent.

$$RCOOH + XeF_2 \xrightarrow[-Xe, \, -HF, \, -CO_2]{} RF$$

$R = n\text{-}C_9H_{19}\,(54\%),\ n\text{-}C_{15}H_{31}\,(62\%),\ PhCH_2\,(76\%),$
$\qquad PhCH_2CH_2\,(76\%),\ PhCH_2CH_2CH_2\,(60\%),\ Ph_2CHCH_2\,(63\%),$
$\qquad PhOCH_2\,(64\%),\ 2,4\text{-}Cl_2C_6H_3OCH_2\,(84\%),$
$\qquad CH_2BrCH_2CH_2\,(91\%)$

It should be noted that no aromatic fluorination proceeds here.

Dicarboxylic acids are converted easily to difluoroalkanes.

$$PhCH_2CH(COOH)_2 + 2XeF_2 \rightarrow PhCH_2CHF_2$$
$$68\%$$

Tertiary carboxylic acids are decarboxylated easily with XeF_2. For example,
1-adamantanoic acid is converted to 1-fluoroadamantane (82% yield), and
fluorotriphenylmethane (65%) is obtained from triphenylacetic acid and XeF_2.
However 3-phenylbicyclo[1.1.1]pentan-1-oic acid reacts with XeF_2 to form a
dimer, and no fluoro product is obtained here.

Several secondary carboxylic acids tested for fluorodecarboxylation give
only small amount of fluoro products.

$$(CH_3)_2CHCOOH + XeF_2 \rightarrow (CH_3)_2CHF$$
$$10\%$$

$$(CH_2)_5CHCOOH + XeF_2 \rightarrow (CH_2)_5CHF$$
$$3\%$$

Benzoic acid gives small amount of benzoyl fluoride (see Sect. 1.5).

The ketone function of levulinic acid is unaffected in the reaction with XeF_2,
and 4-fluoro-2-butanone is obtained (82% yield). But amino acids and cholic
acid give only the unreacted starting material.

In view of these limitations, fluorodecarboxylation of carboxylic acids by
xenon difluoride may be recommended as a convenient method for preparing
small amounts of pure fluoroalkanes and arylfluoroalkanes.

Imides of perfluorinated succinic and glutaric acids react with XeF_2 forming the respective N-fluoroimides in good yields [131].

n=2(55-65%), 3(50-60%)

The fluorination is carried out at 0 to 20 °C without solvent or at 0 °C in CF_2Cl_2 (24 h, sealed tube). It is interesting that such NH-acid as bis(sulfuryl fluoride)amine reacts with XeF_2 in the same conditions, forming $FXeN(SO_2F)_2$ [132].

The reaction of xenon difluoride with 1,1-dinitroethane potassium salt has been described [133]. One of the products of the reaction is 1-fluoro-1,1-dinitroethane, but its yield is small and strongly depends on the reaction conditions (solvent, temperature, reagent ratio).

Finally, the reaction of xenon difluoride with the CF_3 radicals should be mentioned, which afforded bis(trifluoromethyl)xenon, $Xe(CF_3)_2$—the first and as yet the only organoxenon compound [134].

1.9 Conclusions

We give here some practical recommendations on fluorination of organic compounds with xenon difluoride.

In the first works on fluorination with XeF_2 the reactions were carried out in metal vacuum lines. Later it appeared that XeF_2 may be handled in open systems. Most convenient for fluorination are the reactors of quartz, poly-chlorotrifluoroethylene or polytetrafluoroethylene; the pyrex reactor may also be used, but only in cases when it does not react with a catalyst. Radical fluorination under heating or irradiation should be carried out in a quartz reactor.

The recommended solvents for fluorination in the condensed phase are CH_2Cl_2, $CHCl_3$, CH_3CN, SO_2FCl, and ether. The most widely used of them is dichloromethane, though xenon difluoride is slightly soluble in it. Acetonitrile dissolves XeF_2 very well but reacts with it in the presence of Lewis acids. The most convenient solvent for the fluorinations catalysed by such acids as SbF_5, is SO_2FCl. It easily dissolves xenon difluoride, most organic compounds and the compounds initiating the reactions with XeF_2. Being a low-boiling liquid (b.p. 8 °C), SO_2FCl is easily isolated from the reaction mixture. The preparation of SO_2FCl from sulfuryl chloride and ammonium fluoride is described in [135].

In the cases when xenon difluoride does not directly react with a substrate, the catalytic amount of Lewis acid is added (10 to 20 mol %). The most efficient of these is SbF_5, but usually used in practice are boron trifluoride, BF_3OEt_2, HF and CF_3COOH. Fluorination in the presence of trifluoroacetic acid proceeds with trifluoroacetoxylation and trifluoromethylation of a substrate, therefore CF_3COOH as the catalyst of fluorination is not recommended. A reasonably active fluorination catalyst is tungsten hexafluoride, but it is not as easily available as BF_3 or BF_3 etherate.

The Lewis acid-catalysed fluorination of alkenes, alkynes, aromatic and polyfluoroaromatic compounds with xenon difluoride proceeds as a rule at room temperature. The reaction of XeF_2 with aniline, alkoxybenzenes, polycyclic arenes and pyridines are carried out in milder conditions (low temperature, diluted solutions, no catalyst). This increases the yields of fluorine-containing products by suppressing polymerisation of substrates. The presence in the molecule of aromatic compound of such substituents as F, Cl, Br, alkyl, hydroxyalkyl, CN, COOR, NO_2 does not hinder the fluorine introduction in the aromatic ring. But phenols, aryl iodides, arylcarboxylic acids, 1-arylolefins and 1-arylalkynes treated with XeF_2 do not form the respective aryl fluorides, so for the successful synthesis of the latter these functional groups must be protected.

1.10 Preparations

1. Xenon difluoride [4]
A copper load was charged with 5 g of MnF_3 distributed in a thin layer, and placed into a heated horizontal copper reactor. The heated zone should exceed the load length by 30 to 50%, and a trap under cooling should be fixed as near the heated zone as possible to avoid concentration of XeF_2 in the reactor. The temperature in the reactor was kept at 320 to 345 °C. A mixture of xenon and chlorine trifluoride (1 : 0.7 to 1.0, mol) was passed at the rate of 3.8 to 7.2 l h^{-1}. Xenon difluoride condensed in the trap as colourless crystals. Yield 60–80%.
2. Xenon difluoride [6].
A 2 l Pyrex glass flask was filled with the gaseous mixture of 350 ml of xenon and 374 ml of fluorine. The total pressure in the flask was 724 mm at 25°C. The flask was kept for 3 weeks at room temperature exposed to daylight. The formation of fine crystals was noticed on the second day of standing. The initial rate of formation of XeF_2 was about 35 mg per day. The total amount of product obtained during 3 weeks varied from 0.5 to 0.75 g.
3. cis, trans-1,2-Difluoroacenaphthene [35]
In a Kel-F reactor was placed 1 mmol of acenaphthylene in 50 ml of CH_2Cl_2, then anhydrous HF (traces) was added and 1 mmol of XeF_2 was introduced with stirring. The colourless solution became dark blue, and xenon evolved. After 5 min the mixture was diluted with 15 ml of CH_2Cl_2, washed with 5% solution of sodium carbonate (10 ml), with water and dried. The solvent was distilled off and the residue subjected to TLC (silica gel, eluent–petroleum ether (40 to 60 °C)) to obtain cis-1,2-difluoroacenaphthene (8%, m.p. 103 to 105 °C) and trans-1,2-difluoroacenaphthene (60%, m.p. 41 to 43 °C).

In a similar way, cis- and trans-1,2-difluorotetralins were prepared (yields 10 and 60% respectively) [35], cis- and trans-1,2-difluoro-1-phenylcyclopentane (20 and 74%), -cyclohexane (40 and 40%), and -cycloheptane (59 and 32%) [32], with the only difference that the starting alkene was dissolved in 5 ml of dichloromethane and fluorination was carried out for 20–30 min.

4. 1-R-Heptafluoro-1,4-cyclohexadienes

A [89] In a Kel-F reactor was placed a solution of 2 mmol of substituted pentafluorobenzene C_6F_5R in 4 ml of dichloromethane, 2 mmol of xenon difluoride, then boron trifluoride was introduced with stirring. The mixture was stirred for 10–30 min at room temperature, then treated as described above, and the residue, after distilling off the solvent, was purified by the preparative GLC to give 1-R-heptafluoro-1,4-cyclohexadiene with approx. 80% yield (R = H, F, Cl, Br, C_6F_5).

B [90] A solution of 10 mmol of hexafluorobenzene in 20 ml of HF and 20 ml of SO_2FCl was placed in a Kel-F reactor, then cooled to -70 to $-80\,°C$, whereupon 11 mmol of $XeFSbF_6$ was added to it in portions, with stirring (preparation of $XeFSbF_6$ see in [136]). The mixture was further stirred for 25–30 min, then poured onto ice cooled with liquid nitrogen, the organic layer was separated, washed with water and dried. SO_2FCl (b.p. 8 °C) was distilled off to give perfluoro-1,4-cyclohexadiene in a 91% yield. In a similar way, 1-nitroheptafluoro-1,4-cyclohexadiene and 1-nitro-4-trifluoromethylhexafluoro-1,4-cyclohexadiene were prepared from nitropentafluorobenzene and 4-nitro-heptafluorotoluene respectively (yields 13 to 30%).

5. Fluorination of anisole [62]

A solution of 37 mmol of anisole in 12 ml of dichloromethane was placed in a Kel-F reactor and degassed at $-196\,°C/5 \cdot 10^{-6}$ mm Hg. Then it was added to 12.2 mmol of XeF_2 placed in a similar reactor. The mixture was slowly heated to evolution of xenon, which was accompanied by darkening of the solution. The reaction usually proceeds at -10 to 25 °C for several minutes. Distillation gave fluoroanisoles in a 71.5% yield ($o:m:p = 10:1:8$).

6. (Difluoroidodo)benzene [120]

In a quartz reactor connected with an apparatus for measuring the amount of xenon evolved, in a dry nitrogen atmosphere at 20 °C, was placed 10 mmol of iodobenzene and 11 mmol of xenon difluoride. After the theoretical amount of xenon had evolved, the product was distilled in vacuum, in a quartz apparatus. Yield of (difluoroiodo)benzene 80%, m.p. 30 to 40 °C, b.p. 93 °C/0.2 mm Hg.

In a similar way, (difluoroiodo)pentafluorobenzene was obtained from iodopentafluorobenzene, but to finish the reaction, the mixture was heated to 40 °C. Yield 82%, m.p. 48 to 49 °C, b.p. 90 °C/0.2 mm Hg.

7. General procedure for fluorodecarboxylation of carboxylic acids [130]

To a solution of carboxylic acid (1 mmol) in 15 ml of dichloromethane contained in a polyethylene bottle, was added xenon difluoride (1 mmol, 170 mg). The solution was stirred magnetically at 22 °C for 10 h during which the colourless solution became slightly yellow. The resulting mixture was washed with 3% sodium bicarbonate (50 ml) solution. The organic solution was dried ($MgSO_4$) and concentrated to yield the pure product.

1.11 References

1. Neiding AB, Sokolov VB (1974) Usp. Khim. 43:2146; (1975) Chem. Abs. 82:67498
2. Sladky F (1973) In: Gutmann V (ed) Main group elements group VII and noble gases. Butterworth, London, p 2
3. Williamson SM (1968) Inorg. Synth. 11:147
4. Mit'kin VN, Zemskov SV (1981) Izv. Akad. Nauk SSSR. Neorg. mater. 17:1897; (1982) Chem. Abs. 96:58511
5. Hudlicky M (1976) Comprehensive chemistry of organic fluorine compounds. Wiley, New York
6. Streng LV, Streng AG (1965) Inorg. Chem. 4:1370
7. Pepekin VI, Lebedev YuA, Apin AYu (1969) Zh. Fiz. Khim. 43:1564; (1969) Chem. Abs. 71:74947
8. Legasov VA, Marinin AC (1972) Zh. Neorg. Khim. 27:2408; (1973) Chem. Abs. 78:51974
9. Eisenberg M, Des Marteau D (1980) Inorg. Nucl. Chem. Lett. 6:29
10. Kiselev YuM, Goryachenkov SA, Martynenko LI, Spitsyn VI (1984) Dokl. Akad. Nauk SSSR 278:881; (1985) Chem. Abs. 102:105129
11. Zajc B, Zupan M (1986) Bull. Chem. Soc. Jap. 59:1659
12. Podkhalyuzin AT, Nazarova MP (1975) Zh. Org. Khim. 11:1568; (1975) Chem. Abs. 83:96539
13. Kloter G, Pritzkov H, Seppelt K (1980) Angew. Chem. 92:954
14. Zupan M (1976) J. Fluor. Chem. 8:305
15. Marat RK, Janzen AF (1977) Can. J. Chem. 55:3031
16. Forster AM, Downs AJ: J. Chem. Soc. Dalton Trans. 1984:2827
17. Janzen AF, Wang PM, Lemire AE (1983) J. Fluor. Chem. 22:557
18. Fehér I, Sempteg M (1970) Magy. Kem. Foly. 76:141; (1970) Chem. Abs. 73:24576
19. Shien TC, Chernick CL (1964) J. Amer. Chem. Soc. 86:5021
20. Shien TC, Feit E, Chernick CL, Yang N (1970) J. Org. Chem. 35:4020
21. Paprott G, Lentz D, Seppelt K (1984) Chem. Ber. 117:1153
22. Shackelford SA, McGuire RR, Pflug JL: Tetrahedron Lett. 1977:363
23. Shellhamer DF, Conner RJ, Richardson RF, Heasley VL (1984) J. Org. Chem. 49:5015
24. Zupan M, Pollak A: J. Chem. Soc. Chem. Commun. 1973:845
25. Zupan M, Pollak A: Tetrahedron Lett. 1974:1015
26. Zupan M, Pollak A (1977) Tetrahedron 33:1071
27. Gregorcic A, Zupan M (1979) J. Org. Chem. 44:4120
28. Gregorcic A, Zupan M (1979) J. Org. Chem. 44:1255
29. Zupan M, Pollak A (1976) J. Org. Chem. 41:4002
30. Stavber S, Zupan M (1977) J. Fluor. Chem. 10:271
31. Shellhamer DF, Ragains ML, Gipe BT, Heasley VL, Heasley GE (1982) J. Fluor. Chem. 20:13
32. Zupan M, Sket B (1978) J. Org. Chem. 43:698
33. Gregorcic A, Zupan M (1984) J. Org. Chem. 49:333
34. Zupan M, Pollak A (1977) J. Org. Chem. 42:1559
35. Sket B, Zupan M: J. Chem. Soc. Perkin Trans. I. 1977:2169
36. Korytnyk W, Valentekovie-Horvath S, Petrie CR (1982) Tetrahedron 38:2547
37. Zupan M, Gregorcic A, Pollak A (1977) J. Org. Chem. 42:1562
38. Shackelford SA: Tetrahedron Lett. 1977:4215
39. Shackelford SA (1979) J. Org. Chem. 44:3489

40. Gregorcic A, Zupan M (1980) Bull. Chem. Soc. Jap. 53:1085
41. Gregorcic A, Zupan M (1984) J. Fluor. Chem. 24:291
42. Hildreth RA, Druelinger ML, Shackelford SA (1982) Tetrahedron Lett. 23:1059
43. Gregorcic A, Zupan M (1977) Tetrahedron 23:3243
44. Shellhamer DF, Curtis CM, Dunham RH, Hollingsworth DR, Ragains ML, Richardson RE, Heasley VL, Shackelford SA, Heasley GE (1980) J. Org. Chem. 50:2751
45. Yurasova TI (1974) Zh. Obshch. Khim. 44:956; (1974) Chem. Abs. 81:25624
46. Zupan M, Pollak A (1976) J. Fluor. Chem. 8:275
47. Cantrell GL, Filler R (1985) J. Fluor. Chem. 27:35
48. Zajc B, Zupan M (1982) J. Org. Chem. 47:573
49. Tsushima T, Kawada K, Tsuji T (1982) Tetrahedron Lett. 23:1165
50. Zupan M, Pollak A (1974) J. Org. Chem. 39:2646
51. MacKenzie D, Fajer J (1970) J. Amer. Chem. Soc. 92:4994
52. Shaw MJ, Human HH, Filler R (1969) J. Amer. Chem. Soc. 91:1563
53. Shaw MJ, Human HH, Filler R (1970) J. Amer. Chem. Soc. 92:6498
54. US Pat 3833581 (1974); (1974) C. A. 81:135684
55. Shaw MJ, Human HH, Filler R (1970) J. Org. Chem. 36:2917
56. Turkina MYa, Gragerov IP (1975) Zh. Org. Khim. 11:340
57. Stavber S, Zupan M (1983) J. Org. Chem. 48:2223
58. Zupan M (1976) Chimia 30:305
59. Sket B, Zupan M (1978) J. Org. Chem. 43:835
60. Sket B, Zupan M (1981) Bull. Chem. Soc. Jap. 54:279
61. Filler R (1978) Isr. J. Chem. 17:71
62. Anand SR, Quaterman LA, Human HH, Migliorese KG, Filler R (1975) J. Org. Chem. 40:807
63. Zupan M, Pollak A (1976) J. Fluor. Chem. 7:443
64. Goncharov AA, Kozlov YuN, Purmal AP (1977) Zh. Fiz. Khim. 51:2939; (1978) Chem. Abs. 88:36967
65. Goncharov AA, Kozlov YuN (1978) Zh. Fiz. Khim. 52:945; (1978) Chem. Abs. 89:5703
66. Koundstaal H, Olieman C (1981) Rec. Trav. Chim. Pays Bas. 100:246
67. Firnau G, Chirakal R, Sood S, Garnett ES (1980) Can. J. Chem. 58:1449
68. Firnau G, Chirakal R, Sood S, Garnett ES (1981) J. Label. Compounds Radiopharm. 18:7
69. Nikolenko LN, Shustov LD, Bocharova TN, Yurasova TI, Legasov VA (1972) Dokl. Akad. Nauk SSSR 204:1369; (1972) Chem. Abs. 77:101059
70. Bocharova TN, Marchenkova NG, Shustov LD, Prokof'eva TYu, Nikolenko LN (1973) Zh. Obshch. Khim. 43:1325; (1973) Chem. Abs. 79:65953
71. Shustov LD, Tel'kovskaya TD, Nikolenko LN (1974) Zh. Obshch. Khim. 44:2564; (1975) Chem. Abs. 82:97795
72. Shustov LD, Tel'kovskaya TD, Nikolenko LN (1975) Zh. Org. Khim. 11:2137; (1976) Chem. Abs. 84:30028
73. Shustov LD, Semenova MN, Nikolenko LN (1978) Zh. Obshch. Khim. 48:1903; (1979) Chem. Abs. 90:22739
74. Shustov LD, Nikolenko LN (1983) Zh. Obshch. Khim. 53:2408; (1984) Chem. Abs. 100:139008
75. Anand SR, Filler R (1976) J. Fluor. Chem. 7:179

76. Volkova SA, Sinyutina ZM, Nikolenko LN (1974) Zh. Obshch. Khim. 44:2592; (1975) Chem. Abs. 82:31070
77. Gibson JA, Janzen AF (1971) Can. J. Chem. 49:2168
78. Butin KP, Kiselev YuM, Magdesieva TV, Reutov OA (1982) Izv. Akad. Nauk SSSR. Ser. Khim.: 716; (1982) Chem. Abs. 97:39056
79. Butin KP, Kiselev YuM, Magdesieva TV, Reutov OA (1982) J. Organometal. Chem. 235:127
80. Anand SP, Quaterman LA, Christian PA, Human HH, Filler R (1975) J. Org. Chem. 40:3796
81. Rabinovitz M, Agranat I, Selig H, Lin CH (1977) J. Fluor. Chem. 10:159
82. Zupan M, Pollak A (1975) J. Org. Chem. 40:3794
83. Agranat I, Rabinovitz M, Selig H, Lin CH: Chem. Lett. 1975:1271
84. Bergmann ED, Selig H, Lin CH, Rabinovitz M, Agranat I (1975) J. Org. Chem. 40:3793
85. Agranat I, Rabinovitz M, Selig H, Lin CH (1976) Experientia 32:417
86. Stephenson MT, Shine HI (1981) J. Org. Chem. 46:3139
87. Stavber S, Zupan M: J. Chem. Soc. Chem. Commun. 1978:969
88. Bardin VV, Furin GG, Yakobson GG (1980) In: Vsesoyuznaya Konferenciya pamyati AE Favorskogo, Feb 1980. Leningrad, p 110
89. Stavber S, Zupan M (1981) J. Org. Chem. 46:300
90. Bardin VV, Furin GG, Yakobson GG (1982) Zh. Org. Khim. 18:604; (1982) Chem. Abs. 97:72000
91. Yakobson GG, Bardin VV, Furin GG (1983) In: The Third Regular meeting of Soviet-Japanese fluorine chemists. Tokyo, p 14.1
92. Zajc B, Zupan M (1982) Bull. Chem. Soc. Jap. 55:1617
93. Nikolenko LN, Yurasova TI, Man'ko AA (1970) Zh. Obshch. Khim. 40:938; (1970) Chem. Abs. 73:34956
94. Avramenko AA, Bardin VV, Karelin AI, Krasil'nikov VA, Tushin PP, Furin GG, Yakobson GG (1985) Zh. Org. Khim. 21:822; (1985) Chem. Abs. 103:141551
95. Bardin VV, Stennikova IV, Furin GG, Leshina TV, Yakobson GG (1988) Zh. Obshch. Khim. 58:2580
96. Bartlett N, Sladky F: J. Chem. Soc. Chem. Commun. 1968:1046
97. Burns RC, MacLeod ID, O'Donnell TA, Peel TE, Phillips KA, Waugh AB (1977) J. Inorg. Nucl. Chem. 39:1737
98. Janzen AF, Vaidya OC (1973) Can. J. Chem. 51:1136
99. Gibson JA, Marat RK, Janzen AF (1975) Can. J. Chem. 53:3044
100. Alam K, Janzen AF (1987) J. Fluor. Chem. 36:179
101. Forster AM, Downs AI (1985) Polyhedron 4:1625
102. Cowley AH, Chung Yi, Lee R (1979) Inorg. Chem. 18:60
103. Zupan M, Zajc B: J. Chem. Soc. Perkin Trans. I 1978:965
104. Gregorcic A, Zajc B, Zupan M 1979) Phosphorus Sulfur 6:107
105. Yagupol'skii YuL, Savina TI (1979) Zh. Org. Khim. 15:438; (1979) C. A. 90:203618
106. Varwig I, Mews R: J. Chem. Res. (Synops) 1977:245
107. Steinbeisser H, Mews R (1981) J. Fluor. Chem. 17:505
108. Geisel M, Mews R (1982) Chem. Ber. 115:2135
109. Leidinger W, Sundermeyer F (1982) Chem. Ber. 115:2892
110. Bergman I, Engman L (1981) J. Amer. Chem. Soc. 103:2715
111. Alam K, Janzen AF (1985) J. Fluor. Chem. 27:467

112. Naumann D, Herberg S (1982) J. Fluor. Chem. 19:205
113. Klein G, Naumann D (1985) J. Fluor. Chem. 30:259
114. Gibson JF, Janzen AF: J. Chem. Soc. Chem. Commun. 1973:739
115. Zupan M, Pollak A: Tetrahedron Lett. 1975:3525
116. Zupan M, Pollak A (1976) J. Org. Chem. 41:2179
117. Zupan M, Pollak A: J. Chem. Soc. Perkin Trans. I 1976:1745
118. Zupan M: Synthesis 1976:473
119. Stavber S, Zupan M (1978) J. Fluor. Chem. 12:307
120. Maletina II, Orda VV, Aleinikov NN, Korsunskii BL, Yagupol'skii LM (1976) Zh.
     Org. Khim. 12:1371; (1976) Chem. Abs. 85:77778
121. Zupan M, Pollak A (1976) J. Fluor. Chem. 7:445
122. Gregorcic A, Zupan M (1977) Bull. Chem. Soc. Jap. 50:517
123. Zupan M, Pollak A: J. Chem. Soc. Chem. Commun. 1975:715
124. Zupan M: J. Chem. Soc. Chem. Commun. 1977:266
125. Nesmeyanov AN, Lisichkina IN, Tolstaya TP (1978) Dokl. Akad. Nauk SSSR
     246:1463; (1978) Chem. Abs. 88:169693
126. Yagupol'skii LM, Popov VI, Kondratenko NV, Korsunskii BL, Aleinikov VP
     (1975) Zh. Org. Khim. 11:459; (1975) Chem. Abs. 82:171158
127. Bulgakov RG, Maistrenko GYa, Tolstikov GA, Yakovlev VN, Kazakov VP: Izv.
     Akad. Nauk SSSR. Ser. Khim. 1984:2644; (1985) Chem. Abs. 102:122365
128. Bulgakov RG, Yakovlev VN, Maistrenko GYa, Tolstikov GA: Izv. Akad. Nauk
     SSSR. Ser. Khim. 1986:490; (1987) Chem. Abs. 106:84699
129. Patrick TB, Johri KK, White DH (1983) J. Org. Chem. 48:4158
130. Patrick TB, Johri KK, White DH, Bertrand WS, Mokhtar R, Kilbourn MR, Welck
     MJ (1986) Can. J. Chem. 64:138
131. Yagupol'skii YuL, Savina TI (1981) Zh. Org. Khim. 17:1330; (1981) Chem. Abs.
     95:168500
132. LeBlond R, DesMarteau DD: J. Chem. Soc. Chem. Commun. 1974:555
133. Celinskii IV, Mel'nikov AA, Varyagina LG, Trubizyn AE (1985) Zh. Org. Khim.
     21:2490. Celinskii IV, Mel'nikov AA, Trubizyn AE (1987) Zh. Org. Khim. 23:1657
134. Turbini L, Aikman R, Lagow R (1979) J. Amer. Chem. Soc. 101:5833
135. Woyski M (1950) J. Amer. Chem. Soc. 72:919
136. Gillespie R, Netzer A, Schrobilgen G (1974) Inorg. Chem. 13:1455

2 Some "Electrophilic" Fluorination Agents

Georgii Georgievich Furin

Institute of Organic Chemistry, Pr. Lavrent'eva 9, 630090 Novosibirsk, USSR

Contents

2.1 Introduction

Advances in studies of the properties of fluoroorganic compounds continuously extend the field of their practical applications. At first, only for fluorinated polymers and freons could a use be found, whereas now the range of applications of fluoroorganic compounds is so wide and they are so varied that this has become a separate field of chemistry [1,2].

The new field was based on the classic fluorination methods worked out at the beginning of the 1960s. Their advantages were described in a monograph [3], which had a great effect on the further development of fluoroorganic chemistry. The progress in this field, which was achieved in the 1980s was largely determined by the fundamental approach of the pioneers of fluoroorganic chemistry [4]. Today, the main demerit of fluoroorganic compounds from the practical viewpoint is their high cost owing to the high toxicity and corrosion activity of fluorine and hydrogen fluoride, and the difficulty of introducing the required number of fluorine atoms into the strictly definite positions of organic molecules. With the advance of science and technology these demerits are being gradually overcome. This is due to the unique physical and chemical properties of some polyfluorinated compounds, which in their turn stimulate the development of new synthetic methods and improvement of the existing technologies. Thus the possibility of perfluorinated tertiary amines, ethers, and paraffins being used as the components of artificial blood [5–7] led to significant advances in the development of electrochemical and gas-phase fluorination methods.

Serious difficulties encountered in the synthesis of fluoroorganic compounds with elemental fluorine called for the development of other methods of fluorination [8–11]. Among these is the method based on the use of higher transition metal fluorides as the fluorinating agents. However, along with the advantages (simplicity and technological convenience), the method has some disadvantages, such as low productivity, complex equipment, and incompleteness of fluorination leading to complex mixtures, which require further purification to yield the target product.

All this encourages researchers to look for new fluorinating agents. Among the numerous agents used for the fluorination of organic molecules, is a group of inorganic fluorides whose reactions may formally be regarded as "electrophilic" fluorination. An indication of such processes is orientation in the reactions with benzene derivatives, substituted alkenes, and some organoelement compounds. Of course, the direct generation of fluoronium cation (or even polarised structures $\overset{\delta+}{F} - \overset{\delta+}{X}$) is hardly possible in condensed phase because of thermodynamic reasons [12]. As a matter of fact, the mechanisms of these reactions are complex and in many respects not clear as yet. However this does not prevent the use of "electrophilic" fluorination agents for the synthesis of fluoroorganic compounds. The "electrophilic" fluorination agents are most frequently the fluorides of elements with high electronegativity (binary or more complex). Among these are XeF_2, $FClO_3$, ClF_3, BrF_3, and some fluorides with the N–F and O–F bonds. They are partially described in the other chapters of this book. Here we shall analyse synthetic applications of some "electrophilic" fluorination agents with the N–F and O–F bonds.

2.2   Fluorinating Agents with the N–F Bond

An ideal agent for the aromatic electrophilic substitution of hydrogen by fluorine would be the salt containing the F^+ cation. However, no such salt has been obtained as yet. As an alternative, a complex salt $XF_{n+1}^+ Y^-$ may be suggested, which may be formally regarded as a complex of the F^+ cation with the neutral molecule XF_n. Such salts are strong electrophiles as the cation here has a high electronegativity. But high electronegativity gives rise to high oxidative ability. How do these two properties show themselves in the reactions with aromatic substrates? The reactions of salts $XeF^+ SbF_6^-$, $ClF_2^+ SbF_6^-$, $BrF_2^+ SbF_6^-$, and $BrF_4^+ SbF_6^-$ with polyfluoroaromatic compounds proceed not by the electrophilic aromatic substitution but the one-electron oxidation of the organic substrate with subsequent transformation of the radical cation to the cationic σ-complex which then forms the 1,4-difluoro-adduct [13]. At the same time, cation NF_4^+ has a high electronegativity (the oxidation number of nitrogen is $+5$) and a high kinetic stability (NF_4^+ is isoelectronic to CF_4) [14,15]. Christe and his co-workers [14] have shown that in anhydrous HF, benzene and its derivatives are fluorinated by $NF_4^+BF_4^-$ with substitution of hydrogen by fluorine.

$$C_6H_5R + NF_4^+BF_4^- \xrightarrow[-78\,°C]{HF} C_6H_4FR + NF_3 + BF_3$$

$R = H, CH_3, NO_2$

Benzene gives di-, tri-, and tetrafluorobenzenes along with fluorobenzene, their yields being small. It is interesting that in the fluorination of nitrobenzene, the main product is 2-fluoronitrobenzene (yield 62%), whereas the yields of *ortho*- and *para*-isomers are much lower (14 and 6% respectively) [15,16]. At the same time, tetra-, penta-, and hexafluorobenzenes react with $NF_4^+BF_4^-$ in more rigid conditions, giving exclusively the adducts of two fluorine atoms with the aromatic ring. Hydrogen atoms in tetra- and pentafluorobenzene in this case remain intact.

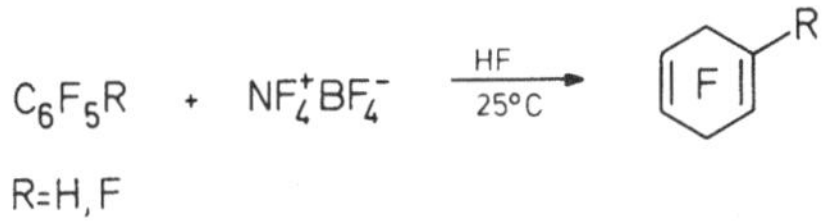

$$C_6F_5R \quad + \quad NF_4^+BF_4^- \quad \xrightarrow[25°C]{HF} \quad$$

R = H, F

Regretfully, there has been no further work on this rather interesting agent.

It should be noted that fluorodiazonium hexafluoroantimonate also reacts with aromatic compounds, forming aryl fluorides, but their yields are small [17].

$$ArH + N_2F^+AsF_6^- \xrightarrow[-78\,°C]{HF} ArF + N_2 + AsF_5$$

As the fluorinating agents, some perfluorinated dialkyl-*N*-fluoro-amines, -amides, and sulfonyl amides may be used. Perfluoropiperidine *1* reacts with *N,N*-dimethylaniline, substituting the *ortho*-hydrogen atom by fluorine. The same reaction gives the demethylation and condensation products [18]. A similar reaction proceeds with *N,N*-diethylaniline.

$$C_6H_5NR_2 + 1 \rightarrow 2\text{-}FC_6H_4NR_2 + C_6H_5NHR$$

$R = Me, Et$

Fluorination of lithium 2,4,6-tris-*tert*-butylphenoxide by perfluoropiperidine proceeds stepwise. Initially, fluorodienone *2* is formed, together with the stable phenoxyl radical. The latter is slowly fluorinated by *1*, giving fluorodienone *3* [18].

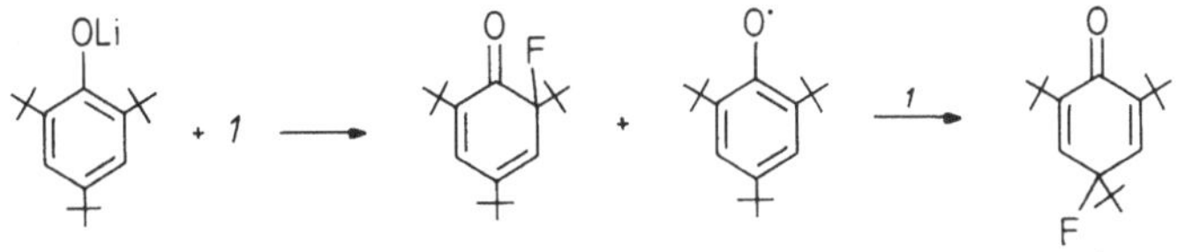

Diethyl phenylmalonate and its sodium salt have been shown [18] to undergo the substitutive fluorination by perfluoropiperidine and perfluorinated polymers 4. The fluorination product is diethyl phenyl-2-fluoromalonate, 1 being a more efficient fluorinating agent than the N–F-containing fluoropolymers 4.

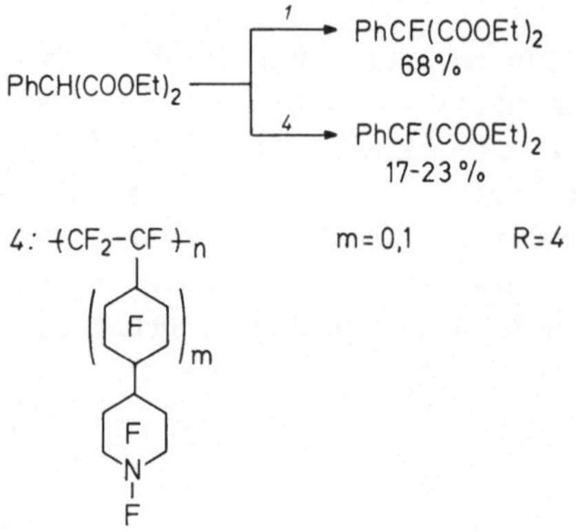

An efficient fluorinating agent is N-fluoro-2-pyridone obtained by the reaction of fluorine with 2-trimethylsiloxypyridine [19].

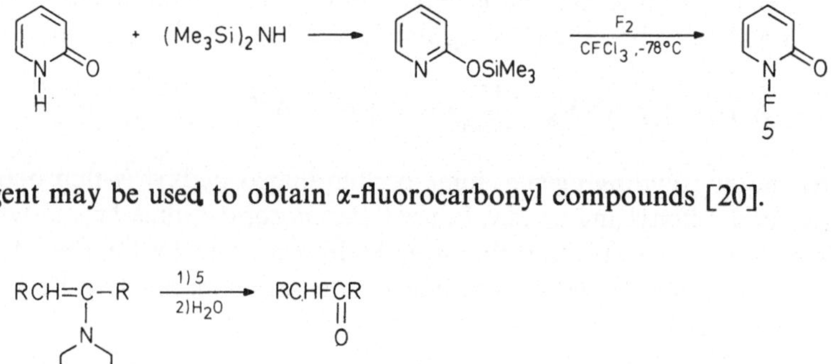

This agent may be used to obtain α-fluorocarbonyl compounds [20].

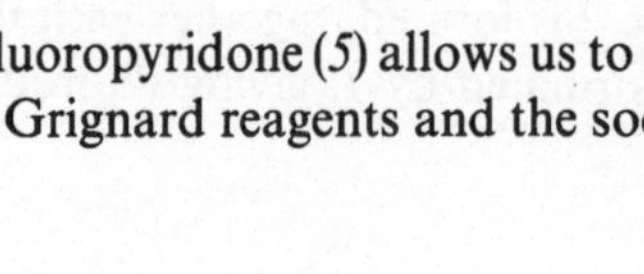

R = Ph (11–33%), RR = cyclo-C_4H_8 (36–44%)

N-Fluoropyridone (5) allows us to carry out the regiospecific fluorodemetall-ation of Grignard reagents and the sodium salts of diethyl malonate derivatives [19,20].

RMgBr + 5 → RF

R = Ph(15%), cyclo-C_6H_{11}(11%), 2-C_8H_{17} (5%)

$R\bar{C}(COOEt)Na^+$ + 5 → $RCF(COOEt)_2$

R = H(9%), Me(17%), Ph(39%)

In the latter case, with $R = PhCH_2$, the fluorinated ester is easily decarboxylated to give ethyl 2-fluoro-3-phenylpropanoate (30–39%) [19].

Regretfully, there are few examples of fluorination by N-fluoro-2-pyridone, so it is difficult to evaluate its synthetic utility.

An interesting compound is formed upon treatment of pyridine with fluorine at a low temperature [21–23]. It is suggested to have the structure of N-fluoropyridinium fluoride [23]

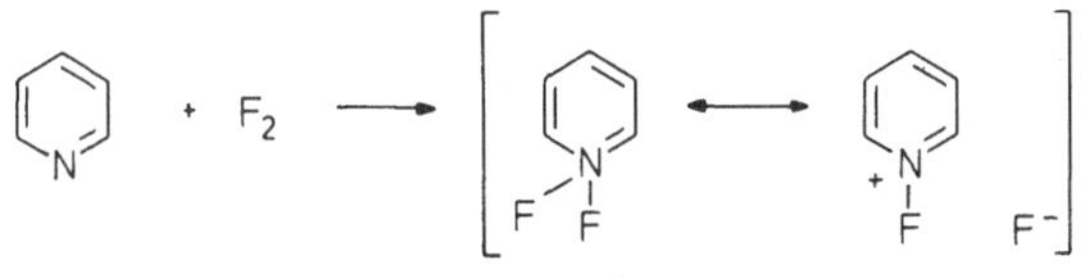

This compound was used for the fluorination of uracyl and some chloroolefins [23,24]. However, the adduct is of little use because of its violent decomposition above $-2\,°C$.

An important achievement of the Japanese scientists is the synthesis of N-fluoropyridinium triflate which, as opposed to the fluoride, is safe, and stable against hydrolysis [24–26]. This method is rather universal, it allows one to obtain a wide range of substituted N-fluoropyridinium triflates and study their fluorinating ability [25,26]. It is interesting to note that introduction of various substituents into the pyridine ring allows variation of the fluorinating ability of the salts to be made.

Pathway A:

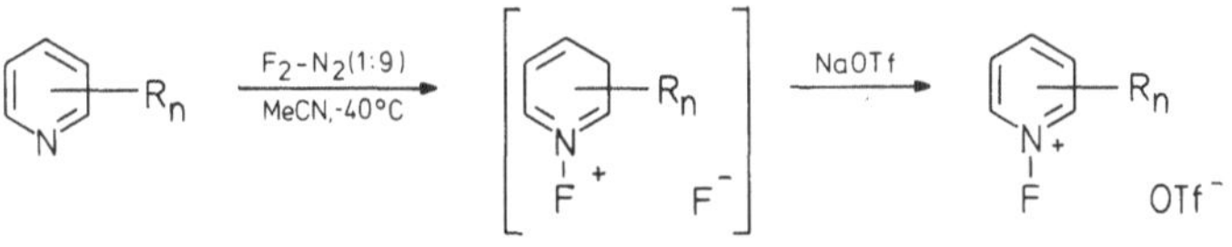

$R = H(6, 80\%)$, 2-Me(60%), 2,6-Me$_2$(73%), 2,4,6-Me$_3$(7,49%), 2-OMe(73%), 2,6-(COOMe)$_2$(9, 72%)

Pathway B:

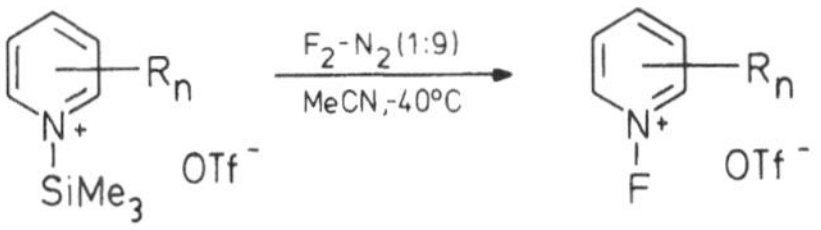

$R = H(6, 78\%)$, 3-Cl(79%), 3,5-Cl$_2$(8, 55%), 3-COOMe(69%), 2,6-(COOMe)$_2$(9, 68%)

The substituted N-fluoropyridinium triflates were used for substitution of hydrogen by fluorine in aromatic compounds [26–28]. Thus benzene was transformed to fluorobenzene (yield 56%, triflate 9), N-(carboxyethyl)aniline gave 2- and 4-fluoro-N-(carboxyethyl)anilines (yields 60 and 27% respectively, triflate 6), phenylurethane also gave the products of fluorination at position 2 (47%) and 4 (32%) of the aromatic ring (triflate 9). In the case of fluorination of anisole, the *ortho*:*para*-fluoroanisole ratio was shown to depend on a substituent in the pyridine ring, and the reactivity of *N*-fluoropyridinium triflates to vary in the series: *7*<*6*<*8*<*9*. At the same time, in the fluorination of thioanisoles, triflate *7* was more reactive, and triflate *8* did not react [29].

$$4\text{-ClC}_6\text{H}_4\text{SCH}_3 + 7 \rightarrow 4\text{-ClC}_6\text{H}_4\text{SCH}_2\text{F}$$

Organosodium and -magnesium compounds easily react with *N*-fluoropyridinium triflates, forming fluoroorganic compounds in good yields.

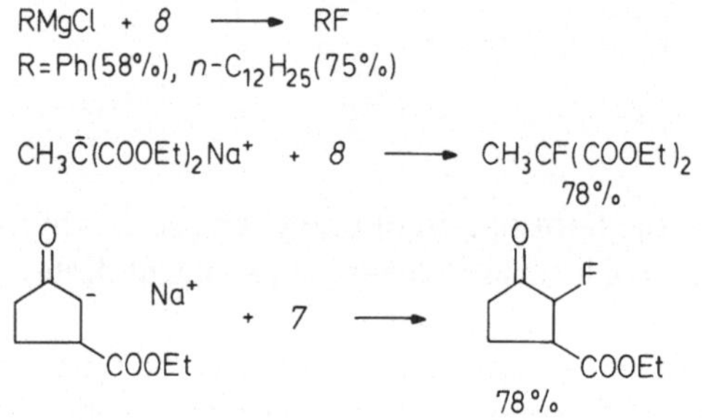
RMgCl + 8 → RF
R=Ph(58%), n-C12H25(75%)
CH3C̄(COOEt)2Na+ + 8 → CH3CF(COOEt)2
78%
Na+
+ 7 →
COOEt
O
F
COOEt
78%

Enol ethers are regiospecifically fluorinated to form α-fluorocarbonyl compounds. It is possible to synthesize in this way fluorinated ketosteroids, as the C=C bonds not activated by the $OSiMe_3$ (or OAc) groups are not fluorinated [26].

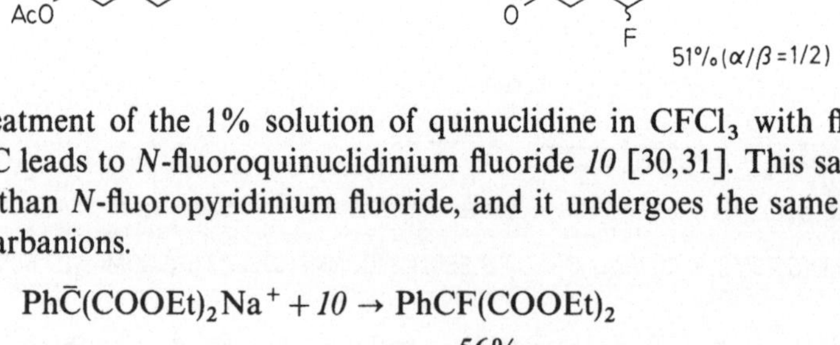
PhCH=C(OSiMe3)OEt + 6 → PhCHFCOOEt
n-C7H15CH=C(OSiMe3)CH3 + 6 → n-C7H15CHFCOCH3
OAc
AcO
+ 6 →
OAc
O
F
51% (α/β = 1/2)

Treatment of the 1% solution of quinuclidine in $CFCl_3$ with fluorine at −78 °C leads to *N*-fluoroquinuclidinium fluoride *10* [30,31]. This salt is more stable than *N*-fluoropyridinium fluoride, and it undergoes the same reactions with carbanions.

$$\text{Ph}\bar{\text{C}}(\text{COOEt})_2\text{Na}^+ + 10 \rightarrow \text{PhCF(COOEt)}_2 \quad 56\%$$

$$(\text{CH}_3)_2\bar{\text{C}}\text{NO}_2\text{Na}^+ + 10 \rightarrow (\text{CH}_3)_2\text{CFNO}_2 \quad 47\%$$

It is interesting that phenyltrichlorosilane is transformed by fluoride *10* to fluorobenzene (22%). The reaction seems to proceed via the intermediate formation of the fluorosiliconium salt which is readily desilylated by electrophiles.

Synthesis of fluoroorganic compounds by the reaction of *N*-fluoro-*N*-alkylsulfonamides with alkaline salts of the carbanions was also suggested [32,33]. These fluorination agents are stable compounds obtained by treatment of the CFCl₃–CHCl₃ solutions of *N*-alkylsulfonamides by fluorine [32,34] or

Table 1. Fluorination with *N*-fluorosulfonylamides [32,33]

Compound	Reagent	Product	Yield (%)
PhC⁻(COOEt)₂Na⁺	A	PhCF(COOEt)₂	81
MeC⁻(COOEt)₂Na⁺	A	MeCF(COOEt)₂	53
C₆H₅MgCl	B	C₆H₅F	50
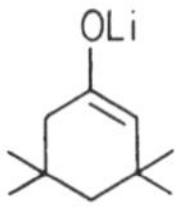 (3,5,5-trimethylcyclohex-1-enyl lithium enolate, OLi)	C	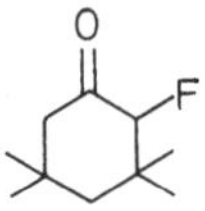 (3,3,5-trimethyl-2-fluorocyclohexanone)	35
PhCOCH⁻CH(CH₃)₂K⁺	C	PhCOCHFCH(CH₃)₂	81
CH₃(CH₂)₁₃MgBr	C	CH₃(CH₂)₁₃F	15
(CH₃)₂C⁻NO₂Bu₄N⁺	B	(CH₃)₂CFNO₂	83–87
Ph₂CCOO²⁻ 2Li⁺	C	Ph₂CFCOOH	69
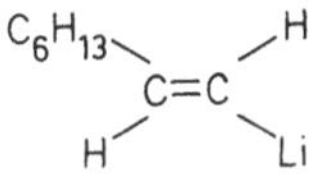 C₆H₁₃CH=CHLi (cis)	D	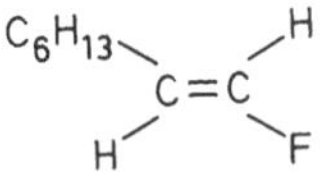 C₆H₁₃CH=CHF (cis)	71
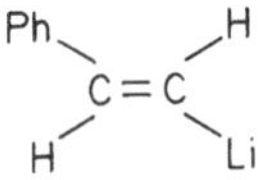 PhCH=CHLi	D	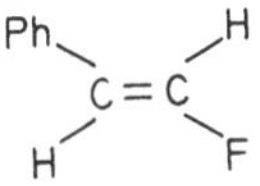 PhCH=CHF	76
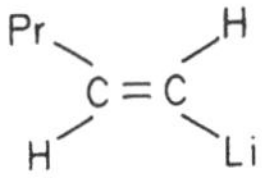 PrCH=CHLi	D	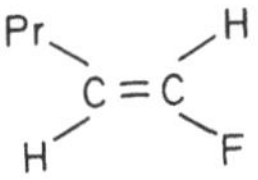 PrCH=CHF	85
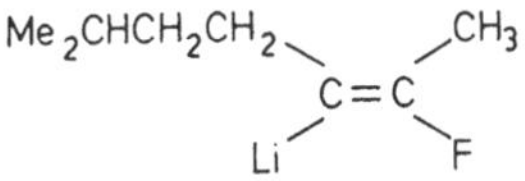 Me₂CHCH₂CH₂C(Li)=C(CH₃)F	D	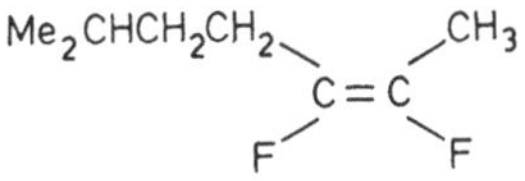 Me₂CHCH₂CH₂C(F)=C(CH₃)F	75
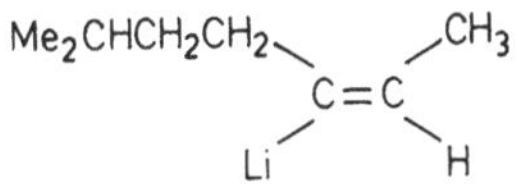 Me₂CHCH₂CH₂C(Li)=C(CH₃)H	D	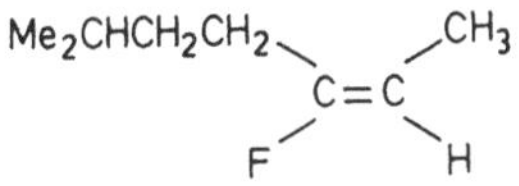 Me₂CHCH₂CH₂C(F)=C(CH₃)H	75
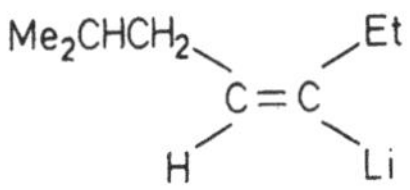 Me₂CHCH₂C(H)=C(Et)Li	D	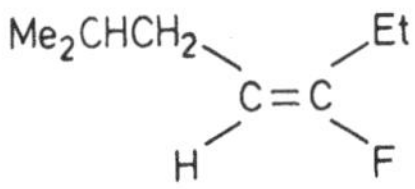 Me₂CHCH₂C(H)=C(Et)F	88

Table 1. (*Continued*)

Compound	Reagent	Product	Yield (%)
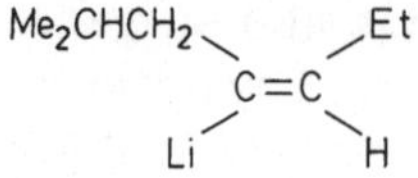	D	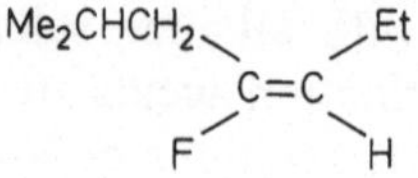	83
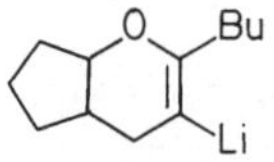	D	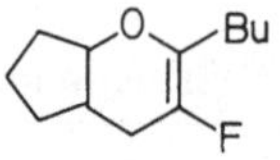	74
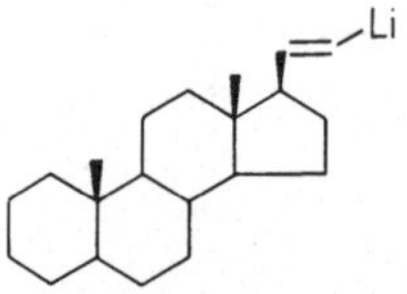	D	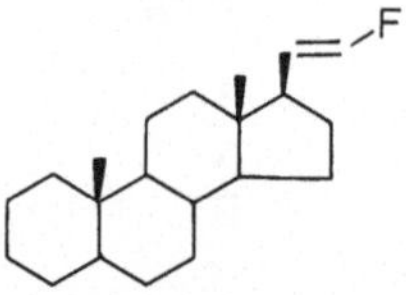	80

A: *N*-Fluoro-*N*-neopentyl-*p*-toluenesulfonamide; B: *N*-Fluoro-*N*-*tert*-butyl-*p*-toluene-sulfonamide; C; *N*-Fluoro-*N*-*exo*-2-norbornyl-*p*-toluenesulfonamide; D: *N*-Fluoro-*N*-*tert*-butylbenzenesulfonamide

Table 2. Fluorination of aromatic compounds with $(CF_3SO_2)_2NF$ (CDCl$_3$, 22 °C) [38]

Compound[a]	Time (h)	Products (%)
Nitrobenzene[b]	12	no reaction
Acetophenone	12	no reaction
Chlorobenzene	24	no reaction
Benzene[b]	18	fluorobenzene, 50
Toluene[b]	10	2-fluorotoluene, 74; 3-fluorotolune, 4; 4-fluorotoluene, 22
Anisole[b]	2	2-fluoroanisole, 69; 4-fluoroanisole, 24; polyfluoroanisole, 7
Phenol[c]	12	2-fluorophenol, 60; 4-fluorophenol, 40
m-Cresol[c]	12	2-fluoro-5-methylphenol, 44; 4-fluoro-3-methylphenol, 56
p-Cresol[c]	12	2-fluoro-4-methylphenol, 80; other, 20
m-Xylene[c]	12	2-fluoro-5-methyltoluene, 2-fluoro-3-methyltoluene (1:2 mixture)
Naphthalene[c]	12	1-fluoronapthalene, 80; 2-fluoronapthalene, 7; other, 13

[a] ArH: $(CF_3SO_2)_2NF = (\geqslant 2)$: 1, mol
[b] Without solvent
[c] An NMR yield (a percent of fluorinated products)

CF_3OF [35–37].

$$RSO_2NHR' \xrightarrow{F_2-N_2} RSO_2NFR'$$

$R' = R = $ alkyl (C_{1-30}), cyclo-alkyl(C_{3-30}), aryl, aralkyl
$R = $ 4-tolyl, $R' = $ Me, t-Bu, cyclo-C_6H_{11}, t-BuCH$_2$, 2-norbornyl
$R = $ Bu, $R' = t$-BuCH$_2$

Table 1 represents some examples of fluorination by these compounds.

It is worthwhile paying attention to the high yields (71–88%) of the fluorodemetallation products of alkenyllithium. This reaction shows the stereo- and regiospecificity of fluorination and the absence of polymers, due to which it may be recommended as a preparative method for the synthesis of fluoroolefins.

$$R_1R_2C=CLiR_3 + PhSO_2NF(t\text{-Bu}) \rightarrow R_1R_2C=CFR_3$$

The enhanced electron-accepting ability of substituents at the nitrogen atom in nitrogen-fluorine containing compounds raises the reactivity of the latter. These fluorides are obtained by the reaction of 1% F_2 in N_2 with bis(perfluoroalkylsulfonyl)imides in $CFCl_3$ ($-78\,°C$) with 60–96% yields [38].

Among N-fluoro-bis(perfluoroalkylsulfonyl)imides, the reactions of $(CF_3SO_2)_2NF$ are the most extensively studied ones. This fluoride reacts with benzene, toluene, xylene, and other aromatic compounds with electron-donating substituents, but is inert with chlorobenzene and nitrobenzene (Table 2). The orientation in this fluorination is formally that of an electrophilic attack, though the reaction gives an unusually large amount of *ortho*-isomer. There are no polyfluorinated compounds among the products, which seems to result from the low nucleophilicity of monofluorinated arenes as compared with their precursors. The selectivity of fluorination observed here is especially important, because the partially fluorinated benzenes are used for the synthesis of biologically active compounds, though their use is restricted by the absence of simple and convenient methods of synthesis [39].

Literature contains no data on the reaction of perfluorinated N-fluoro-sulfonylamides with unsaturated compounds.

2.3 Alkaline Metal Fluorooxysulfates

Since Appelman reported, in 1979, the first synthesis of caesium fluorooxysulfate by the reaction of fluorine with caesium sulfate in water [40], there has been constant growth of interest in this reagent. As a rule, caesium and, more rarely, rubidium salts are used. These salts are potential oxidants and fluorinating agents. The standard electrode potential of $SO_4F^- - HSO_4^-$ is 2.5 V [41]. Detailed structural analysis of these salts led to a conclusion that caesium fluorooxysulfate is a hypofluorite of ionic structure. It is already clear that such salts are synthetically and analytically convenient in organic and inorganic chemistry. Caesium fluorooxysulfate is a unique example of an anionic electrophile having two potential electrophilic centres.

Let us consider the main results of fluorinations of aromatic compounds with $CsSO_4F$. A common condition for such reactions is the presence of a polar solvent (acetonitrile) and a catalyst—strong protic acid or BF_3. But even in these conditions satisfactory results have been obtained only for the aromatic compounds with electron-donating substituents. Thus benzotrifluoride reacts slowly with $CsSO_4F$, forming 3-fluorobenzotrifluoride with only a 2–3% yield [42]. Nitrobenzene is transformed to a mixture of 2-, 3-, and 4-fluoronitrobenzenes [43]. At the same time, benzene gives fluorobenzene in a good yield [44]. Toluene and acetanilide are easily converted to the respective monofluorinated derivatives [42].

$$C_6H_5R + CsSO_4F \xrightarrow[MeCN]{BF_3} 2\text{-}FC_6H_4R + 3\text{-}FC_6H_4R + 4\text{-}FC_6H_4R$$

R=H		30–35	
NHAc	75	—	11
CH_3	31	4	8
NO_2	6	16	3

Upon treatment of phenol and alkoxybenzenes with $CsSO_4F$ in the presence of BF_3 (MeCN, 20 °C), monofluoroderivatives are formed with a 70–80% yield. An interesting feature of these reactions is the predominant *ortho*-fluorination, but increased amount of a substituent leads to the increased yield of *para*-isomer [45].

$$C_6H_5OR + CsSO_4F \xrightarrow[MeCN]{BF_3} 2\text{-}FC_6H_4OR + 4\text{-}FC_6H_4OR$$

R=H	6.2	:	1
CH_3	2.8	:	1
n-Bu	1.8	:	1
i-Bu	1.2	:	1

A similar picture is observed in the naphthalene series. 1-Naphthol and 1-alkoxynaphthalenes are chiefly fluorinated at the 2- and 4-positions, the dependence of isomer ratio on the size of group R being similar to that for alkoxybenzenes [42]. Yields of alkoxyfluoronaphthalenes are about 50%.

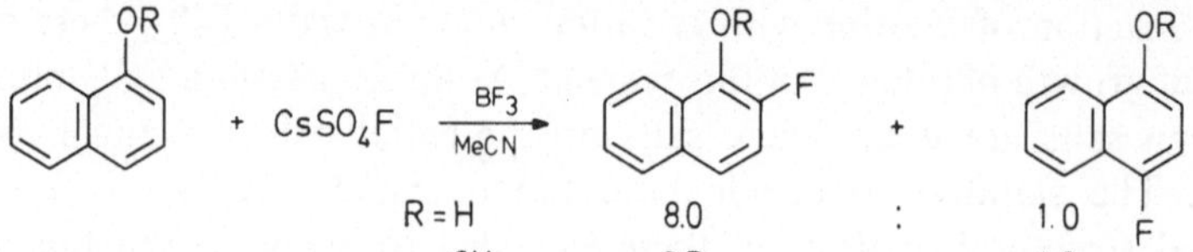

R = H	8.0	:	1.0
CH_3	3.5	:	1.0
C_2H_5	3.0	:	1.0
$i\text{-}C_3H_7$	1.8	:	1.0

2-Naphthol and 2-alkoxynaphthalenes react with $CsSO_4F$ and BF_3 giving a mixture of 1-fluoro-2-alkoxynaphthalene and the product of its further reaction with $CsSO_4F$–1,1-difluoro-2-oxo-1,2-dihydronaphthalene with a total yield of 60–80% [45]. Steric hindrance induced by the branched substituent R decreases the quantity of the 4-fluoro-substituted derivative. Increased amount of the fluorinating agent raises the proportion of the fluorination product which now contains the difluoroderivatives.

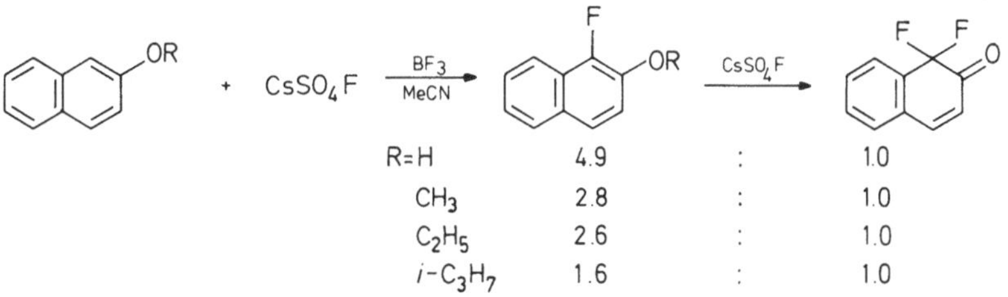

R = H	4.9	:	1.0
CH$_3$	2.8	:	1.0
C$_2$H$_5$	2.6	:	1.0
i-C$_3$H$_7$	1.6	:	1.0

Along with the main products, the reaction mixture contains the products of deeper fluorination. Thus 1-RONf produces 1-alkoxy-2,4-difluoronaphthalene and 2,2-difluoro-1-oxo-1,2-dihydronaphthalene. These products are formed as a consequence of the increased amount of fluorinating agent and elevated temperature.

Caesium fluorooxysulfate reacts with polycyclic aromatic compounds. As a rule, the yield of monofluorination products is 40–70%, therefore these reactions represent a rather convenient method of synthesis of such substances. They are as convenient for synthesis as the reactions of XeF_2 [44].

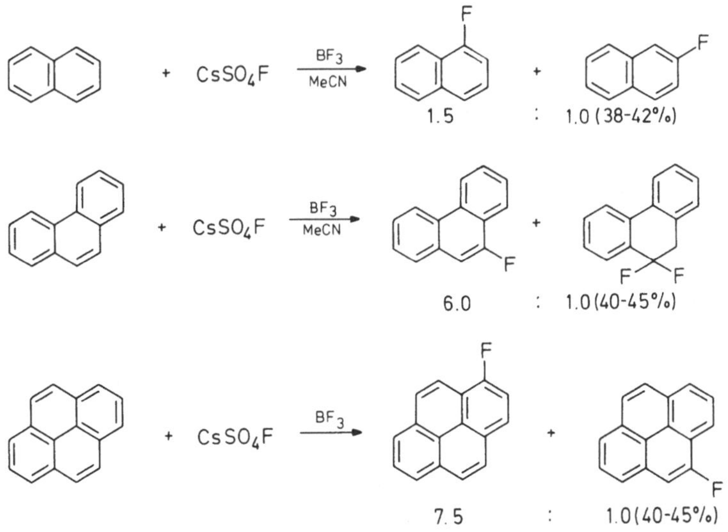

The relative reactivity of the benzene derivatives in the reaction with caesium fluoroxysulfate is determined by the substituted electronic effect. Table 3 shows

Table 3. Relative reactivities of C_6H_5R toward
SO_4F^- (MeCN, 25 °C) [41]

R	Reactivity
NO$_2$	0.02
CN	0.07
COOMe	0.17
F	0.55
H	1.00
Ph	11
Me	13–29
OMe	190
OH	440

the electron-donating substituents to increase the fluorination rate, and the electron-accepting ones—to decrease it [48]. The ρ^+ value found from the data of Table 3 equals -3.5. This indicates that this reagent is more selective than the elementary fluorine for which the ρ^+ value is -2.45.

The effect of catalysts: HF, H_2SO_4, CF_3SO_3H, HSO_3F, BF_3, and SbF_5–HSO_3F in the fluorinations of toluene, nitrobenzene, and naphthalene by fluorooxysulfates has been studied in detail [43]. The acid catalysts accelerate the reaction, and the efficiency of the catalyst parallels the H_0 function of the respective acid. In particular, in the fluorination of naphthalene with $CsSO_4F$, H_2SO_4 was found to be a more efficient catalyst than HSO_3F. It is worth giving here some considerations on the mechanism of fluorination of aromatic compounds by caesium fluorooxysulfate. It should be noted that no detailed mechanistic studies have been carried out, therefore these considerations are rather speculative.

The key intermediate is supposed to be complex A. Elimination of SO_4^{2-} produces the benzenonium cation, whereas elimination of SO_4^- and F^- leads to radical cation C, implying the occurrence of the one-electron oxidation of the aromatic compound. The benzenonium ion B gives the product of fluorination of the aromatic substrate, whereas radical cation C undergoes further transformation, forming a lot of by-products. The latter prevail in the case of non-activated or deactivated aromatic substrates.

Protic and Lewis acids promote the transformation of intermediate A to ion B, thus raising the yield of the products of fluorination in the benzene ring. The leaving group here is HSO_4^-, and in the case of the BF_3 catalyst—$BF_3OSO_3^-$.

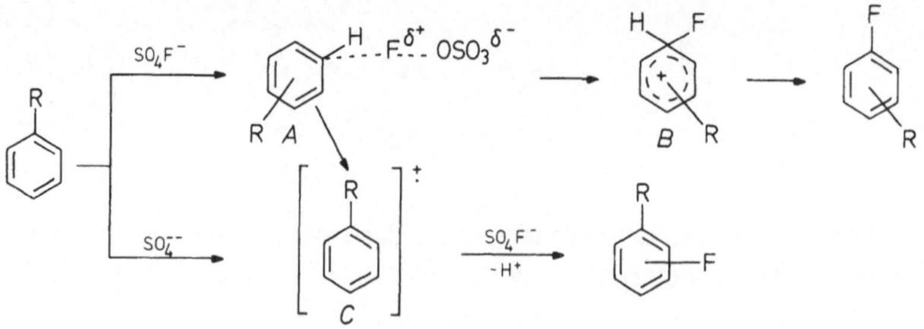

To explain formation of the hydroxy-derivatives as by-products in these reactions, a mechanism has been suggested involving a rearrangement in the intermediate A induced by the catalyst.

$$ArH + SO_4F^- + BF_3 \longrightarrow ArH + OSO_3BF_4^- \longrightarrow ArOH + BF_3 + FSO_3^-$$

The phenol products undergo fluorination or condensation leading to the dimer and polymers. An alternative mechanism of the catalysis, protonation of the SO_4F^- ion, should not be excluded from consideration.

Appelman and his co-workers consider that fluoroaromatic compounds are chiefly formed via intermediate B, whereas the polyfluorinated by-products, and benzyl fluoride (from toluene) are produced by the reactions involving radical cation C. However this does not account for the reason of pronounced *ortho*-fluorination of anisole, toluene, and biphenyl. Even the reaction with fluorobenzene gives a great amount of 1,2-difluorobenzene, though in the electrophilic reactions (halogenation, sulfonation, nitration, etc.) substitution of hydrogen in C_6H_5F proceeds for more than 80% at the *para*-position. Fluorination of benzene derivatives with *meta*-orienting substituents gives about equal or slightly different (NO_2) yields of *ortho*- and *meta*-isomers [41]. In our opinion, this indicates a more complex mechanism of fluorination of aromatic compounds than this is considered by the authors of [41,43]. Thus the isomer ratio here may be affected by the substituent 1,2-shift in the arenonium cation.

Fluorination of alkenylaromatic and related compounds by caesium fluoro-oxysulfate has been studied. The reactivity of the $C=C$ bond was found to exceed that of the aromatic ring, and $CsSO_4F$ always reacts with the alkenyl fragment, irrespective of its position in a molecule. In the case of 1,1-diphenylethylene, the difluoro-derivative is formed when the reaction is carried out in methylene chloride (yield 74%). In other media, especially in the nucleophilic ones, chiefly the monofluoro-derivatives are formed [46].

$$Ph_2C=CH_2 + CsSO_4F \text{ —}$$

$\xrightarrow{CH_2Cl_2, 35\ °C} Ph_2C=CF_2$
74%

$\xrightarrow{HOAc} Ph_2C=CFOAc$
48%

$\xrightarrow{MeOH} Ph_2C=CFOMe$
36%

$\xrightarrow{HF} Ph_2CH–CHF_2 + Ph_2CH–CH_2F$
80%   20%

This reaction is illustrative of the effect of the nucleophilic solvents involved in it.

It is interesting to note that in α-methylstyrene, the allyl hydrogens are formally substituted, but as a matter of fact the reaction obviously proceeds with isomerisation of the intermediate carbocation.

$$PhMeC=CH_2 + CsSO_4F \rightarrow PhC(CH_2F)=CH_2 + PhC(CHF_2)=CH_2$$
$$ 30\% 32\%$$

In [47], the stereochemical results have been analysed in the "electrophilic" fluorination of acenaphthene, stilbene, indene, and 1-phenylindene by various fluorinating agents, including $CsSO_4F$.

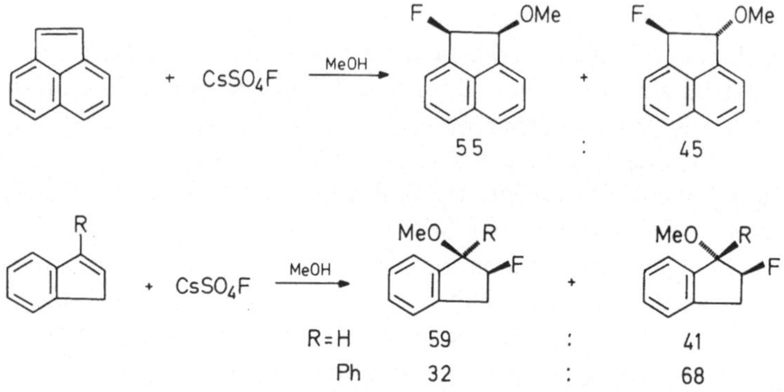

The information on the reactivity of a reagent is usually obtained by carrying out comparative studies on the reactions of such reagents with a model compound. In respect of the fluorinating agents, it would be interesting to compare the reactivity of $CsSO_4F$ with F_2, CF_3OF, CF_3COOF, and XeF_2. As a model, the authors of [42] chose 1,2-diphenylacetylene which had earlier been fluorinated by all these agents.

At a low temperature, the reaction with fluorine in methanol [49] leads to three products: 1,1,2,2-tetrafluoro-1,2-diphenylethane, 1,1,2-trifluoro-2-methoxy-1,2-diphenylethane (main product), and 1,1-difluoro-2,2-dimethoxy-1,2-diphenylethane, whereas CF_3OF at these conditions gives 1,2,2-trifluoro-1-trifluoromethoxy-1,2-diphenylethane as a main product [50]. A similar reaction with CF_3COOF leads to 2-fluoro-1,2-diphenylethanone and dibenzoyl [51], and at room temperature fluorination with XeF_2 (depending on the catalyst used) leads to 1,1,2,2-tetrafluoro-1,2-diphenylethane (in the presence of HF [52]) or to six products (in the presence of trifluoroacetic acid [53]).

The authors of [48] showed that the reaction of 1,2-diphenylacetylene with $CsSO_4F$ in methanol yields two products: 1,1-difluoro-2,2-dimethoxy-1,2-diphenylethane (11) and 2,2-difluoro-1,2-diphenylethanone (12). In this case compound 12 is formed as a result of sequential transformations of compound 11 in the reaction conditions. This distinguishes $CsSO_4F$ from the above-mentioned fluorinating agents. The authors of [48] studied the regioselectivity of fluorination of other acetylene derivatives by this reagent and found that addition of the methoxy group follows the Markownikov rule for 1-phenylace-tylene and 1-phenyl-2-tert-butylacetylene, whereas 1-phenyl-1-pentyne leads to

Table 4. Fluorination of phenylacetylenes $PhC \equiv CR$ with $CsSO_4F$ (MeOH, 22 °C) [48]

R	Relative yield (%)							
	11	*12*	*13*	*14*	*15*	*16*	*17*	*11+12*
Ph	40	60						100
H	50	50						100
t-Bu	54	46						100
Pr	31	25		25	5	10	4	56
Pr[a]	57	35		8	1	1	1	92
Me	20	19	3	26	6	17	9	39
Me[a]	27	28	2	14	4	16	9	55
Me[b]	41	41	1	12	1	1	5	82

[a] Acetylene: nitrobenzene = 10:1, mmol
[b] Acetylene: nitrobenzene = 1:1, mmol

1-phenyl-1,1-difluoropentan-2-one (*14*), 1-phenyl-2-fluoropentan-1-one (*15*), 1-phenyl-1-fluoropentan-2-one (*16*), and 1-phenylpentan-1,2-dione (*17*). Similar products are formed from 1-phenyl-1-propyne (together with a small amount of 1-phenyl-1,1-difluoro-2,2-dimethoxypropane (*13*)) (Table 4). The regioselectivity may be explained by the generation of intermediate carbocations, whereas formation of by-products may be attributed to the formation of by-products is confirmed by the experiments of the authors of [48]—they added nitrobenzene and 2,4,6-tri-*tert*-butylphenol as the free-radical traps (Table 4).

$$PhC \equiv CR + CsSO_4F \xrightarrow[22\,°C]{MeOH} PhC(OMe)_2CF_2R + PhC(O)CF_2R$$

	11	*12*
R = H	25%	26%
Ph	23%	32%
t-Bu	23%	19%

Enol acetates are readily transformed by $CsSO_4F$ to α-fluoroketones [52].

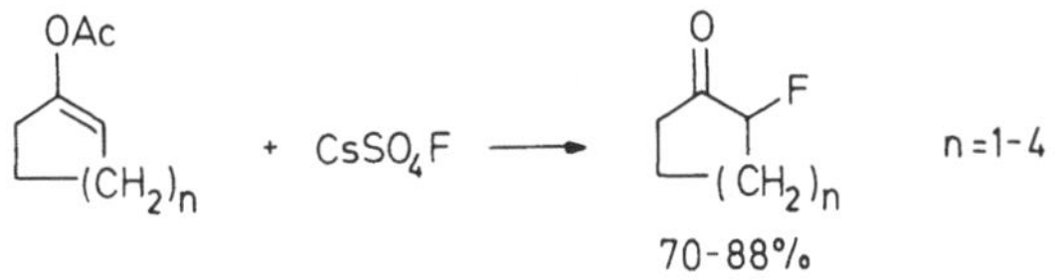

There are only few examples of such reactions, among which is successful synthesis of 2α-fluorinated vitamin D_3 [54].

$CsSO_4F$ was used for the synthesis of some fluoro-derivatives of uracyl and uridine [55]. Thus 1,3-dimethyluracyl is transformed to *cis, trans*-5-fluoro-6-methoxy-derivatives to produce 5-fluoro-1,3-dimethyluracyl.

Table 5. Fluorination of aryltrialkylstannanes
$4\text{-}XC_6H_4SnR_3$ with $CsSO_4F$ or fluorine [56]

X	R	Yield (%)	
		$CsSO_4F$	Fluorine
H	Me	69	30
Me	Me	86	57
OMe	Me	79	60
Cl	Me	87	67
OMe	Bu	43	—
Me	Bu	11	—
H	C_6H_{13}	0	47
OMe	C_6H_{13}	0	—
H	Bu	—	41

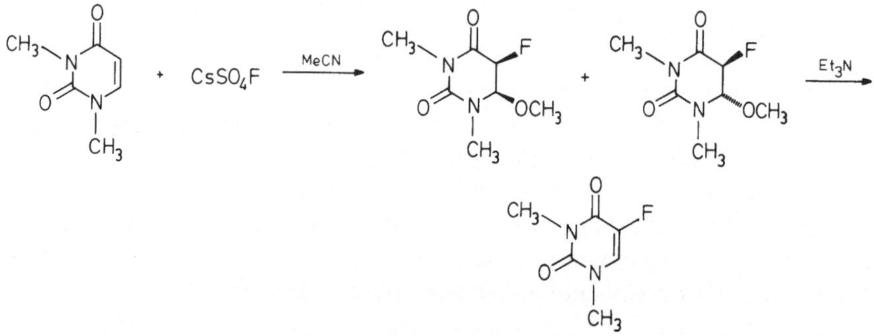

It is necessary to mention the regioselective fluorodestannylation of aryl-trialkylstannanes to give aryl fluorides in good yields, carried out in [56].

$$4\text{-}XC_6H_4SnR_3 + CsSO_4F \rightarrow 4\text{-}XC_6H_4F$$

The yields of aryl fluorides here are significally higher than in the reactions with fluorine (Table 5) [56]. With increased size of group at tin, the yield of fluorobenzene decreases, whereas in the reaction with fluorine the case is just the opposite.

The possibilities of caesium fluorooxysulfate are not exhausted in this treatment, and we hope that subsequent investigations will open up its wide potentialities as a fluorinating agent. This agent may be a substrate for the synthesis of monofluoro-derivatives of the aromatic series.

2.4 Acyl Hypofluorites

Fluorination of organic compounds by caesium fluorooxysulfate has analogues in the reactions of acetyl and trifluoroacetyl hypofluorites. The behaviour of organic hypofluorites as fluorinating agents is considered in another chapter, but it would be reasonable to discuss here the results of fluorinations by CH_3COOF and CF_3COOF to compare them with those for $CsSO_4F$. At the same time, these reagents are interesting in their own right. Thus during the last 5 years, acetyl hypofluorite has shown itself to be useful and has been employed for various purposes [57–60]. The need for the samples with radioisotope ^{18}F extended the field of applications of this fluorinating agent [61–66].

Acetyl hypofluorite is prepared by passing elemental fluorine diluted by nitrogen to 5–10%, through a mixture of sodium acetate in anhydrous acetic acid or $CFCl_3$ at $-75\,°C$ [67–69]. It is also synthesized from ammonium [65] or potassium [70] acetates in HOAc at room temperature or by passing fluorine over the solid salt $KOAc\cdot2HOAc$ [71,72]. But fluorination of potassium alkanoates in $CFCl_3$ (or $CF_2ClCFCl_2$, perfluoro-2-butyltetrahydrofuran) leads to a mixture of fluorooxy-compounds. Thus, the potassium perfluorooctanoate yields $CF_3(CF_2)_7OF$, $CF_3(CF_2)_6CF(OF)_2$, and $CF_3(CF_2)_6COOF$. From sodium trifluoroacetate, a mixture of CF_3COOF and CF_3CF_2OF may be obtained [73]. The best results in this reaction (exclusive formation of CF_3COOF) are achieved in HF or H_2O.

In the liquid state, AcOF is an explosive and toxic product, but at $-80\,°C$ and in quantities of less than 0.2 mmol (~15 mg) it is quite safe. Physical and spectral data for AcOF are given in [70,74]. If a substrate used for fluorination with AcOF is insoluble in $CHCl_3$, it is possible to use HOAc, $MeNO_2$, and MeCN as solvents [76].

It should be noted that many "electrophilic" fluorination reactions proceed with the formation of by-products via radicals. Therefore the mechanistic conclusions should be made carefully. For example, the reaction of cyclohexane with AcOF in acetic acid gives a mixture of products containing both the products of radical reactions and those of carbocationic reactions. The ratio of these products depends on the reaction conditions [77].

In 1981, Rozen and his co-workers [68] reported fluorination of aromatic compounds with electron-donating substituents by acetyl hypofluorite. Good yields are produced in the fluorination of aryl alkyl ethers, substituted phenols, and acetanilide derivatives, with fluorine substituted for the hydrogen atoms in the *ortho*- and, to a less extent, *para*-position to the electron-donating substituent. Such an unusual isomer ratio is explained by the authors by the addition of AcOF at the C_{ipso}–C_{ortho} bond with subsequent elimination of HOAc.

Likewise, 1-methoxynaphthalene gives a mixture of 1-methoxy-2-fluoro- and 1-methoxy-4-fluoronaphthalenes in the ratio of 6:1 (yield 70%).

The hypothesis about the addition-elimination has been experimentally supported. Thus piperonal (*18*) gives compound *19* which was isolated and its structure was determined by spectral methods.

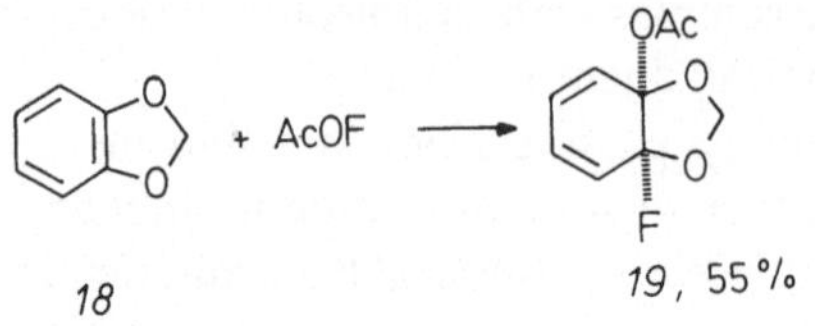

Benzene, chlorobenzene, toluene, and aniline produce low yields of fluoro-derivatives [78,79] (Table 6).

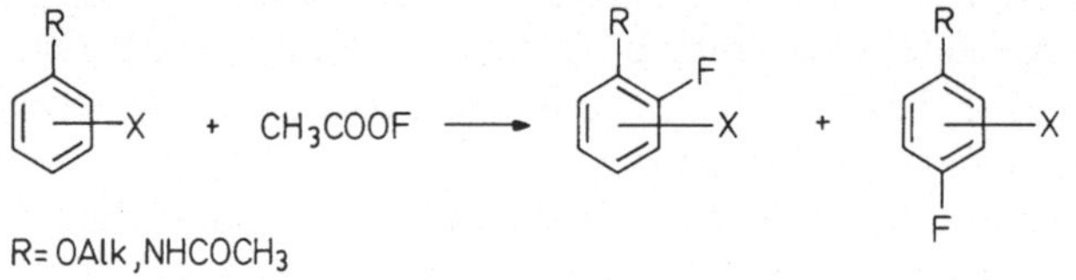

R= OAlk, NHCOCH₃

The data presented in Table 6 allow one to make two interesting conclusions. First, successful fluorination of an aromatic compound requires the presence of the OAlk or NHCOCH$_3$ substituent. Second, the electronic nature and orienting effect of another substituent in the aromatic ring are unimportant. Thus toluene is transformed to isomeric fluorotoluenes in a total yield of 14%, but 3-trifluoromethylacetanilide gives the products of fluorination at positions 2 and 6, in a 62% yield.

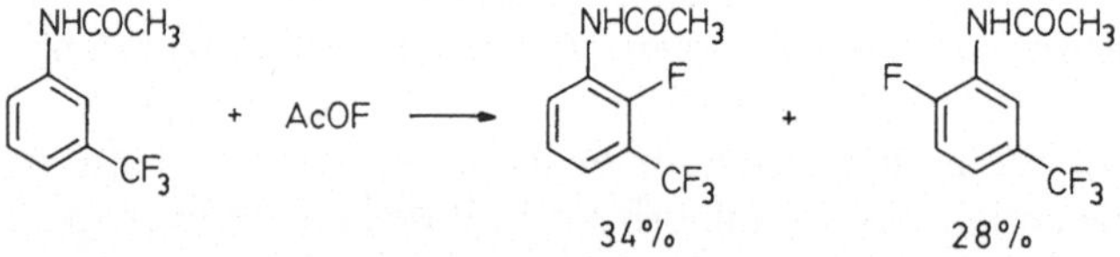

Table 6. Reactions of aromatic compounds with acetyl hypofluorite

Compound	Product	Yield (%)	Ref.
PhOCH$_3$	2-FC$_6$H$_4$OCH$_3$	77	59
	4-FC$_6$H$_4$OCH$_3$	8	
3-CH$_3$OC$_6$H$_4$OCH$_3$	3-CH$_3$O-4-FC$_6$H$_3$OCH$_3$	39	59
	3-CH$_3$O-4,6-F$_2$C$_6$H$_2$OCH$_3$	55	
PhOEt	2-FC$_6$H$_4$OEt	46	59
	4-FC$_6$H$_4$OEt	6	
2-CH$_3$OC$_6$H$_4$NO$_2$	2-CH$_3$O-3,5-F$_2$C$_6$H$_2$NO$_2$	42	59
4-CH$_3$OC$_6$H$_4$NO$_2$	4-CH$_3$O-3-FC$_6$H$_3$NO$_2$	47	59
4-HOC$_6$H$_4$NO$_2$	4-HO-3-FC$_6$H$_3$NO$_2$	62	59

Table 6. (*Continued*)

Compound	Product	Yield (%)	Ref.
2-HOC$_6$H$_4$COOMe	2-HO-3-FC$_6$H$_3$COOMe	9	59
	2-HO-4-FC$_6$H$_3$COOMe	14	
PhNHCOCH$_3$	2-F-C$_6$H$_4$NHCOCH$_3$	55	59
	4-FC$_6$H$_4$NHCOCH$_3$	8	
PhNHCOCF$_3$	2-FC$_6$H$_4$NHCOCF$_3$	57	59
PhNHCOBu-*t*	2-FC$_6$H$_4$NHCOBu-*t*	52	59
2-CF$_3$C$_6$H$_4$NHCOCH$_3$	2-CF$_3$-6-FC$_6$H$_3$NHCOCH$_3$	62	59
3-CH$_3$C$_6$H$_4$NHCOCH$_3$	3-CH$_3$-6-FC$_6$H$_3$NHCOCH$_3$	32	59
	3-CH$_3$-2-FC$_6$H$_3$NHCOCH$_3$	28	
	3-CH$_3$4,6-F$_2$C$_6$H$_2$NHCOCH$_3$	11	
	3-CH$_3$-2,4-F$_2$C$_6$H$_2$NHCOCH$_3$	10	
3-BrC$_6$H$_4$NHCOCH$_3$	2-F-3-BrC$_6$H$_3$NHCOCH$_3$	25	59
	6-F-3-BrC$_6$H$_3$NHCOCH$_3$	47	
3-CF$_3$C$_6$H$_4$NHCOCH$_3$	2-F-3-CF$_3$C$_6$H$_3$NHCOCH$_3$	34	59
	6-F-3-CF$_3$C$_6$H$_3$NHCOCH$_3$	28	
3,5-Me$_2$C$_6$H$_3$NHCOCH$_3$	2-F-3,5-Me$_2$C$_6$H$_2$NHCOCH$_3$	67	59
4-CH$_3$C$_6$H$_4$NHCOCH$_3$	2-F-4-CH$_3$C$_6$H$_3$NHCOCH$_3$	85	59
4-CF$_3$C$_6$H$_4$NHCOCH$_3$	2-F-4-CF$_3$C$_6$H$_3$NHCOCH$_3$	72	59
4-BrC$_6$H$_4$NHCOCH$_3$	2-F-4-BrC$_6$H$_3$NHCOCH$_3$	65	59
3,5-(MeO)$_2$C$_6$H$_3$SnBu$_3$	3,5-(MeO)$_2$C$_6$H$_3$F	68	59
C$_6$H$_5$CH$_3$	2-FC$_6$H$_4$CH$_3$	8	78, 79
	3-FC$_6$H$_4$CH$_3$	1	
	4-FC$_6$H$_4$CH$_3$	4	
	C$_6$H$_5$CH$_2$F	1	
C$_6$H$_5$OH	2-FC$_6$H$_4$OH	45	78, 79
	4-FC$_6$H$_4$OH	30	
C$_6$H$_5$OMe	2-FC$_6$H$_4$OMe	64	78, 79
	4-FC$_6$H$_4$OMe	21	
C$_6$H$_5$NHCOCH$_3$	2-FC$_6$H$_4$NHCOCH$_3$	44	78, 79
	4-FC$_6$H$_4$NHCOCH$_3$	22	
C$_6$H$_6$	C$_6$H$_5$F	18	78, 79
C$_6$H$_5$NH$_2$	2-FC$_6$H$_4$NH$_2$	3.5	78
	4-FC$_6$H$_4$NH$_2$	2.5	
C$_6$H$_5$Cl	2-FC$_6$H$_4$Cl	5	78
	4-FC$_6$H$_4$Cl	5	
C$_6$H$_5$NHCOCH$_3$	2-FC$_6$H$_4$NHCOCH$_3$ $\Big\}$ 7:1	55	68
	4-FC$_6$H$_4$NHCOCH$_3$		
C$_6$H$_5$OMe	2-FC$_6$H$_4$OMe $\Big\}$ 9:1	85	68
	4-FC$_6$H$_4$OMe		
3-CH$_3$OC$_6$H$_4$OCH$_3$	2,4-(OMe)$_2$C$_6$H$_3$F $\Big\}$ 1:2	99	68
	2,4-(OMe)$_2$-1,5-F$_2$C$_6$H$_2$		
4-CH$_3$OC$_6$H$_4$NO$_2$	4-CH$_3$O-3-FC$_6$H$_3$NO$_2$	43	68

This apparently follows from the reaction mechanism which seems to be more complex than suggested by the authors of [59,78].

The field of applications of AcOF as a fluorinating agent for aromatic compounds is rather wide. Though the yields of monofluoroderivatives are not very high, AcOF which is similar in its fluorinating ability to elemental fluorine, seems to exceed it in respect of the mild conditions and selectivity of the fluorinating process.

Apart from the reactions with benzene derivatives, fluorination of methoxy-naphthalenes by acetyl hypofluorite has been studied [59].

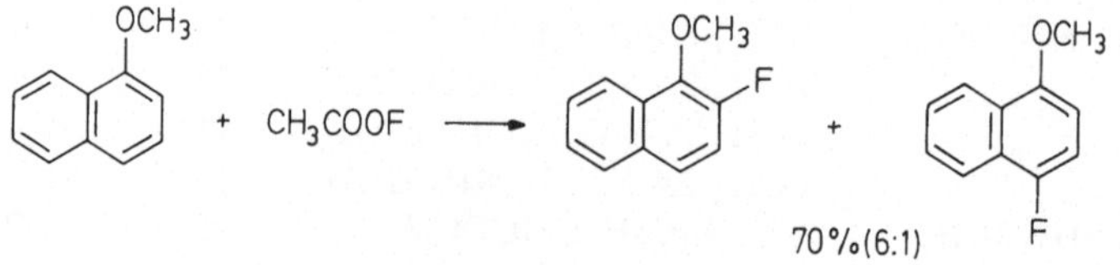

2-Methoxynaphthalene gives 65% of 1-fluoro-2-methoxynaphthalene, and 6-methoxyquinoline was converted in a 75% yield to 5-fluoro-6-methoxy-quinoline; 6-methoxy-1-tetralone affords 5-fluoro-6-methoxy-1-tetralone (53%) and 5,7-difluoro-6-methoxy-1-tetralone (13%) [59].

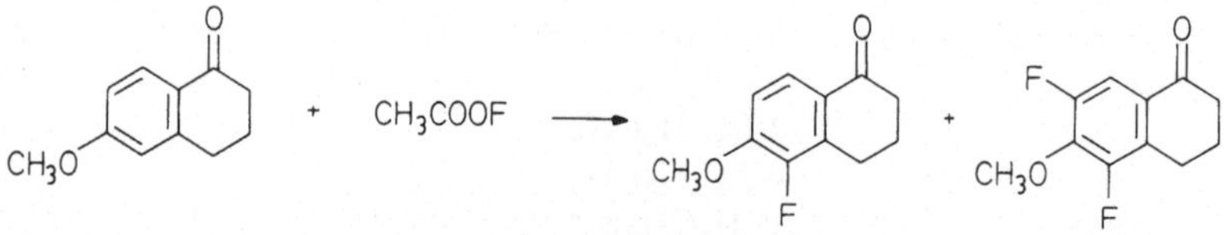

Meso-hexaestrol is an important hormone. Its fluorination with AcOF in $CHCl_3$ at $-75\,°C$ affords 40% of the monofluoroderivative and a small amount of the difluoro-derivative.

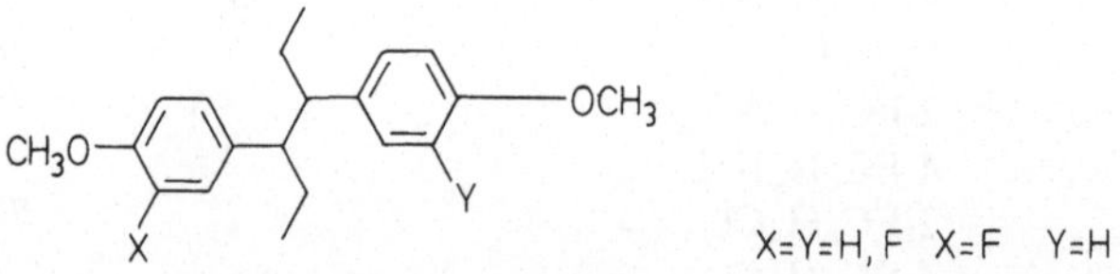

X=Y=H, F   X=F   Y=H

The interaction of acetyl hypofluorite with unsaturated compounds proceeds in mild conditions (-70 to $-80\,°C$) in chloroform or the HOAc–CFCl$_3$ mixture and leads to the fluoroacetoxylation products. Thus cyclohexene reacts with AcOF to form 1-acetoxy-2-fluorocyclohexane (yield 60%), and 1-dodecadecene is transformed to 2-acetoxy-1-fluorododecane (yield 30%) [50,58].

The stereochemistry of addition of acetyl and trifluoroacetyl hypofluorite to olefins has been studied in the case of the reactions with aryl and diarylolefins. The reaction of AcOF with *cis*- and *trans*-stilbenes proceeds stereoselectively, whereas trifluoroacetyl hypofluorite gives the product of stereospecific fluoro-acylation of these olefins (with $R = CF_3$ the reaction gives after hydrolysis the respective hydroxy-compounds).

cis-PhCH=CHPh + RCOOF → PhCH(OR)–CHFPh

	erythro : threo (%)	
R = CH₃	11	51
CF₃	58	—

trans-PhCH=CHPh + RCOOF → PhCH(OR)–CHFPh

	erythro : threo (%)	
R = CH₃	7	50
CF₃	—	62

According to the data reported in [73], the presence of electron-accepting groups COMe, COOMe, and Cl in the *para*-position of one of the aromatic rings does not affect the stereoselectivity of the addition of CF_3COOF at a multiple bond. In the case of methylcarbonyl- and methylcarboxyl-substituted stilbenes the addition proceeds regioselectively, whereas the chlorine-containing stilbene forms an equimolar mixture of regioisomers. 2-Carboxymethyl-*trans*-stilbene adds CF_3COOF non-specifically, though the *threo*-conformer is predominant.

trans-RC₆H₄–CH=CHPh + CF₃COOF → RC₆H₄CHF–CH(OH)Ph

R = 2-COOMe	threo-40% erythro-14%
4-COOMe	threo-80%
4-COMe	threo-28%
4-Cl*	threo-32%
4-Me*	threo-8%

* + threo-4-RC₆H₄CH(OH)CHFPh (R = Cl, 32%; R = Me, 48%)

It is interesting that treatment of 4-methoxystilbenes both with acetyl and trifluoroacetyl hypofluorites leads to the fluoroacetoxylation products which also contain the fluorine atom in the 3-position.

trans-4-MeOC₆H₄CH=CHPh $\xrightarrow{\text{CH}_3\text{COOF or CF}_3\text{COOF}}$

4-MeOC₆H₄CH(OAc)CHFPh + 3-F-4-MeOC₆H₄CH(OR)CHFPh

erythro-15%	erythro-14% (R = Ac)
threo-42%	threo-5%
	erythro-57% (R = H, after hydrolysis
	threo-14% of trifluoroacetate)

Reaction of 1,2-difluoro-1,2-diphenylethylene with AcOF gives the product of fluoroacetoxylation [64].

$$PhCF=CFPh + CH_3COOF \rightarrow PhCF_2-CFPh$$
$$\overset{|}{O}COCH_3$$

The α,β-unsaturated carbonyl compounds easily undergo fluoroacetoxylation by AcOF, forming α-fluoro-β-acetoxy carbonyl compounds [58]. Elimination of HOAc leads to the fluorine-containing unsaturated ketones [80].

$$trans\text{-}PhCH=CHCOR + CH_3COOF \rightarrow threo\text{-}PhCH(OAc)CHFCOR$$

R=OEt	57%
Ph	70%

Acetyl hypofluorite is effectively used for the synthesis of α-fluorocarbonyl compounds from enol acetates. As opposed to many reactions considered above, these processes are characterised by high yields and regiospecificity of fluorination. The presence of an aromatic ring in the molecule of the starting compounds produces no effect on the reaction results (Table 7).

$$RCH=CR' + CH_3COOF \rightarrow RCHFCR'$$
$$\overset{|}{O}Ac \qquad\qquad\qquad \overset{\|}{O}$$
$$40\text{–}80\%$$

Similar reactions yielded some acetoxyfluorosteroids [80].

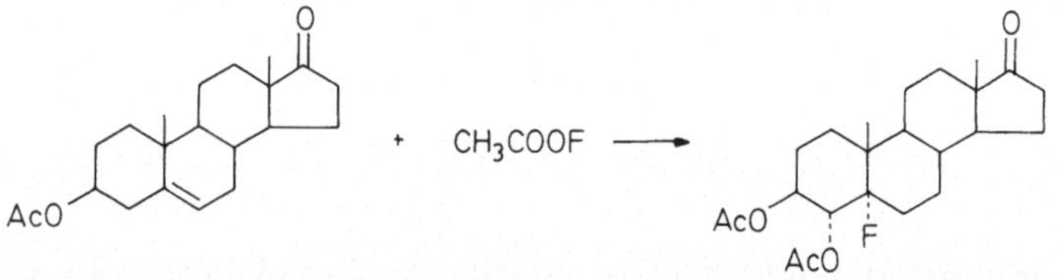

Δ^4-3-Ketosteroids are fluorinated by acetyl hypofluorite at the 6-position [84].

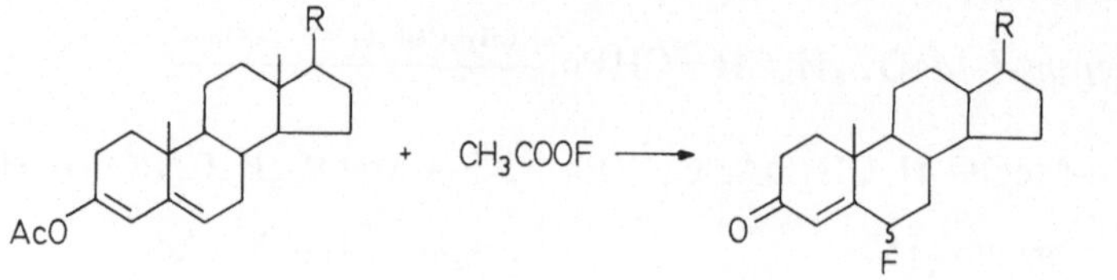

R	Yield(%)	α/β
COCH$_3$	48	2/5
OCOCH$_3$	50	3/5

Table 7. Reactions of enol acetates with acetyl hypofluorite

Compound	Product	Yield (%)	Ref.
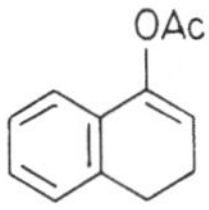 OAc	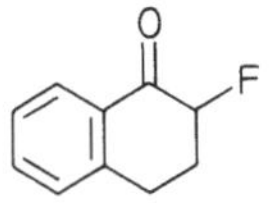	85	81
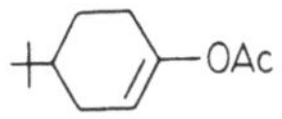 —OAc	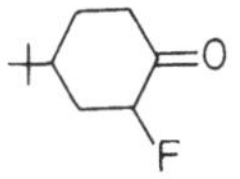	43-*trans* 29-*cis*	82,83
2-NfC=CH₂ with OAc (2-NfC=CH$_2$, ȮAc)	2-NfCCH$_2$F (2-NfCCH₂F, O)	45	82,83
4-C$_6$H$_5$C$_6$H$_4$C=CH$_2$ ȮAc	4-C$_6$H$_5$C$_6$H$_4$CCH$_2$F O	62	82
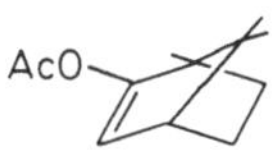 AcO	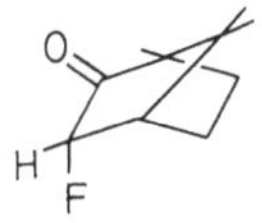 H F	65	82
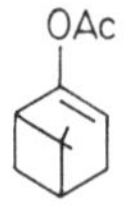 OAc	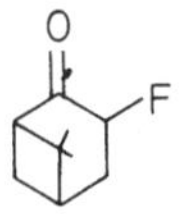 O F	80	82
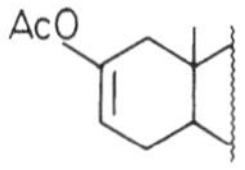 AcO	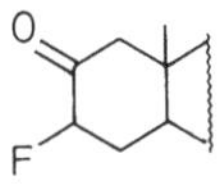 O F	40–50	81
PhCH=CCH$_2$Ph ȮAc	PhCHFCCH$_2$Ph O	50	82
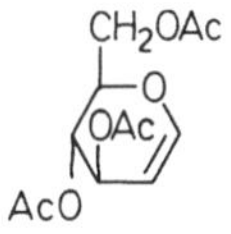 CH$_2$OAc AcO	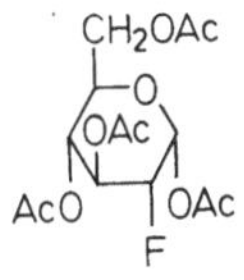 CH$_2$OAc AcO OAc F	78	62, 63

Compounds *20* and *21* are also fluorinated at the 6-position without cleavage of the epoxy cycle to form compounds *22* and *23* as α-isomers.

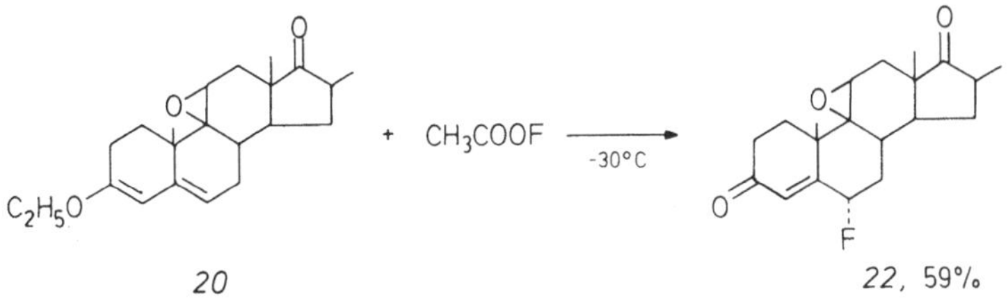

+ CH$_3$COOF $\xrightarrow{-30°C}$

20   *22*, 59%

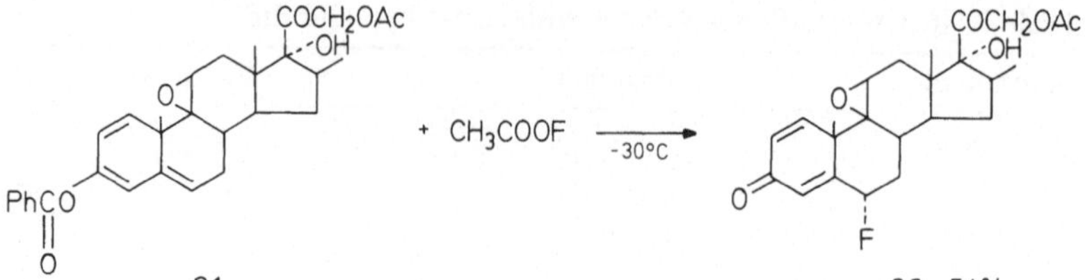

21   23, 54%

Treatment of 1,4-androsten-3,17-dione with acetyl hypofluorite leads to the acetoxyfluoro-derivative which eliminates HOAc to give the new fluorodienone steroid [80].

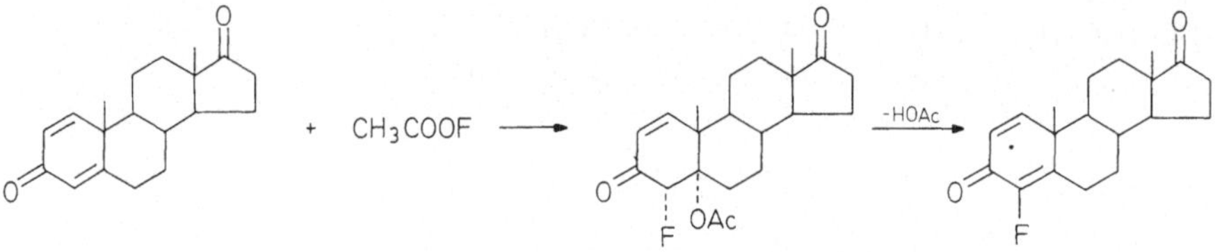

The reaction of AcOF with uracyl and citosine has been investigated [85,86]. In HOAc the reaction gives the fluoroacetoxylation product, whereas in water, hydroxy difluoride is formed. More recently, the fluorination in HOAc with subsequent elimination of the HOAc molecule was used for the synthesis of ^{18}F-labelled uracyl- and citosyl nucleosides [87].

The antipyrine molecule contains the ethylene fragment, and one could expect its fluoroacetoxylation under the action of AcOF. Indeed, this reaction leads to 4-fluoroantipyrine in a 82% yield [80]. In a similar way proceeds fluorination of bimane 24 [88]. If the reaction is carried out in the $CHCl_3$–$MeNO_2$ (2:1) mixture, it gives fluoropyrazolinone 26, along with the mono- and difluorobimanes [76].

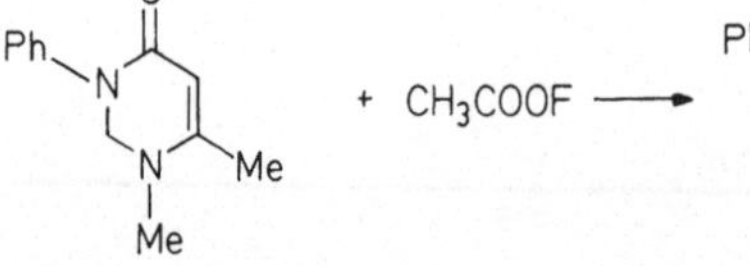

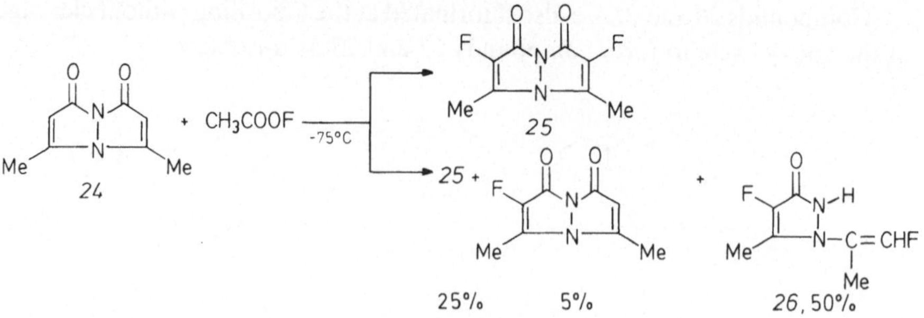

25   25   26, 50%
25%   5%

The "electrophilic" fluorination of diazepam by AcOF produces 3-fluoro-diazepam whose yield is almost independent of the solvent used. The reaction is supposed to involve the enol form of diazepam [89]. Indeed, the reaction of CF$_3$COOF with 3-trimethylsiloxydiazepam afforded fluorodiazepam in a 80% yield [90].

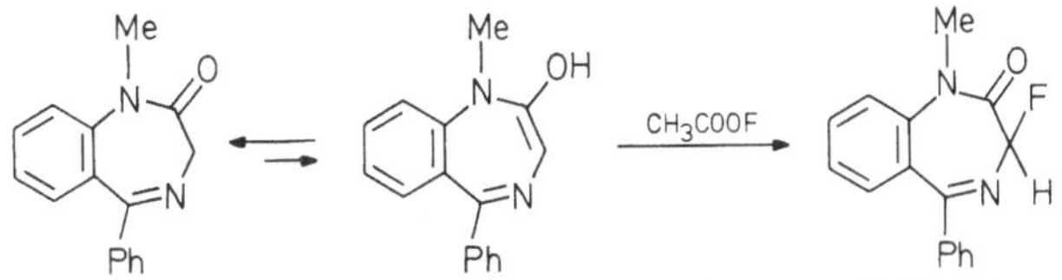

Solvent	Yield (%)
HOAc	15
CHCl$_3$	21
CFCl$_3$	18

The ability of acetyl hypofluorite to add at the C=C bond opened up a simple route to fluorinated carbohydrates. Thus, treatment of 3,4,6-tri-O-acetyl-D-glucal with acetyl hypofluorite leads to two isomeric 2-desoxy-2-fluoro-D-glucoses [63,74,91,92].

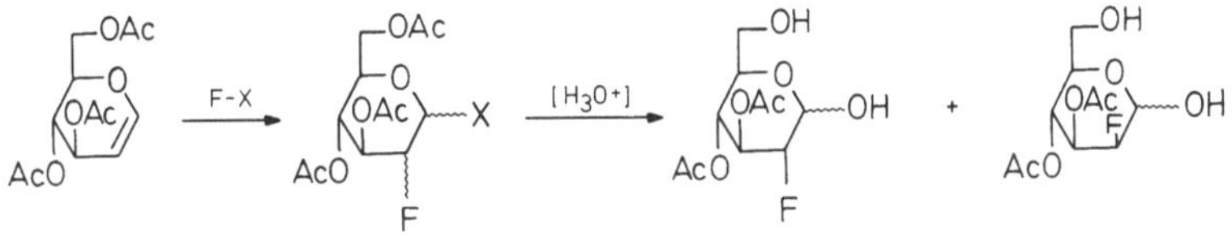

X = F, OAc, OR$_F$, XeF

Table 8 gives the isomer ratio depending on the fluorinating agent. If AcOF is used in non-polar solvents (CFCl$_3$, CCl$_4$), the yield of products is only 4%, whereas in polar solvents (HOAc, MeOH, DMF), the yield is about 20% [93]. The size of the substituent at the hydroxy groups does not affect the yield of product [94]. Detailed analysis of the reaction products revealed compounds that are the typical products of radical reactions [95].

The stereoselectivity of the reaction of AcO ^{18}F with sugars has been studied in [64,95,96].

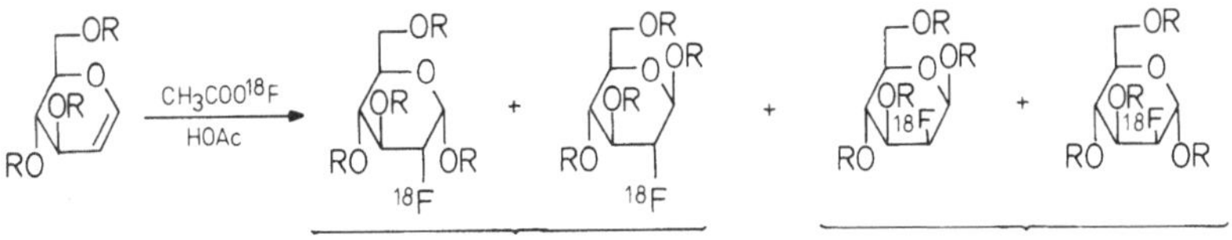

A remarkable feature of acetyl hypofluorite is the regiospecific fluorode-metallation of organoelement compounds. The most well-studied are the

Table 8. Relative yield of fluorocarbohydrates *28* and *29* obtained from *27* by the action of fluorinating agents [91]

Fluorinating agent	Solvent	Temperature (°C)	Relative yield (%)	
			28	*29*
CH_3COOF	$CFCl_3$	-78	95	5
CH_3COOF	HOAc	20	82	18
F_2	$CFCl_3$	-78	80	20
CF_3OF	$CFCl_3$	-78	81	19
XeF_2-BF_3	$Et_2O-C_6H_6$	20	93	7

reactions of aromatic derivatives of mercury, tin, silicon (Table 9). The aromatic ring may contain the HO, AlkO, CH_3, F, Br, and OAc substituents, but 3- and 4-fluoroanilines are obtained from the respective arylmercury acetates and AcOF in low yields. The presence of electron-accepting groups (NO_2, COOH) also produces a negative effect on the yields of fluoro-derivatives.

The examples of the reactions with alkylmercury compounds are not numerous [101].

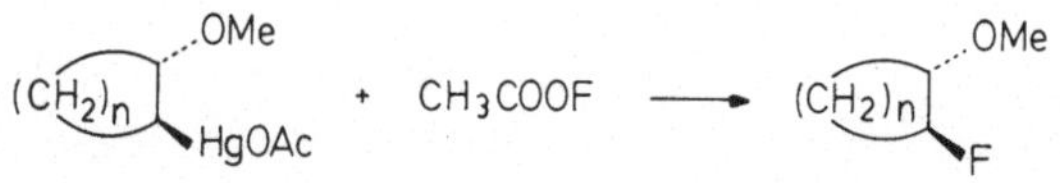

The fluorination of aryltributylstannanes gives higher yields than of arylmercuriates. However the latter are more readily available and are therefore better for use in the synthesis of fluoroaromatic compounds.

The solvent produces a significant effect of fluorodemetallation. Thus, in the fluorodestannylation of phenyltrimethylstannane with AcOF, the yield of fluorobenzene at 0 °C is: in $CFCl_3$ 68.2%, in CCl_4 65.5%, in CH_2Cl_2 14.5% [100]. As follows from Table 9, the yield of fluorobenzene decreases in the series: Sn > Ge > Si derivatives. Fluoroaromatic compounds are also obtained in the reaction of acetyl hypofluorite with organogermanium compounds [100], aryltrimethylsilanes [102], and arylpentafluorosilicates [99].

Despite the large volume of information on synthetic aspects of fluorination of organic molecules by acetyl hypofluorite, the data on its chemical behaviuor in various solvents, particularly in acetic acid, are very scarce. Due to this, the mechanistic details of fluorinations by this reagent will have to be cleared up. Most researchers suggest the one-electron oxidation of the aromatic substrate by AcOF in acetic acid [78]. The scheme below shows the possible reaction route. It should be noted that all the products fit into this scheme irrespective of the class of aromatic substrate.

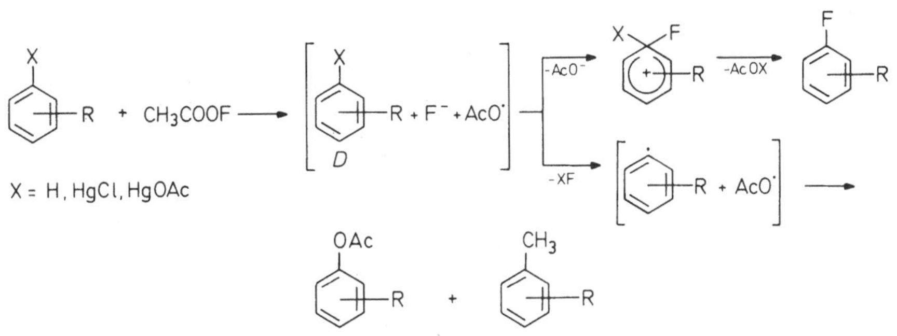

Table 9. Reactions of organoelement compounds with acetyl hypofluorite

Compound	Product	Yield (%)	Ref.
4-MeOC$_6$H$_4$HgOAc	4-MeOC$_6$H$_4$F	65	78, 79
4-MeOCOHgC$_6$H$_4$NHCOCH$_3$	4-FC$_6$H$_4$NHCOCH$_3$	60	78, 79
2-HOC$_6$H$_4$HgCl	2-HOC$_6$H$_4$F	53	78, 79
4-HOC$_6$H$_4$HgCl	4-HOC$_6$H$_4$F	47	78, 79
C$_6$H$_5$HgCl	C$_6$H$_5$F	55	78, 79
C$_6$H$_5$HgOAc	C$_6$H$_5$F	58	78, 79
4-NH$_2$C$_6$H$_4$HgOAc	4-NH$_2$C$_6$H$_4$F	4	78, 79
3-NH$_2$C$_6$H$_4$HgOAc	3-NH$_2$C$_6$H$_4$F	19	78, 79
4-MeOC$_6$H$_4$SnBu$_3$	4-MeOC$_6$H$_4$F	78	97
4-MeC$_6$H$_4$SnBu$_3$	4-MeC$_6$H$_4$F	72	97
3-MeC$_6$H$_4$SnBu$_3$	3-MeC$_6$H$_4$F	71	97
2-MeC$_6$H$_4$SnBu$_3$	2-MeC$_6$H$_4$F	57	97
C$_6$H$_5$SnBu$_3$	C$_6$H$_5$F	72	97
4-ClC$_6$H$_4$SnBu$_3$	4-ClC$_6$H$_4$F	68	97
4-FC$_6$H$_4$SnBu$_3$	1,4-F$_2$C$_6$H$_4$	73	97
C$_6$H$_5$SiMe$_3$	C$_6$H$_5$F	10	98
4-MeC$_6$H$_4$SiMe$_3$	4-MeC$_6$H$_4$F	13	98
4-MeOC$_6$H$_4$SiMe$_3$	4-MeOC$_6$H$_4$F	9	98
4-ClC$_6$H$_4$SiMe$_3$	4-ClC$_6$H$_4$F	15	98
4-BrC$_6$H$_4$SiMe$_3$	4-BrC$_6$H$_4$F	14	98
4-MeCOC$_6$H$_4$SiMe$_3$	4-MeCOC$_6$H$_4$F	6	98
4-AcOC$_6$H$_4$SiMe$_3$	4-AcOC$_6$H$_4$F	16	98
4-Me$_3$SiC$_6$H$_4$SiMe$_3$	4-FC$_6$H$_4$SiMe$_3$	16	98
K$_2$[C$_6$H$_5$SiF$_5$]	C$_6$H$_5$F	20	99
K$_2$[C$_6$H$_5$CH$_2$SiF$_5$]	C$_6$H$_5$CH$_2$F	6	99
K$_2$[4-MeC$_6$H$_4$SiF$_5$]	4-MeC$_6$H$_4$F	18	99
4-MeOC$_6$H$_4$SnMe$_3$	4-MeOC$_6$H$_4$F	66	100
4-MeC$_6$H$_4$GeMe$_3$	4-MeC$_6$H$_4$F	16	100
4-MeC$_6$H$_4$SiMe$_3$	4-MeC$_6$H$_4$F	9	100
C$_6$H$_5$SnMe$_3$	C$_6$H$_5$F	68	100
C$_6$H$_5$GeMe$_3$	C$_6$H$_5$F	9	100
C$_6$H$_5$SiMe$_3$	C$_6$H$_5$F	4	100

Acetyl hypofluorite oxidizes the aromatic substrate to form complex D which involves the aryl cation radical, the acetoxyl radical, and the fluoride ion. In the case of organometallic derivative, the metal stabilises the positive charge on the carbon atom and regioselectively gives the fluorodemetallation product. The interaction of the fluoride ion with metal leads to elimination of metal fluoride, and the aryl radical formed is regiospecifically transformed to the aryl acetate, apparently as a result of radical recombination. The acetoxyl radical decomposes to CO_2 and CH_3 and recombinates to form the toluene derivatives.

Though this route requires further investigations, it allows one to predict the character of products in the fluorinations by acetyl hypofluorite.

A short and simple synthesis of AcOF via elemental fluorine opened up a route to ^{18}F-labelled fluoroorganic compounds. These factors are especially important in view of the half-life period of $^{18}F \sim 110$ min. 3-Methoxy-4-hydroxy-L-phenylalanine reacts with AcO ^{18}F at 20 °C, forming a mixture of 2-, 5-, and 6-fluoroderivatives [103].

$$HO-\underset{MeO}{\langle\rangle}-CH_2CH(NH_2)COOEt \xrightarrow[2)48\% HBr]{1)AcO^{18}F} HO-\underset{HO}{\langle\rangle}_F-CH_2CH(NH_2)COOH$$

Acyl hypofluorites synthesized from perfluorinated carboxylic acids proved to be efficient catalysts and initiators of polymerisation of fluoroolefins [104,105]. Thus polymerisation of tetrafluoroethylene catalysed by perfluoro-octanoyl hypofluorite to give perfluoro-2-butyltetrahydrofurane affords poly(tetrafluoroethylene) in a 89% yield, which is stable up to 372 °C [104].

Thus within a short period of time CH_3COOF became a popular fluorinating agent used inter alia for synthesis of reagents for use in biology. The use of ^{18}F stimulated the development of methods for the synthesis of AcOF and expanded the range of accessible fluoroorganic compounds.

2.5   Conclusion

Choosing a fluorinating agent, a chemist is usually interested in its availability, selectivity of fluorination, equipment for the process, and, of course, the safety of working with it. This review allows one to evaluate these factors, but it seems reasonable to give a general overview of the possibilities of these fluorinating agents.

Among the nitrogen-fluorine-containing compounds the most promising are, apparently, *N*-fluoropyridinium triflate and its derivatives. These fluorinating agents are relatively easily synthesized, may be stored, and are easy to handle in synthesis. Fluorinations are usually carried out in dichloromethane, dichloro- and trichloroethane at 20 to 100 °C, during several hours (5 to 24 h). This agent may be used to fluorinate aromatic compounds with electron-donating substituents, olefins, enol ethers, and carbanion salts; the processes may be conducted in glassware, as usual. The Japanese researchers [29] conducted fluorination using approx. 1 mmol of *N*-fluoropyridinium triflate. Other nitrogen-fluorine-containing compounds are either less available or dangerous to handle, or insufficiently studied as fluorinating agents.

Another accessible fluorinating agent is caesium fluorooxysulfate. Zupan and Stavber [42] reported that they synthesized more than 200 g of $CsSO_4F$ (approx. 0.8 g-atom of active fluorine) at a time and stored it in polyethylene flask at $0\,°C$ for 14 days without decrease of its activity. We believe the most remarkable feature of $CsSO_4F$ (or other fluorooxysulfates) to be the predominant *ortho*-fluorination of alkoxyarenes and acetanilides, as the classical methods, such as the Balz-Schiemann reaction, give low yields of such compounds. Using caesium fluorooxysulfate, it is possible to carry out substitution of vinyl hydrogen by fluorine—a rather rare reaction type. It is advisable to use BF_3 as a catalyst. Strong protic acids may cause side reactions, for example, isomerisation.

The use of acyl hypofluorites is certainly justified when it is necessary to synthesize vicinal fluoroacetoxy- (from AcOF) or fluorohydroxy- (from CF_3COOF) containing compounds. α-Fluorocarbonyl compounds are easily obtained in high yields from enol acetates, but fluorination of aromatic compounds with acetyl hypofluorite has no advantages over the use of $CsSO_4F$ or XeF_2. At the same time, the regiospecificity of fluorodemercuration of arylmercury derivatives is an advantage of acetyl hypofluorite, and its use here is preferable. A common disadvantage of CH_3COOF and CF_3COOF is their instability, the necessity of using them immediately after preparation and also restricting the fluorination degree by selecting the amount of acyl hypofluorite (~ 15–100 mg).

The data reviewed allow one to make the following conclusions. First, interest in the development of the method of direct substitution of hydrogen by fluorine in various organic molecules remains as high as ever, and we can expect new results in the future. Second, in addition to the direct fluorination method, new fluorinating agents are being found, which have some advantages over the traditionally used ones. They are widely used for the fluorination of natural and biologically active compounds. Third, the industrial need for new materials with specific features calls for extensive research into such substances among organic compounds. Some fluoroorganic agents have already been found to satisfy these requirements. The development of fluorination methods and search for new fluorinating agents will remain an urgent task in the future.

2.6 Preparations

1. Caesium fluoroxysulfate [42]
In a 100 ml polyethylene vessel containing 10 g of Cs_2SO_4 in 16 ml of water, was introduced a 20% mixture of F_2 in nitrogen at $0\,°C$ for 5 h (complete amount of F_2 introduced was approximately 40 mmol (1.5 g)). After half an hour the insoluble $CsSO_4F$ was filtered off and washed with water (1 ml) and combined

precipitates were dried under vacuum at room temperature. Dry $CsSO_4F$ (4 g) was obtained, which must be stored in a polyethylene vessel at $0\,°C$. The compound is stable for at least 14 days, but any contact with a metallic spatula or mechanical pressure must be avoided, since decomposition or even an explosion may take place.

2. Fluorination of acetanilide [42]

$CsSO_4F$ (1 mmol) and 1.5 ml of acetonitrile were stirred at room temperature for 5 min, and after the introduction of BF_3 (0.5–1 mmol) over the reaction mixture, 1 mmol of acetanilide in 0.5 ml of acetonitrile was added. The reaction mixture was stirred at room temperature for 30 min, then 10 ml of dichloromethane was added, the insoluble residue was filtered off, the filtrate washed with water and dried over Na_2SO_4, and the solvent was evaporated under vacuum. The crude reaction mixture (130 mg, 71% conversion of acetanilide) was separated by preparative TLC (SiO_2, dichloromethane : methanol = 9.5 : 0.5). The yield of 2-fluoro-1-(acetylamino)benzene was 80 mg (74.5%), m.p. 76 to 78 $°C$, and the yield of 4-fluoro-1-(acetylamino)benzene was 12 mg (11%), m.p. 152 to 153 $°C$.

3. Acetyl hypofluorite in solution [58]

A fluorine-nitrogen mixture ($\sim 10\%$ F_2) was bubbled slowly through a suspension of sodium acetate–acetic acid mixture (14 g; from equimolar amounts of both components) in $CFCl_3$ (450 ml) at $-75\,°C$, which was agitated by an efficient vibromixer. The oxidizing power of the solution was determined by treating aliquots with acidic potassium iodide solution and subsequent titration of the liberated iodine with thiosulfate.

4. ω-Fluoroacetophenone [58]

Diisopropylamine (1.70 ml, 12 mmol) and THF (10 ml) were cooled to $-78\,°C$. Butyllithium as a 15% solution in hexane (7.32 ml, 12 mmol) was added in one portion to produce lithium diisopropylamide and the resultant solution was stirred at $-78\,°C$ for 10 min. Then, a solution of acetophenone (1.202 g, 10 mmol) in THF (2 ml) was added in portions and stirring of the lithium enolate solution was continued at $-78°C$ for 15 min and at $25°C$ for further 15 min. Diisopropylamine and solvent were then distilled off under reduced pressure at $25\,°C$ and THF (10 ml) was added to the oily residue. The resultant solution was cooled to $-78\,°C$ and added dropwise, quickly, to the stirred acetyl hypofluorite solution (~ 18 mmol) at $-78\,°C$. After 1 min, the mixture was poured into 5% sodium thiosulfate solution (400 ml). The organic layer was washed with conc. $NaHCO_3$ solution (150 ml) and then with water (2×100 ml) until neutral, dried with $MgSO_4$, and evaporated. The crude product was purified by flash vacuum chromatography (silica gel, ethyl acetate : petroleum ether = 1 : 10). The yield of ω-fluoroacetophenone was 0.94 g (68%), m.p. 25 to 26 $°C$; the recovery of acetophenone was 0.12 g (10%).

5. Fluorination of meso-hexestrol dimethyl ether [80]

To the solution of acetyl hypofluorite (15 mmol) was added 2.1 g (7 mmol) of meso-hexestrol dimethyl ether dissolved in 40 ml $CHCl_3$. After 10 min the reaction was stopped and worked up as usual. The crude reaction mixture was chromatographed using 20% EtOAc in petroleum ether. Two compounds were obtained and purified by HPLC using 20% EtOAc in cyclohexane. The first one proved to be the mono-fluoro derivative (40% yield), m.p. 140 °C (from EtOAc). The second fraction proved to be the difluoro derivative (20% yield), m.p. 160 °C (from EtOAc).

6. Preparation of N-fluoropyridinium triflates [25]

Method A. The adduct Py·F$_2$ prepared in CFCl$_3$ at −75°C was allowed to react with sodium triflate in dry acetonitrile at −40 °C for 2 h to give N-fluoropyridinium triflate (6) as very stable white crystals in 67% yield, m.p. 185 to 187 °C.

Method B. It was found that 6 was synthesized conveniently by the reaction of F$_2$/N$_2$ (1:9) with pyridine in MeCN at −40 °C followed by treatment with sodium triflate (71%).

Method C. Triflate 7 was prepared by bubbling F$_2$/N$_2$(1:9) into a solution of 2,4,6-trimethylpyridine in MeCN at −40 °C in the presence of sodium triflate (49%, m.p. 164 to 166 °C).

7. Fluorination of sulfides (typical procedure) [29]

4-Chlorophenyl methyl sulfide (158 mg, 1 mmol) was added into a mixture of N-fluoro-2,4,6-trimethylpyridinium triflate 7 (290 mg, 1 mmol) and dry dichloromethane (3 ml) at room temperature under argon atmosphere and the reaction mixture was stirred for 8 h. The reaction was followed by checking the oxidation power with aq. KI solution. After triflate 7 completely disappeared, anhydrous sodium carbonate (0.5 g) was added and the organic layer was thin-layer chromatographed on silica gel by using a mixture of hexane and Et$_3$N (100:1) as an eluent to give 4-chlorophenyl fluoromethyl sulfide as an oily product (133 mg, 76%).

2.7 References

1. Ishikawa N (ed) (1984) Novoe v tekhnologii soedinenii ftora. Russ. per. Mir, Moscow, p 591
2. Ishikawa N, Kobayashi E (1982) Ftor. Khimiya i primenenie. Russ. per. Mir, Moscow, p 91
3. Sheppard WA, Sharts CM (1969) Organic fluorine chemistry. Benjamin, New York
4. Adv. Fluorine Chem. (1961) Butterworth, Washington, vol 2
5. Leidinger S (1984) Bull. Soc. Quim. Peru 50:344
6. Banks RE, Tatlow JC (1986) J. Fluor. Chem. 33:71
7. Riess JC (1987) J. Chem. Phys. 84:1119
8. Haas A, Gerstenberger MRC (1981) Angew. Chem. Int. Ed. Engl. 20:647
9. Haas A, Lieb M (1985) Chimia 39:134
10. Purrington ST, Kagan BS (1986) Chem. Rev. 86:997
11. Zupan M (1984) Vestn. Slov. Kem. Drus. 31:151
12. Cartwright MM, Woolf AA (1984) J. Flour. Chem. 25:263; Christe KO (1984) J. Fluor. Chem. 25:269
13. Yakobson GG, Furin GG (1984) Soviet Sci. Rev. Sect. B. Chem. Rev. 5:255
14. Christe KO, Wilson RD, Goldberg IB (1979) Inorg. Chem. 18:2578
15. Christe KO, Schack CJ, Wilson RD (1976) J. Fluor. Chem. 8:541
16. Schack CJ, Christe KO (1981) J. Fluor. Chem. 18:363

17. Olah GA, Laali K, Farnia M, Shih J, Singh BP, Schack CJ, Christe KO (1985) J. Org. Chem. 50:1338
18. Banks RE, Tsiliopoulos E (1986) J. Fluor. Chem. 34:281
19. Purrington ST, Jones WA (1983) J. Org. Chem. 48:761
20. Purrington ST, Jones WA (1984) J. Fluor. Chem. 26:43
21. Simons JH (1950) In: Simons JH (ed) Fluorine Chemistry. vol 1, Academic, New York, p 420
22. Umemoto T (1988) in: 5th Regular Meeting of Soviet-Japanese Fluorine Chemists, 25–26 Jan 1988, Tokyo, p XI-1
23. Meinert H (1965) Z. Chem. 5:64
24. Meinert H, Cech D (1972) Z. Chem. 12:292
25. Umemoto T, Tomita K (1986) Tetrahedron Lett. 27:3271
26. Umemoto T, Tomita K (1986) Tetrahedron Lett. 27:4465
27. Umemoto T, Onodera K, Tomita K (1986) In: 52nd Nat. Meeting of Japanese Chem. Soc., 1 Apr 1986. Kyoto, abstr. N 1L15
28. Kawada K, Tomita K, Umemoto T (1986) In: 52nd Nat. Meeting of Japanese Chem. Soc., abstr. N 1Z05
29. Umemoto T, Tomizawa G (1986) Bull. Chem. Soc. Jap. 59:3624
30. Banks RE, Boisson R, Tsiliopoulos E (1987) J. Fluor. Chem. 35:13
31. Banks RE, Boisson R, Tsiliopoulos E (1986) J. Fluor. Chem. 32:461
32. Barnette WE (1984) J. Amer. Chem. Soc. 106:452
33. Lee SH, Schwartz J (1986) J. Amer. Chem. Soc. 108:2445
34. US Pat 4479901 (1984); (1985) Chem. Abs. 102:113537
35. Barton DHR, Hasse RH, Pechet MM, Toh HT: J. Chem. Soc. Perkin Trans. I 1974:732
36. US Pat 3917688 (1975); (1976) Chem. Abs. 84:30714
37. Seguin M, Adenis JC, Michand C, Basselier JJ (1980) J. Fluor. Chem. 15:201
38. Singh S, DesMarteau DD, Zuberi SS, Witz M, Huang Hsu-Nan (1987) J. Amer. Chem. Soc. 109:7194
39. Banks RE, Tatlow JC (1986) J. Fluor. Chem. 33:71
40. Appelman EH, Basile LJ, Thompson RC (1979) J. Amer. Chem. Soc. 101:3384
41. Ip DP, Arthur CD, Winans RE, Appelman EH (1981) J. Amer. Chem. Soc. 103:1964
42. Stavber S, Zupan M (1985) J. Org. Chem. 50:3609
43. Appelman EH, Basile LJ, Hayatsu R (1984) Tetrahedron 40:189
44. Stavber S, Zupan M (1981) J. Fluor. Chem. 17:597
45. Visser GWM, Halteren BV, Herscheid JDM, Brinkman G, Hoekstra A (1984) J. Labell. Compounds Radiopharm. 21:1185
46. Stavber S, Zupan M: J. Chem. Soc. Chem. Commun. 1981:795
47. Stavber S, Zupan M (1987) J. Org. Chem. 52:919
48. Stavber S, Zupan M (1987) J. Org. Chem. 52:5022
49. Merrit RF (1987) J. Org. Chem. 32:4124
50. Barton DHR, Danks LJ, Ganguly AK, Hesse RH, Tarzia G, Pechet MM: J. Chem. Soc. Perkin Trans. I 1976:101
51. Stavber S, Zupan M: J. Chem. Soc. Chem. Commun. 1981:148
52. Zupan M, Pollak A (1974) J. Org. Chem. 39:2646
53. Gregorcic A, Zupan M (1979) J. Org. Chem. 44:4120
54. Kobayashi Y, Nakazawa M, Kumadaki I, Taguchi T, Ohshima E, Ikekawa N, Tanaka Y, Deluca HF (1986) Chem. Pharm. Bull. 34:1568
55. Stavber S, Zupan M: J. Chem. Soc. Chem. Commun. 1983:563

56. Bryce MR, Chambers RD, Mullins ST, Parkin A: Bull. Soc. Chim. Fr. 1986:55
57. Lerman O, Rozen S (1983) J. Org. Chem. 48:724
58. Rozen S, Brand M: Synthesis 1985:665
59. Lerman O, Tor Y, Hebel D, Rozen S (1984) J. Org. Chem. 49:806
60. Chirakal R, Firnau G, Gause J, Garnett ES (1984) Int. J. Appl. Radiat. Isotop. 35:651
61. Shine CY, Salvadori PA, Wolf AP (1972) J. Nucl. Med. 23:108
62. Adam MJ: J. Chem. Soc. Chem. Commun. 1983:730
63. Adam MJ, Pate BD, Nesser JR, Hale LD (1983) Carbohyd. Res. 124:215
64. Fowler JS, Shine CY, Wolf AP, Salvadori PA, MacGregor RR (1982) J. Labell. Compounds Radiopharm. 19:1634
65. Shine CY, Salvadori PA, Wolf AP, Fowler JS, MacGregor RR (1982) J. Nucl. Med. 23:899
66. Adam MJ, Ruth TJ, Javan S, Pate BD (1984) J. Fluor. Chem. 25:329
67. Rozen S, Lerman O, Kol M: J. Chem. Soc. Chem. Commun. 1981:443
68. Lerman O, Tor Y, Rozen S (1981) J. Org. Chem. 46:4629
69. Hebel D, Lerman O, Rozen S (1985) J. Fluor. Chem. 30:141
70. Diksic M, Jolly D (1983) Int. J. Appl. Radiat. Isotop. 34:893
71. Jewett DM, Potocki JF, Ehronkaufer RE (1984) Synth. Commun. 14:45
72. Yewett DM, Potocki JF, Ehronkaufer RE (1984) J. Fluor. Chem. 24:477
73. Rozen S, Lerman O (1980) J. Org. Chem. 45:672
74. Appelman EH, Mendelsohn MH, Kim H (1985) J. Amer. Chem. Soc. 107:6515
75. Bida GT, Satyamurthy N, Barrio JR (1984) J. Nucl. Med. 25:1327
76. Kosower EM, Hebel D, Rozen S, Radkowski AE (1985) J. Org. Chem. 50:4152
77. Rozen S, Lerman O, Kol M, Hebel D (1985) J. Org. Chem. 50:4753
78. Visser GWM, Bakker CNM, Halteren BW, Herscheid JDM, Brinkman GA, Hoekstra A (1986) J. Org. Chem. 51:1886
79. Visser GWM, Halteren BW, Herscheid JDM, Brinkman GA, Hoekstra A: J. Chem. Soc. Chem. Commun. 1984:655
80. Hebel D, Lerman O, Rozen S: Bull. Soc. Chim. Fr. 1986:861
81. Barnette WE, Wheland RC, Middleton WJ, Rozen S (1985) J. Org. Chem. 50:3698
82. Rozen S, Menahem Y (1980) J. Fluor. Chem. 16:19
83. Rozen S, Menahem Y: Tetrahedron Lett. 1979:725
84. Shimokawa K (1988) In: 5th Regular Meeting of Soviet-Japanese Fluorine Chemists, 26–27 Jan 1988. Tokyo, p X-1
85. Diksic M, Farrokhzad S, Colebrook LD (1986) Can. J. Chem. 64:424
86. Visser GWM, Boele S, Halteren BW, Knops G, Herscheid JDM, Brinkman GA, Hoekstra A (1986) J. Org. Chem. 51:1466
87. Kosower EM, Hebel D, Rozen S, Radkowcki AE (1985) J. Org. Chem. 50:4152
88. Visser GWM, Noordhuis P, Zwaagstra O, Herscheid JDM, Hoekstra A (1986) Int. J. Appl. Radiat. Isotop. 37:1074
89. Luxen A, Barrio JR, Satyamurthy N, Bida GT, Phelps ME (1987) J. Fluor. Chem. 36:83
90. Middleton WJ, Bingham EM (1980) J. Amer. Chem. Soc. 102:4845
91. Mulholland GK, Ehronkaufer RE (1986) J. Org. Chem. 51:1482
92. Vyplel H (1985) Chimia 39:305
93. Shine CY, Wolf AP (1985) J. Nucl. Med. 26:129
94. Shine CY, Wolf AP (1986) J. Fluor. Chem. 31:255
95. Cornelis JS, Herscheid JDM, Visser GWM, Hoekstra A (1985) Int. J. Appl. Radiat. Isotop. 36:111

96. Adam MJ, Ruth TJ, Javan S, Pate BD (1984) J. Labell. Compounds Radiopharm. 21:11
97. Adam MJ, Ruth TJ, Javan S, Pate BD (1984) J. Fluor. Chem. 25:329
98. Speranza M, Shine CY, Wolf AP, Wilbur DS, Angeloni G (1985) J. Fluor. Chem. 30:97
99. Speranza M, Shine CY, Wolf AP, Wilbur DS, Angeloni G: J. Chem. Soc. Chem. Commun. 1984:1448
100. Coenen HH, Moerlein SM (1987) J. Fluor. Chem. 36:63
101. Habel D, Rozen S (1987) J. Org. Chem. 52:2588
102. Speranza M, Shine CY, Wolf AP, Wilbur DS, Angeloni G (1984) J. Labell. Compounds Radiopharm. 21:1189
103. Chirakal R, Firnau G, Couse J, Garnett ES (1984) Int. J. Appl. Radiat. Isotop. 35:651
104. US Pat 4535136 (1985); (1985) Chem. Abs. 103:178795
105. US Pat 4588796 (1986)

3 Hypofluorites and their Application in Organic Synthesis

Farid Mubarakshevich Mukhametshin

Institute of Applied Chemistry, 614034 Perm, USSR

Contents

3.1 Introduction

Hypofluorites are represented by a wide-spread class of compounds with the O–F functional group bonded to an inorganic or organic residue.

The most well-known representatives of inorganic hypofluorites are oxygen difluoride OF_2, and pentafluorosulfur hypofluorite SF_5OF. Less known are the analogues of the latter—pentafluoroselenium hypofluorite SeF_5OF, pentafluorotellurium hypofluorite TeF_5OF, higher oxygen fluorides such as FOOF and FOOOF, and the derivatives of fluorosulfonic FSO_2OF, chloric O_3ClOF, and nitrous O_2NOF acids.

The relative stability of inorganic hypofluorites is explained by the fact that their elements are in the highest oxidation state. At a partial oxidation state, the strong electron-accepting nature of the fluorine atom bonded with oxygen produces a destabilising effect in the element-oxygen-fluorine system. For that reason such compounds as, e.g. F_3SOF and F_2NOF are hypothetical and exist only as the respective oxides $O=SF_4$ and $O{\leftarrow}NF_3$.

As opposed to inorganic hypofluorites, the organic ones are much more widely spread, this being again due to their structural peculiarities. Such compounds may be synthesized and may exist in a free form, provided that the OF-bonded carbon substituent is fully or partially (but to a sufficient extent) fluorinated. Their stability is explained by the fluorine screening of the molecular carbon frame, a certain inertness of the C–F bond, and the electron-accepting nature of the perfluoro (or polyfluoro)alkyl substituents. The hydrogen-containing alkyl hypofluorites do not occur in normal conditions. For example, methyl hypofluorite CH_3OF whose formation has been postulated in the reaction of methanol with xenon difluoride [1], was only fixed in situ, as a product of addition to the C=C multiple bond at a low temperature.

The wider occurrence of organic hypofluorites is due to the variety of poly- and perfluorinated compounds. Owing to this, the organic hypofluorites involve not only monofunctional derivatives but also compounds containing two geminal OF fragments, for example difluoromethane-bis-hypofluorite and tetra-fluoroethane-bis-hypofluorite $CF_3CF(OF)_2$. The large capabilities of the per-fluorinated organic compounds are illustrated by several instances of the selective synthesis of hypofluorites with the OF group bonded to the acyl or alkoxy substituents of the type of $CF_3C(O)OF$ and CF_3OOF.

Despite the large number of hypofluorites synthesized up to now, their chemistry is a relatively new field of science, which is not older than 25 to 30 years. Nevertheless, there have been frequent attempts at reviewing the chemistry of hypofluorites, indicating a constant interest in it.

Inorganic hypofluorites were reviewed in [2–8] discussed chiefly inorganic reactions, which are beyond the scope of this review and are not considered here.

Organic hypofluorites were first separated as a subject in the review [9] published in 1980 and devoted to the critical analysis of the current views on the reaction mechanisms and the reactivity towards organic compounds.

This review follows the same trend, developing the previous ideas and covering the literature up till 1987. An attempt is made to explain the available experimental data on the synthesis and chemical properties of organic hypo-fluorites in terms of novel approaches.

3.2 Synthesis of Hypofluorites

The first representative of organic hypofluorites—trifluoromethyl hypofluorite CF_3OF—was synthesized in 1948 by the reaction of fluorine with carbon oxide [10].

$$CO + 2F_2 \rightarrow CF_3OF$$

The reaction was found later to proceed more efficiently in a two-section apparatus with a separate fluorine inlet [11]. One section was intended for the synthesis of carbonyl difluoride: $CO + F_2 \rightarrow COF_2$, in another its further fluorination took place: $COF_2 + F_2 \rightarrow CF_3OF$. The highest temperature for the fluorination of CO is 670 K. Above this temperature, formation of tetra-fluoromethane was reported [12]:

$$CO + 2F_2 \rightarrow CF_4 + 1/2\,O_2$$

Kinetic studies [13] have shown carbon oxide to slowly react with fluorine even at room temperature, the activation barrier being insignificant ($55\,kJ\,mol^{-1}$). This is a typical radical–chain reaction, with chain generation proceeding by the bimolecular mechanism: $CO + F_2 \rightarrow {}^\cdot COF + F$. Due to its radical-chain nature, the reaction is explosive and hazardous and requires the appropriate safety measures.

It is an important condition for the reaction of $CO + F_2$ that the starting reagents should be oxygen-free. Otherwise, a series of transformations $^\cdot COF + O_2 \rightarrow {}^\cdot OOC(O)F \rightarrow {}^\cdot OC(O)F \rightarrow FC(O)OOC(O)F$ leads to the formation of bis-fluoroformyl peroxide as a by-product, which at elevated temperatures yields such by-products as carbon dioxide, oxygen, and carbonyl difluoride [14].

However, the presence of carbon dioxide in the reaction mixture should not be a constraint, as in an excess of fluorine, CO_2 is known [15] to be fluorinated to trifluoromethyl hypofluorite: $CO_2 + 2F_2 \rightarrow CF_3OF + 1/2\,O_2$. This reaction proceeds smoothly in the temperature range of 473 to 523 K, is explosion-safe unlike fluorination of CO, and therefore may be regarded as a separate method for the preparation of CF_3OF.

It is important to know that fluorination of carbon dioxide as distinct from fluorination of CO requires the preliminary generation of atomic fluorine. The fluorine atoms generated, e.g. photochemically, initially add to the C=O multiple bond, forming carbon- and oxygen-centred radicals, as shown in [16,17].

$$F_2 \xrightarrow{h\nu} 2F; \qquad CO_2 + F \rightarrow O{=}CFO^\cdot + {}^\cdot C(O)OF$$

The same beginning is expected for the thermal fluorination of CO_2 described in [18]. Thus the mechanism of fluorination of this reagent reminds one of fluorination of carbon oxide in the presence of oxygen.

One of the questions inevitably arising in efforts to rationalise the mechanism of the transformation of CO_2 to CF_3OF is the question of the mechanism of formation of COF_2, which alone can be the source of trifluoromethyl hypofluorite. In the literature this matter has not yet been discussed, but it seems to be solvable in terms of the general approach.

First, only one radical of those formed at the stage of initiation is thermodynamically favourable—the oxygen-centred one (the heats of formation vary by about $270\,kJ\,mol^{-1}$). Second, it is reasonable to admit that this radical further reacts with fluorine to give fluorooxyformyl fluoride which is converted by atomic fluorine to the fluorooxydifluoromethoxyl radical. The latter undergoes fragmentation to form carbonyl difluoride and the fluoroxyl.

$$O{=}CFO^\cdot + F_2 \rightarrow O{=}CFOF + F \rightarrow {}^\cdot OCF_2OF \rightarrow O{=}CF_2 + {}^\cdot OF$$

$$2FO^\cdot \rightarrow O_2 + F_2$$

Such a view of the mechanism of transformation of CO_2 to COF_2 is in agreement with the close reactivities of these reagents towards fluorine. Fluorination of carbonyl difluoride, the best reagent for the synthesis of CF_3OF, proceeds smoothly in the same temperature range as for CO_2. The reaction is an equilibrium, and above 523 K trifluoromethyl hypofluorite slowly decomposes to the starting reagents [19].

When initiated photochemically [20], fluorination of COF_2 is irreversible, but leads to substantial amounts of bis-trifluoromethyl peroxide (up to 20%).

Due to the apparent possibility of the fluorine attack on the carbonyl carbon and oxygen, there are also two viewpoints on the mechanism of fluorination of COF_2. According to one of them, CF_3OF is generated from the oxygen-centred radical $CF_3O^.$ [20], according to another—from the carbon-centred radical $^.CF_2OF$ [21]. But again, the radical $CF_3O^.$ is thermodynamically more favourable.

$$F_2 \xrightarrow{hv \text{ or } \Delta} 2F$$

$$F + COF_2 \rightarrow CF_3O^.$$

$$CF_3O^. + F_2 \rightarrow CF_3OF + F$$

$$2CF_3O^. \rightarrow CF_3OOCF_3$$

The intermediate formation of radical $CF_3O^.$ is indicated by the above experimental data on fluorination of carbon oxide by fluorine excess where the reaction produces tetrafluoromethane in accordance with the scheme.

$$CF_3O^. + CO \rightarrow CF_3OC^.(O) \rightarrow {}^.CF_3 + CO_2$$

$$^.CF_3 + F_2 \rightarrow CF_4 + F$$

The latter scheme was confirmed by studies of the thermal reaction of bis-trifluoromethyl peroxide and carbon oxide, where the main products are carbon dioxide and hexafluoroethane [22].

$$CF_3OOCF_3 + CO \rightarrow CO_2 + C_2F_6$$

It should be noted that the above methods of thermal and photochemical generation of fluorine may only serve to obtain CF_3OF. The higher organic hypofluorites may not be obtained by this method, as the starting compounds are converted by fluorine at high temperature or under irradiation to the respective perfluoroalkanes (see, e.g. [23]). The process may be directed as desired by using certain catalysts.

The catalytic method plays a leading role in the synthesis of perfluoroalkyl hypofluorites. The recommended heterogeneous catalysts are the sodium, potassium, rubidium, caesium, magnesium, calcium, strontium, barium, brass, iron, copper, nickel, silver, and cobalt fluorides [24]. The homogeneous catalysts are represented by nitrosyl fluoride [25].

The highest catalytic activity in the above series of fluorides is shown by caesium fluoride. It is effective for the whole series of perfluorinated carbonyl-containing reagents (excluding carbon oxide).

Thus perfluoroacyl fluorides and perfluoroketones in the presence of CsF give the primary and secondary perfluoroalkyl hypofluorites respectively [26–32].

$$R_FCF_2C(O)F + F_2 \xrightarrow{CsF} R_FCF_2CF_2OF$$

$$(R_F)_2C=O + F_2 \xrightarrow{CsF} (R_F)_2CFOF$$

The extremely high catalytic activity and selectivity of CsF make it effective even with such unstable compounds as peroxides [31,32].

$$FC(O)OOC(O)F + F_2 \xrightarrow{CsF} FOCF_2OOCF_2OF$$

$$CF_3OOC(O)F + F_2 \xrightarrow{CsF} CF_3OOCF_2OF$$

Another example is the synthesis of difluoromethane-bis-hypofluorite with a high yield, by fluorination of carbon dioxide in mild conditions, which has only become feasible due to the high catalytic activity of CsF [27,34–36].

$$CO_2 + 2F_2 \xrightarrow{CsF} CF_2(OF)_2$$

The preparative syntheses of hypofluorites with the use of caesium fluoride show nearly quantitative yields of target products. But to achieve the maximal efficiency, the catalyst should be carefully dried. This procedure is generally carried out in the reactor, by heating the catalyst to 500 to 550 K, with simultaneous evacuation of the reactor for many hours. Fluorine and the carbonyl-containing reagent are then mixed in the reactor at 77 K and are allowed to slowly warm to room temperature. Strict dosing of the reagents removes the need for additional purification of the products.

The catalytic action of CsF is generally recognised to be based on its ability to form alkoxides with perfluoroacyl fluorides [37,38].

$$R_FC(O)F + CsF \rightleftarrows R_FCF_2O^- Cs^+$$

Until recently, the transformation of the alkoxides to hypofluorites had been considered (see, e.g. [39]) to result from the nucleophilic S_N2 substitution of fluorine in a molecule by the alkoxide.

$$R_FCF_2O^- + F_2 \rightarrow R_FCF_2OF + F^-$$

This opinion is now considered erroneous. As shown in [40], in reality the system $COF_2 + F_2 + CsF$ generates the trifluoromethoxyl radicals, which may be recorded by adding carbon oxide into the reaction mixture to obtain tetrafluoromethane and carbon dioxide in accordance with the above scheme. The generation of the $CF_3O^.$ radicals is considered in [40] to be best explained by the one-electron transition.

$$CF_3\overset{\frown}{O}Cs^+ + F\text{--}F \xrightarrow{-e} [CF_3O^. \ldots Cs^+ \ldots F^- \ldots F^.] \rightarrow CF_3O^. + CsF + F$$

The radical mechanism of the reaction of fluorine with anionoid species is also indicated by the results of fluorination of carbon dioxide in the presence of CsF. This reaction almost always produces insignificant amounts of trifluoromethyl hypofluorite along with difluoromethane-bis-hypofluorite, and may only be rationalised on the assumption that the system generates carbonyl difluoride.

The initial stage of the catalytic fluorination of CO_2 is formation of the alkoxide, which has been strictly proved [41].

$$O=C=O + CsF \rightleftarrows O=CFO^-Cs^+$$

Subsequent interaction of the alkoxide and fluorine by the one-electron transition mechanism leads to generation of the oxygen-centred radical in the system.

$$O=CFO^-Cs^+ + F_2 \xrightarrow{-e} O=CFO^. + CsF + F$$

Further on this radical is transformed, by the same route as in the thermal fluorination of CO_2, to the fluorooxydifluoromethoxyl radical, which either reacts with fluorine or undergoes minor fragmentation, giving carbonyl difluoride. Apart from the gas-phase route of the reaction, a considerable role is presumably played by the heterophase route which involves the intermediate formation of a complex of CsF with fluorooxyformyl fluoride.

$$O=CFO^. + F_2 \rightarrow O=CFOF + F$$

$$O=CFOF + CsF \rightleftarrows Cs^{+-}OCF_2OF$$

$$Cs^{+-}OCF_2OF + F_2 \rightarrow {}^.OCF_2OF + CsF + F$$

$$^.OCF_2OF \underset{F_2}{\overset{\longrightarrow CF_2O + FO^.}{\underset{\longrightarrow FOCF_2OF + F}{\left[\rule{0pt}{18pt}\right.}}}$$

A shift of the reaction towards synthesis of difluoromethane-bis-hypofluorite is achieved by conducting it at a low temperature (heating of the mixture of the reagents from 77 to 195 K).

In comparison to caesium fluoride (and, possibly, potassium fluoride), the catalysis by other metal fluorides is not so effective. In any case, the literature contains no data on the use of other heterogeneous catalysts in the synthesis of higher perfluoroalkyl hypofluorites, and their activity was estimated in the case of fluorination of carbonyl difluoride.

The fluorides of alkaline earth and transition metals are essentially much less active than caesium fluoride, and fluorinations in their presence require temperatures from 300 to 400 K [24]. The fluorides of transition metals show a greater activity than those of alkaline earth metals.

The catalytic activity of transition metal fluorides is directed to a different object of catalysis. As opposed to CsF activating the carbonyl-containing reagent, they activate fluorine. According to [42], the role of a catalyst consists in atomisation of fluorine, with the subsequent radical (but not chain) process described by the scheme.

$$F_2 \xrightarrow{\text{Ct}} 2F_{gas}$$

$$F_{gas} + COF_2 \rightarrow CF_3O^{\cdot}$$

$$CF_3O^{\cdot} + F_{gas} \rightarrow CF_3OF$$

$$2CF_3O^{\cdot} \rightarrow CF_3OOCF_3$$

The principal arguments in support of this mechanism are the observed inability of transition metal fluorides to form alkoxides and the presence of bis-trifluoromethyl peroxide among the reaction products. Nevertheless there was another viewpoint suggesting formation of an intermediate complex from, e.g. silver difluoride, with participation of carbonyl difluoride [43].

$$AgF_2 + COF_2 \rightarrow Ag(OCF_3)_2 \xrightarrow{F_2} [Ag(OCF_3)_2F_2]$$

$$\rightarrow AgF_2 + CF_3OOCF_3$$

The existence, for more than 15 years, of two conflicting opinions on the mechanism of catalysis by metal fluorides seems to be an obvious flaw in elemental fluorine chemistry, which stimulated the author of this review to present his own view, which seems to be well substantiated by the following facts.

The fluorides of transition metals, such as cobalt, silver, and nickel are known to react readily with fluorine, forming higher fluorides: CoF_3 [44], AgF_2 [44], and $NiF_{2.22}$ [45]. To cause the reverse reaction of fluorine elimination, the above higher fluorides should be heated to a temperature at least as high as 500 K. But fluorination of carbonyl difluoride on a catalyst, e.g. AgF_2 [43], proceeds at a temperature 100 K lower than that. It means that the catalyst does not generate the atomic fluorine into the gas phase, but only chemosorbs it, thus causing the activation (weakening or cleavage) of the F–F bond. In the case of NiF_2, the chemosorption of fluorine has been revealed by special experiments, and starts at the temperature of > 373 K. As for silver difluoride, it can fluorinate, at 373 K, not only carbonyl difluoride but also carbon oxide.

In the light of the above facts, it is reasonable to suggest that the radical species are formed in the system "transition metal fluoride–fluorine–reagent" as a result of the interaction of the reagent with the centre on the catalyst surface, at which fluorine is sorbed, i.e. by a two-step mechanism.

$$F_2 \xrightarrow{\text{Ct}} 2F_{adsorb}$$

$$F_{adsorb} + COF_{2,gas} \rightarrow CF_3O^{\cdot}_{gas}$$

From the moment of the radical species generation, the reaction proceeds in the gas phase via the chain mechanism with the chief stages: $CF_3O^. + F_2 \rightarrow CF_3OF + F$ and $COF_2 + F \rightarrow CF_3O^.$, and ends as a result of the quaternary break: $2CF_3O^. \rightarrow CF_3OOCF_3$.

This explains both the fluorinating ability of silver difluoride in respect of CO and COF_2 and formation of bis-trifluoromethyl peroxide in the presence of this catalyst, as mentioned above.

$$AgF_2 + COF_2 \rightarrow AgF + CF_3O^.$$

As the energy of the F–F bond (155 kJ mol^{-1}) [46] is close to that of the CF_3O–F bond (182 kJ mol^{-1}) [47], another mechanism of the transformations is quite possible—this involves decomposition of trifluoromethyl hypofluorite by the catalysts via the chemosorption mechanism.

$$CF_3OF \xrightarrow{\text{Ct}} CF_3O^._{gas} + F_{adsorb}$$

Thus the catalysis of the reaction of fluorination of COF_2 may be defined as a typical homogeneous–heterogeneous process, where the transition metal fluoride has the positive chain-initiating function, but at the same time suppresses the process by decomposing the end product. The homogeneous process results in the synthesis of trifluoromethyl hypofluorite and bis-trifluoromethyl peroxide.

It should be noted that recombination of the trifluoromethoxyl radicals to the peroxide is not a single way of their reaction. Even for the systems $COF_2 + F_2 + Ct$ (where $Ct = CsF$, NiF_2), formation of small amounts of tetrafluoromethane in addition to the peroxide have been found [48]. The most reasonable explanation to this is the reaction of the $CF_3O^.$ radical with the reactor walls.

$$CF_3O^. \xrightarrow{\text{walls}} CF_3^. \xrightarrow{F_2} CF_4 + F$$

In its nature this process is similar to the oxidation of carbon oxide to CO_2 by the trifluoromethoxyl radicals, which was proved by the experiments of thermal decomposition of CF_3OF (see below).

The comparative analysis of the catalytic synthesis of CF_3OF on caesium fluoride and transition metal fluorides from the preparative point of view allows one to conclude that they do not exclude but rather complement each other. The advantage of CsF, as mentioned above, is in obtaining a sufficiently pure product, the disadvantage is a lengthy procedure and difficulties in handling because of high hygroscopicity. In other words, caesium fluoride may be recommended for the synthesis of small amounts of highly pure CF_3OF samples. When large amounts of trifluoromethyl hypofluorite are to be obtained, it is better to use transition metal fluorides as catalysts. They are more readily available and convenient to handle, and allow a continuous process, which is more productive than a discrete one. However the latter advantage of the method over the use of CsF is reduced by the increased impurity of CF_3OF, which may contain such impurities as O_2, CF_4, CO_2, COF_2, and CF_3OOCF_3.

When choosing a method for the synthesis of hypofluorites, one should also take into account the substantionally larger synthetic possibilities of caesium fluoride, which extend, as mentioned above, to the whole series of carbonyl-containing reagents except carbon oxide. In this respect, among the recognised catalysts of the synthesis of hypofluorites, only nitrosyl fluoride seems to compete effectively with caesium fluoride, its catalytic activity based on the intermediate formation of alkyl nitrites [25].

$$COF_2 + NOF \rightleftarrows CF_3ONO \xrightarrow{F_2} CF_3OF + NOF$$

However the wide use of this method is restricted by the poor availability of nitrosyl fluoride.

The development of the catalytic methods of synthesis of organic hypofluorites from compounds containing the C=O group relates to the last and most fruitful period in the chemistry of these compounds, and no substantial advance is to be expected here in the near future. At an early stage, this work involved numerous efforts to use alcohols as the starting compounds (trichloro- and trifluoroethanol, trichloronitroisopropanol, perfluoro-*tert*-butanol) [49–53], hexafluoroacetone hydrate [49], trichloroacetic acid esters [54], as well as the alkali perfluoroalkanoates [51,53,55–57], trifluoroacetyl nitrite [50,51], and some other compounds [58,59].

These fluorinations were carried out in static or flow systems at temperatures of 200 to 300 K. To avoid profound destruction, fluorine was diluted by an inert gas in concentrations of 5 to 10% at the start and 20 to 50% at the end.

As compared to the CsF-catalysed fluorination of carbonyl-containing reagents, synthesis of hypofluorites by the direct fluorination of alcohols, acids or their derivatives is preparatively not very efficient. The main reason for that is formation of a great number of by-products and/or complex mixtures of hypofluorites. Thus, hexafluoroacetone hydrate, along with the expected heptafluoroisopropyl hypofluorite, gives tetrafluoromethane, trifluoroacetyl fluoride, carbon dioxide, and carbonyl difluoride [49]. The sodium salt of the hydrate is transformed into a mixture of four different hypofluorites [55].

$$(CF_3)_2C(OH)ONa \xrightarrow{F_2/N_2} (CF_3)_2CFOF + (CF_3)_2C(OF)_2$$
$$+ CF_3CF(OF)_2 + C_2F_5OF$$

Sodium trifluoroacetate reacts in a similar way [55].

$$CF_3COONa + F_2/N_2 \rightarrow CF_3CF(OF)_2 + C_2F_5OF + CF_2(OF)_2$$

In the fluorination of trifluoroethanol and trifluoroacetyl nitrite, the yield of pentafluoroethyl hypofluorite does not exceed 20%. Fluorination of sodium oxalate also gives by-products [56,60].

$$(COONa)_2 \begin{array}{c} \xrightarrow[295\,K]{} CF_2(OF)_2 + \text{by-products} \\ \xrightarrow[357\,K]{} CF_3OF + \text{by-products} \end{array}$$

Because of by-products and difficulties with subsequent separation and purification of the products, the above fluorination may hardly be recommended for the synthesis of individual hypofluorites. At the same time, they must not be completely ignored. Some of these methods (in particular, fluorination of trifluoroacetates at a decreased temperature) use the fluorinating agents that are milder and more selective than fluorine, and which may fluorinate organic compounds in suitable solvents. Another example of the rejection of the direct fluorination method is the transformation of sodium perfluoro-*tert*-butoxide, giving a high yield [50].

$$(CF_3)_3CONa + F_2 \rightarrow (CF_3)_3COF + NaF$$

The similarity of chemical behaviour, the absence of critical discrepancies between the procedures, and theoretical considerations necessitate discussion in this review of one more interesting class of the derivatives containing the OF group—fluoroperoxides.

The only convenient preparative method for the synthesis of fluoroperoxides is considered to be the caesium fluoride-catalysed reaction, where one of the reagents is a carbonyl-containing compound, another—oxygen difluoride. Thus, with the use of COF_2 as carbonyl containing reagent, trifluoromethyl fluoroperoxide is formed with a high yield [61].

$$COF_2 + OF_2 \xrightarrow{CsF} CF_3OOF$$

Theoretically, it is essential that the reaction be considered in terms of the nucleophilic S_N2 substitution of fluorine in the oxygen difluoride molecule, as in the synthesis of hypofluorites [62]. This is supported by serious arguments. In particular, it has been shown that in the reaction of ^{17}O-labelled carbonyl difluoride with OF_2, the isotope label of the peroxide formed in the reaction is part of the trifluoromethoxyl fragment of the molecule.

$$C^{17}OF_2 + OF_2 + \xrightarrow{CsF} CF_3{}^{17}O\text{--}OF$$

On the other hand, when $^{17}OF_2$ and COF_2 are made to react, the product is the peroxide with the isotope label in the fluorooxy fragment.

$$COF_2 + {}^{17}OF_2 \xrightarrow{CsF} CF_3O\text{--}{}^{17}OF$$

Though the arguments seem to be attractive, the question should be considered open owing to the possibility of the alternative interpretation of the reaction in terms of the one-electron transfer, as may be seen from the following schemes.

$$C^{17}OF_2 + CsF \rightleftarrows Cs^{+\,-}(^{17}O)CF_3 \xrightarrow[-e,\ -CsF]{FOF} {}^{\cdot}(^{17}O)CF_3 + {}^{\cdot}OF$$

$$^{\cdot}(^{17}O)CF_3 + {}^{\cdot}OF \rightarrow CF_3{}^{17}O\text{--}OF$$

or

$$COF_2 + CsF \rightleftarrows CF_3O^-Cs^+ \xrightarrow[-e, -CsF]{F^{17}OF} CF_3O^{\cdot} + {}^{\cdot}(^{17}O)F$$

$$CF_3O^{\cdot} + {}^{\cdot}(^{17}O)F \rightleftarrows CF_3O-^{17}OF$$

Both schemes reproduce the peculiarities of the catalytic synthesis of trifluoromethyl hypofluorite and difluoromethane-bis-hypofluorite from COF_2 and CO_2 respectively. The necessary (and principal) condition for these reactions is the low reaction temperature.

Quite serious arguments in favour of the mechanism involving the fluoro-oxide radicals are suggested [63,64]. The former shows that oxygen difluoride reacts with trifluoromethane to form trifluoromethyl hypofluorite.

$$CHF_3 + OF_2 \rightarrow CF_3OF$$

The latter studies the reaction of O_2F_2 with perfluoropropylene giving the following fluoroperoxides.

$$CF_3CF=CF_2 + FOOF \rightarrow CF_3CF_2CF_2OOF + CF_3CF(OOF)CF_3$$

Participation of fluorooxide radicals in the reaction is also indicated by the results of an investigation [65], where the UV irradiation of a mixture of oxygen, difluoride with dichlorotetrafluoroacetone led to chlorodifluoromethyl hypofluorite.

$$(ClCF_2)_2CO \rightarrow ClCF_2\overset{\cdot}{C}O + ClF_2\overset{\cdot}{C}$$
$$\underset{-CO}{\underline{\qquad\qquad}}$$

$$ClF_2\overset{\cdot}{C} + FOF \rightarrow ClCF_3 + {}^{\cdot}OF$$

$$ClF_2\overset{\cdot}{C} + {}^{\cdot}OF \rightarrow ClCF_2OF$$

The one-electron transfer mechanism is also possible for the fluorination of perfluorocarboxylic acids in the presence of CsF leading to bis-hypofluorites [66].

$$R_FCOOH + F_2 \xrightarrow{CsF} R_FCF(OF)_2$$

$$R_F = CF_3, C_2F_5, C_3F_7$$

3.3 Physical Properties and Structure

The elementary knowledge on the nature of fluoroorganic hypofluorites is provided by the acquaintance with the physical properties and structure of the most fully described and simplest compounds–trifluoromethyl hypofluorite and difluoromethane-bis-hypofluorite.

Trifluoromethyl hypofluorite is a gas (b.p. 178 K). Liquid CF_3OF has a density of 1.9 g cm^{-3} and solidifies only at 63 K or below. The standard heat of formation, evaporation heat and the Trouton constant of CF_3OF are estimated at 756, 15.5, and 20.8 kJ mol^{-1} respectively [2].

The weakest bond in the CF_3OF molecule is the O–F bond; as reported in [47,67], its energy is about 180 kJ mol^{-1}. For identification of this bond, it is convenient to use the IR spectra [68,69], whose characteristic absorption band of the O–F group lies around 945 cm^{-1}. As the number of the electron-accepting groups at the carbon atom (NO_2, CF_3, NF_2) increases, the band is displaced to lower wavenumbers [5].

In the Raman spectrum, the intensity of v(O–F) is higher than of v(C–O), whereas in the IR spectrum the case is just the opposite. The authors of [70] explain this by the more covalent character of the O–F bond as compared to the C–O bond. As indicated by the microwave spectrum in the frequency range of 8 to 40 Hz, CF_3OF has the structure of an elongated asymmetric top [71].

The structure of trifluoromethyl hypofluorite is represented by a distorted tetrahedron with the following bond length and angles: C–F (1.319±0.003A), C–O (1.395±0.006A), O–F (1.421±0.006A). ∠FOF (109.4±1°), ∠COF (104.8 ±0.6°) [72].

Important information on the nature of hypofluorites is presented by the NMR spectral data [25,26,31,33,73–75]. In the ^{19}F NMR spectra the O–F fluorine signal is considerably shifted downfield (140–160 ppm relative to $CFCl_3$), indicating a decrease of the electron density on fluorine. As a consequence, in hypofluorites the C–O bond is more stable than in the structurally similar hydrocarbon derivatives (alcohols, ethers). As shown by mass spectrometry [76], calorimetric measurements [77], and the quantum-chemical calculations [78], the difference is 40 to 60 kJ mol^{-1}. The ^{13}C NMR spectral data for CF_3OF are reported in [79].

Introduction of a second OF group into hypofluorites significantly raises their melting and boiling points. Thus difluoromethane-bis-hypofluorite boils at 209 K, and in liquid nitrogen it vitrifies. The O–F bond energy in $CF_2(OF)_2$ is 30 kJ mol^{-1} lower than in CF_3OF, which is in agreement with the lower standard heat of formation (about 540 kJ mol^{-1}). Mutual influence of the two OF groups results in the appearance of their absorption bands in the IR spectrum at 916 and 933 cm^{-1}. The bis-hypofluorite has a more complex IR spectrum [80,81]. Other data on the physical properties, geometry, and molecular structure of hypofluorites and fluoroperoxides, obtained by spectroscopic measurements and quantum-chemical calculations, are presented in [9,82–93].

3.4 Chemical Properties

The most characteristic tendency in the reactivity of hypofluorites is cleavage of the weakest O–F bond. Hence hypofluorites should be regarded first of all as fluorinating agents, to some extent—elemental fluorine analogues. At the same time, in many cases hypofluorites allow the introduction of perfluoroalkoxy fragments into organic molecules.

Owing to a relatively low energy of the O–F bond cleavage, the most typical reactions of this class of compounds are the radical processes, which are highly exothermal and, as a rule, lead to formation of various by-products.

Inadequate control of the reaction conditions frequently leads to the decomposition of the starting hypofluorites and further reactions of decomposition products with each other and the starting compounds. Careless work with hypofluorites may lead to explosions, as has been stated many times in the literature.

Due to all this, work with hypofluorites requires dilution of the reagents with inert gases or suitable solvents (perfluorocarbons or perchlorofluorocarbons), and/or low temperatures. Hypofluorite is dosed as a diluted gas or solution with vigorous stirring of the reaction mixture.

An important condition of successful and safe synthesis is the use of apparatus units (reactors, tubes, lubricants) made of the materials that are stable against the action of hypofluorites. It is recommended to use fluorocarbon lubricants, Teflon communication tubes and Teflon, Kel-F or metal (stainless steel, nickel or Monel) reactors. At low temperatures, work may be done in the ordinary glass or quartz equipment.

The analysis of literature data on the chemistry of hypofluorites allows one to divide all known reactions into two large groups: addition at multiple bonds and substitutive fluorination. The first group involves the reactions of compounds containing the C=O, C=N, and C=C bonds. The second group involves various transformations of hydrogen-containing derivatives. An insignificant group represents the fluorination reactions, where the compounds containing, for example, sulfur or phosphorus are converted by hypofluorites to the highest oxidation state.

In most cases hypofluorites have a reasonably high reactivity. This is obvious from the fact that addition to the C=C and C=N bonds generally proceeds spontaneously, in the absence of an external source of initiation. This shows unambiguously the similarity of the reactivity of hypofluorites to that of elemental fluorine, and the theoretic treatment of this phenomenon is one of the most important problems of the chemistry of this class of compounds.

Apart from spontaneous reactions, there is a relatively large group of transformations that require preliminary activation of the O–F bond by heating or by the UV irradiation. These transformations are represented by the reactions with carbonyl-containing fluorinated compounds. These are the typical free-radical reactions, and for their profound understanding it is important to know the peculiarities of the thermal and photochemical decomposition of hypofluorites, which are discussed below.

3.4.1 Thermal Decomposition

In the series of fluoroorganic hypofluorites the most thermally stable compound is trifluoromethyl hypofluorite. Decomposition of CF_3OF in a sealed reactor starts only at 473 K, and at temperatures above 523 K it proceeds at a marked rate. Decomposition products depend on the reactor material and heating conditions.

If CF_3OF is heated in a nickel reactor pre-passivated by fluorine, reversible decomposition occurs [94].

$$CF_3OF \rightleftarrows COF_2 + F_2$$

The reaction is accompanied by the formation of considerable amounts of bis-trifluoromethyl peroxide, indicating the initial formation of the trifluoro-methoxyl radicals. Theoretical treatment of this question is presented in works [95,96].

At temperatures above 673 K, decomposition of CF_3OF becomes irreversible due to the reaction of the liberated fluorine with the reactor walls. With the carbonyl difluoride additions, decomposition of CF_3OF proceeds more slowly.

The effect of glow discharge on trifluoromethyl hypofluorite differs in its results from ordinary heating: in a quartz reactor this gives CF_4, CO_2, O_2, and SiF_4; in a metal one—CF_4, CO_2, COF_2, and F_2 [94].

As opposed to CF_3OF, thermolysis of difluoromethane-bis-hypofluorite is irreversible under any conditions. Slow heating of $CF_2(OF)_2$ leads to the formation of the products similar to the products of decomposition of CF_3OF.

$$CF_2(OF)_2 \xrightarrow{\Delta} COF_2 + F_2 + 0.5\,O_2$$

This result is completely in agreement with the above-suggested mechanism of fluorination of carbon dioxide with participation of the fluorooxydifluoromethoxyl radical. Only here the radical is formed as a result of decomposition of the bis-hypofluorite at the first stage of the reaction.

$$CF_2(OF)_2 \xrightarrow{\Delta} FOCF_2O^{\cdot} + F$$

In conditions of explosive decomposition of $CF_2(OF)_2$ [77], radical $FOCF_2O^{\cdot}$ undergoes more complex transformations.

$$CF_2(OF)_2 \xrightarrow{explosion} COF_2 + CF_3OF + CF_3OOCF_3 + CF_3OOF + CF_4$$

A distinction of thermal decomposition of higher perfluoroalkyl hypofluorites is the chain character of the reaction. It raises the tendency to stabilisation of the alkoxy radicals by the homolytic decomposition of the C–C bond with subsequent formation of alkanes [49].

$$CF_3CF_2OF \xrightarrow{\Delta} CF_3CF_2O^{\cdot} + F$$

$$CF_3CF_2O^{\cdot} \rightarrow CF_3^{\cdot} + COF_2$$

$$^{\cdot}CF_3 + CF_3CF_2OF \rightarrow CF_4 + CF_3CF_2O^{\cdot} \quad \text{etc.}$$

In accordance with this tendency, secondary and tertiary hypofluorites are transformed to acyl fluorides and ketones.

$$(CF_3)_2CFOF \xrightarrow{\Delta} (CF_3)_2CFO^{\cdot} + F$$

$$(CF_3)_2CFO^{\cdot} \rightarrow CF_3C(O)F + \overset{\cdot}{C}F_3$$

$$(CF_3)_3COF \xrightarrow{\Delta} (CF_3)_3CO^{\cdot} + F$$

$$(CF_3)_3CO^{\cdot} \rightarrow (CF_3)_2CO + {}^{\cdot}CF_3$$

Thus higher perfluoroalkyl hypofluorites, as well as difluoromethane-bis-hypofluorite, are less thermally stable than CF_3OF. There is a linear correlation between the O–F bond energy and the ^{19}F chemical shift [96]. Due to this, for the tentative estimation of thermal stability of hypofluorites we may use the NMR spectral data.

3.4.2 Photochemical Decomposition

The synthetic purpose in the photolysis of hypofluorites is to generate the perfluoroalkoxyl radicals in mild conditions. As the O–F bond cleavage energy is small, any UV source may be used for this purpose. Though very short waves may produce an undesirable effect. For example, photolysis of difluoromethane-bis-hypofluorite at $\lambda = 254$ nm (~ 460 kJ mol^{-1}) results in cleavage of the O–F and C–OF bonds [97], whereas at long waves the process follows its usual route [98].

$$CF_2(OF)_2 \xrightarrow{h\nu} COF_2 + F_2 + 0.5\,O_2$$

The photolysis of CF_3OF in gas phase yields bis-trifluoromethyl peroxide [20]. In the argon matrix (8 K, $\lambda < 400$ nm), the photolysis is accompanied by the formation of COF_2. The ESR control of irradiation of the mixtures of CF_3OF with NF_3 or CF_4 at 77 K allowed the detection of radical $CF_3OO^{\cdot}$ [98].

On the basis of the authors' suggestion about the presence of an oxygen trace in the mixture, it is possible to admit that the trifluoromethoxyl radical is first converted to the trioxide one, which subsequently recombinates.

$$CF_3OF \xrightarrow{h\nu} CF_3O^{\cdot} + F$$

$$CF_3O^{\cdot} + O_2 \rightarrow CF_3OOO^{\cdot} \rightarrow CF_3OO^{\cdot} + 0.5\,O_2$$

This route of the reaction is supported, first, by the studies of the photosensibilised oxidation of CO in CO_2 by the UV irradiation of a mixture of $CF_3OF + CO + O_2$, where the quantum yield of carbon dioxide reaches

$\sim 10^4$ molecules $(h\nu)^{-1}$ [99], and, second, by the results of [100].

$$CF_3O^{\cdot} + O_2 \rightarrow CF_3OOO^{\cdot}$$

$$CF_3OOO^{\cdot} + CO \rightarrow CF_3OO^{\cdot} + CO_2$$

$$CF_3OO^{\cdot} + CO \rightarrow CF_3O^{\cdot} + CO_2$$

$$CF_3O^{\cdot} + CO \rightarrow {}^{\cdot}CF_3 + CO_2$$

$${}^{\cdot}CF_3 + O_2 \rightarrow CF_3OO^{\cdot} \quad \text{etc.}$$

3.4.3 Reactions with Carbonyl-Containing Compounds

Interaction of hypofluorites with carbonyl-containing compounds is a method for the synthesis of alkyl peroxides, ethers and esters, carbonates, etc. It has been most fully studied in the case of the reaction of CF_3OF and $CF_2(OF)_2$.

Most known reactions of this type require the thermal or photochemical pre-activation of the O–F bond. Besides, some cases of activation by catalysts have been reported, and some of the reactions proceed on their own.

In particular, thermal or photochemical initiation is required for the reaction of CF_3OF with COF_2, leading to the formation of bis-trifluoromethyl peroxide [101–110].

$$CF_3OF + COF_2 \xrightarrow{\Delta} CF_3OOCF_3$$

In conditions of thermal initiation the process is reversible. The optimal temperature for the synthesis of the peroxide both in the static [102,103] and flow [43] conditions is in the range of 500 to 550 K. Up to 520 K the equilibrium is shifted to the right. The system $CF_3OF + COF_2$ obeys the Le Chatelier principle: at the pressure of 10 MPa and the temperature of 550 K, yield of the peroxide reaches 91% [102]. As shown by the kinetic measurements [105–107], the limiting stage of the thermal synthesis of CF_3OOCF_3 is the decomposition of trifluoromethyl hypofluorite.

$$CF_3OF \xrightarrow{slow} CF_3O^{\cdot} + F$$

$$COF_2 + F \xrightarrow{fast} CF_3O^{\cdot}$$

$$2CF_3O^{\cdot} \xrightarrow{fast} CF_3OOCF_3$$

With UV irradiation, the process follows the same route.

The photodecomposition of CF_3OF ($\lambda = 360$ nm) in the presence of carbon oxide leads to the formation of trifluoromethyl fluoroformiate and trifluoro-methyl oxalate [111–113].

$$CF_3O^{\cdot} + CO \rightarrow CF_3O\dot{C}O$$

$$CF_3O\dot{C}O + CF_3OF \rightarrow CF_3OC(O)F + CF_3O^{\cdot}$$

$$2CF_3O\dot{C}O \rightarrow CF_3OC(O)C(O)OCF_3$$

For the analogous thermal reaction at 353–383 K, a mechanism of self-initiation of chains has been suggested on the basis of the fact that in these conditions CF_3OF does not tend to dissociate [114].

$$CF_3OF + CO \xrightarrow{\Delta} CF_3O^{\cdot} + {}^{\cdot}COF$$

A convincing proof of the bimolecular initiation is the low activation energy of the total process (75 kJ mol^{-1}). If CF_3OF decomposed at the O–F bond, the activation energy would be at least as high as the energy of the bond cleavage, i.e. it would be more than twice as high as it really is.

Another route in the reaction of trifluoromethyl hypofluorite with carbon oxide is the synthesis of bis-trifluoromethyl carbonate.

$$CF_3O^{\cdot} + CF_3O\dot{C}O \rightarrow (CF_3O)_2CO$$

The latter may also be obtained in a small yield by the photolysis of a mixture of CF_3OF with hexafluoroacetone [115].

The reaction of CF_3OF with carbonyl-containing compounds indicates both an extremely high reactivity of the trifluoromethoxyl radical and its relative stability, which is not the case with the alkoxyl radical formed from $CF_2(OF)_2$. Under conditions of thermal initiation the reaction of $CF_2(OF)_2$ with CO gives carbon dioxide and carbonyl difluoride in the 1:2 mole ratio [116]. The activation energy value (83 kJ mol^{-1}) and the reaction temperature (383–400 K), which are similar to those for CF_3OF, indicate the same chain initiation stage for both processes [116].

$$CF_2(OF)_2 + CO \rightarrow FOCF_2O^{\cdot} + {}^{\cdot}COF$$

Thus, here again we come across the $FOCF_2O^{\cdot}$ radical, whose transformation to CO_2 and COF_2 for the system $CF_2(OF)_2 + CO$ seems to be quite legal from the point of view of the above-considered schemes.

$$CF_2(OF)_2 + {}^{\cdot}COF \rightarrow FOCF_2O^{\cdot} + COF_2$$

$$FOCF_2O^{\cdot} \rightarrow {}^{\cdot}OF + COF_2$$

$$FO^{\cdot} + CO \rightarrow FO\dot{C}O \rightleftarrows CO_2 + F$$

Proceeding to draw the analogy between the catalytic systems of hypofluorites and their catalysed reactions with carbonyl-containing reagents, it should be noted that the latter reactions also have the radical nature. The main products here are also the alkyl peroxides; the catalytic effect is produced by the silver, copper, nickel, mercury, cobalt, iron [103], and caesium fluorides [43].

The catalytic effect of these fluorides is obviously based on the generation of the $FOCF_2O^{\cdot}$ radicals as a result of chemosorption of atomic fluorine and subsequent fluorination of the carbonyl-containing reagent. Thus synthesis of bis-trifluoromethyl peroxide from CF_3OF and COF_2, in the presence of silver fluoride may be represented by the following scheme.

$$CF_3OF_{gas} + AgF \rightarrow CF_3O^{\cdot}_{gas} + AgF_2(F_{ads})$$

$$COF_2 + AgF_2 \rightarrow CF_3O^{\cdot}_{gas}$$

$$2CF_3O^{\cdot}_{gas} \rightarrow CF_3OOCF_{3\,gas}$$

Consequently, like the similar synthesis of CF_3OF, the total process of the catalytic synthesis of CF_3OOCF_3 is the heterogeneous process.

As shown in [40], for the CsF-catalysed processes, the mechanism of formation of both CF_3OF and CF_3OOCF_3 is again the same. Synthesis of CF_3OOCF_3 here is also preceded by the generation of the $CF_3O^{\cdot}$ radical, which may be detected by adding carbon oxide to the reaction mixture to produce CO_2 and CF_4 (see above).

The most satisfactory explanation of this result is offered by the one-electron transfer theory. The role of electron donor here is played by caesium trifluoromethoxide, that of acceptor—by trifluoromethyl hypofluorite. The electron transfer proceeds with the formation of two $CF_3O^{\cdot}$ radicals, which subsequently recombine [40].

$$CF_3\overset{\frown}{O}^- Cs^+ + \overset{\frown}{F}-OCF_3 \xrightarrow{-e} [CF_3\overset{\uparrow}{O}^{\cdot}\cdots Cs^+ \cdots F^- \cdots \overset{\downarrow}{O}CF_3] \longrightarrow CF_3\overset{\uparrow}{O}^{\cdot}_{gas} + CsF + CF_3\overset{\downarrow}{O}^{\cdot}_{gas}$$

It is interesting that in the transition state free electrons of the CF_3O fragments are in the fields of opposite charge. Owing to this, the electron spins are opposite, and recombination is possible.

It should be noted that the one-electron transfer theory offers a simple explanation of some other experimental facts as well, which seem anomalous at first sight. This concerns, first of all, the reaction of $CF_2(OF)_2$ with COF_2 in the presence of CsF. From the viewpoint of the S_N2 mechanism, the main product in this reaction should be expected to be CF_3OOCF_3. As a matter of fact, the main products are bis-trifluoromethyl trioxide and trifluoromethylfluoroformyl peroxide [117–119].

$$CF_2(OF)_2 + COF_2 \xrightarrow{CsF} \begin{cases} X \longrightarrow CF_2(OOCF_3)_2 \\ \longrightarrow CF_3OOOCF_3 + CF_3OOC(O)F \end{cases}$$

The one-electron transfer accounts for the result observed in the following series of radical transformations.

$$FOCF_2OF + CF_3O^-Cs^+ \xrightarrow{-e} FOCF_2O^{\cdot} + CsF + CF_3O^{\cdot}$$

$$FOCF_2O^{\cdot} \longrightarrow COF_2 + {}^{\cdot}OF \xrightarrow{{}^{\cdot}OF} F + F + O_2$$

$$F + FOC(O)F \xrightarrow{COF_2} CF_3O^{\cdot} + {}^{\cdot}OC(O)F \longrightarrow CF_3OOC(O)F$$

$$CF_3O^{\cdot} + O_2 \longrightarrow CF_3OOO^{\cdot} \longrightarrow CF_3OO^{\cdot} + 0.5O_2$$

$$CF_3O^{\cdot} + CF_3OO^{\cdot} \longrightarrow CF_3OOOCF_3$$

A similar scheme may be presented to explain the formation of the products of the reaction of potassium perfluoro-*tert*-butoxide with trifluoromethyl hypofluorite [120].

$$(CF_3)_3CO^-K^+ + CF_3OF \xrightarrow{-e} (CF_3)_3CO^\cdot + KF + CF_3O^\cdot$$

$$CF_3O^\cdot \rightarrow F + COF_2 \xrightarrow{KF} CF_3O^-K^+$$

$$(CF_3)_3CO^-K^+ + COF_2 \rightarrow (CF_3)_3COC(O)F \xrightarrow{(CF_3)_3CO^-K^+} [(CF_3)_3CO]_2CO$$

$$(CF_3)_3CO^\cdot + F \rightarrow (CF_3)_3COF$$

Lastly, it is necessary to mention two other "anomalous" results. One case is the reaction of CF_3OF with difluorocarbamyl fluoride, where, instead of the expected peroxide, the reaction product is O-trifluoromethyl-N,N-difluoro-hydroxylamine [121].

$$F_2NC(O)F + CF_3OF \xrightarrow{CsF} \begin{cases} -X \longrightarrow F_2NCF_2OOCF_3 \\ \\ \longrightarrow F_2NOCF_3 \end{cases}$$

Another case is the intramolecular cyclization of perfluoroglutaryl difluoride by fluorine [122].

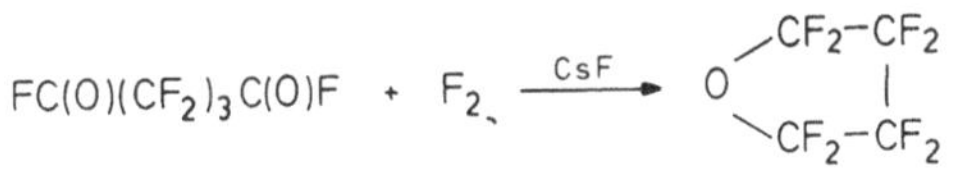

$$FC(O)(CF_2)_3C(O)F \ + \ F_2 \xrightarrow{CsF} O\begin{array}{c} CF_2-CF_2 \\ | \\ CF_2-CF_2 \end{array}$$

It is impossible to explain these results in terms of the ionic mechanism, whereas one-electron transfer gives a quite convincing explanation of them.

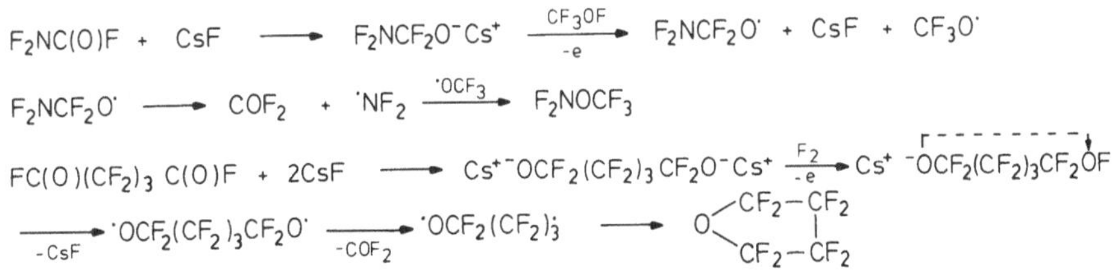

$$F_2NC(O)F + CsF \longrightarrow F_2NCF_2O^-Cs^+ \xrightarrow[-e]{CF_3OF} F_2NCF_2O^\cdot + CsF + CF_3O^\cdot$$

$$F_2NCF_2O^\cdot \longrightarrow COF_2 + {}^\cdot NF_2 \xrightarrow{{}^\cdot OCF_3} F_2NOCF_3$$

$$FC(O)(CF_2)_3C(O)F + 2CsF \longrightarrow Cs^+{}^-OCF_2(CF_2)_3CF_2O^-Cs^+ \xrightarrow[-e]{F_2} Cs^+ {}^-OCF_2(CF_2)_3CF_2OF$$

$$\xrightarrow{-CsF} {}^\cdot OCF_2(CF_2)_3CF_2O^\cdot \xrightarrow{-COF_2} {}^\cdot OCF_2(CF_2)_3^\cdot \longrightarrow O\begin{array}{c} CF_2-CF_2 \\ | \\ CF_2-CF_2 \end{array}$$

Thus the use of CsF to activate the carbonyl-containing reagents leads to significantly milder reaction conditions, but in most cases it fails to give the expected results due to the specific character of the processes.

   Milder reaction conditions may be brought about by the presence of anionoid donor centres in the reactant systems. This is also the case when the starting covalent compounds contain reasonably polarised multiple bonds. This refers to thioanalogues of carbonyl-containing compounds. Thus thiophosgene spontaneously reacts with CF_3OF at temperatures as low as 195 K, forming two products (due to the doubly oriented addition), one of which is unstable [123].

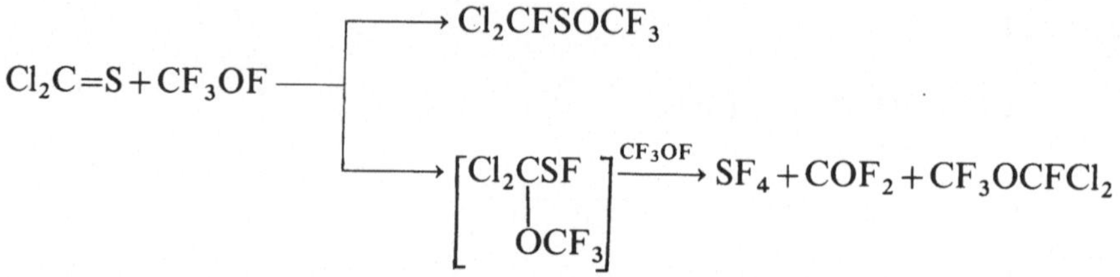

$$Cl_2C=S+CF_3OF \xrightarrow{\hspace{2cm}} \begin{cases} \longrightarrow Cl_2CFSOCF_3 \\[2em] \longrightarrow \left[\underset{\underset{OCF_3}{|}}{Cl_2CSF}\right] \xrightarrow{CF_3OF} SF_4+COF_2+CF_3OCFCl_2 \end{cases}$$

Carbonyl sulfide shows less activity: at 373 K after 3 h, the conversion degree is 75% and the main product is again the S–OCF$_3$ derivative.

$$O=C=S+CF_3OF \rightarrow CF_3OSC(O)OCF_3+(CF_3O)_2CO+CF_3OC(O)F$$

The radical nature of this reaction has been confirmed by the ESR identification of radicals $(CF_3O)_2\dot{S}C(O)OCF_3$ and $(CF_3O)_2\dot{S}C(O)F$, showing that the more nucleophilic sulfur atom is trifluoromethoxylated first. A similar reaction has been reported for CSF$_2$ [124].

3.4.4 Reactions with Compounds Containing The C=N Bond

The reactions of hypofluorites with compounds containing the C=N bond are, with very few exceptions, self-initiated. This type of transformation corresponds with the pseudoelectrophilic character of hypofluorites and the donor properties of the multiple bond nitrogen. For that reason, the initial fluorine attack is directed to the nitrogen.

To obtain the results close to preparative, treatment of the C=N-containing compounds with hypofluorites should be conducted in inert solvents (CH$_2$Cl$_2$, CHCl$_3$, CFCl$_3$), at a low temperature, and with diluted (10–12%) hypofluorite.

In the case of CF$_3$OF it has been found that at the primary stage the reaction proceeds with addition of the CF$_3$O and F fragments at the C=N bond, and its fluorination. Both types of the initial products are generally unstable, and, upon heating, undergo various transformations [125–128].

Thus after fluorination of the double bond, phenylalkyl-(cycloalkyl)imines containing vinylic hydrogen readily release HF [126].

$$4\text{-}X\text{-}C_6H_4CH=NR+CF_3OF \rightarrow 4\text{-}X\text{-}C_6H_4CF=NR$$

$$X=H, NO_2; \quad R=i\text{-}Pr, t\text{-}Bu, cyclo\text{-}C_6H_{11}$$

Subsequent interaction between the fluoroimine formed and an excess of CF$_3$OF leads to the formation of N–F amines [126].

$$4\text{-}X\text{-}C_6H_4CF=NR+CF_3OF \rightarrow 4\text{-}X\text{-}C_6H_4CF_2NFR$$

In the case of hydrazones, the reaction stops at the stage of azaalkanes [129].

$$RNHN=CClR' + CF_3OF \rightarrow [RNHNF-CClFR'] \rightarrow RN=NCClFR'$$

$$R = 2,4,6\text{-}C_6H_2Cl_3; \ R' = i\text{-}Pr, \text{cyclo-}C_6H_{11}$$

The presence of the alkyl, or especially the alkoxy, group at the carbon atom of the C=N bond promotes the destructive character of the process involving cleavage of the carbon-nitrogen bond. This underlies the synthesis of N,N-difluoroamines with yields of up to 70%.

$$4\text{-}X\text{-}C_6H_4\underset{\underset{\text{OEt}}{|}}{C}=NR + CF_3OF \rightarrow 4\text{-}X\text{-}C_6H_4CF_2OEt + RNF_2$$

$$R = \text{adamantyl}; \ X = H, OH, OMe, F$$

The destruction has also been found to be promoted by methanol, but in this case the reaction gives the respective acetals in addition to difluoroamine [128].

$$4\text{-}NaOOC\text{-}C_6H_4CH=NR + CF_3OF$$

$$\xrightarrow{\text{MeOH}} 4\text{-}NaOOC\text{-}C_6H_4CH(OMe)_2 + RNF_2$$

The reason for the formation of acetals is the competition of methanol as a reasonably strong nucleophile with the trifluoromethoxide in the attack at the cationoid reaction centre; the reaction follows the route postulating the electrophilic nature of CF$_3$OF.

$$RN=CHR' + CF_3OF \xrightarrow{\text{MeOH}} RNF\underset{\underset{\text{OMe}}{|}}{-}CHR' \xrightarrow{CF_3OF} R\overset{+}{N}F_2\underset{\underset{\text{OMe}}{|}}{CHR'}$$

$$\xrightarrow[-RNF_2]{\text{MeOH}} (MeO)_2CHR'$$

Although there is no direct proof, such an explanation seems reasonable at first sight. In any case, it is consistent with the fact of synthesis of acetal, unambiguously indicating the existence of intermediate states with cationoid centres. This result would not have been possible in the case of the radical mechanism, but the electrophilic mechanism is also unlikely from the viewpoint of classical knowledge, as the heterolysis of the O–F bond to form the fluorine cation is impossible.

Thus, considering the reactivity of hypofluorites, we again, as in the case of their synthesis, meet with the necessity of searching for a reasonable approach removing all the contradictions. One such approach seems to be the one-electron transfer theory [130–134]. Indeed, in the electron transfer from one reagent (donor) to another (acceptor), the transition state is represented by the pair— radical cation–radical anion. In the system CF$_3$OF-imine, the role of donor is played by the conjugated system of n-electrons and the π-bond, that of acceptor—by the non-bonding orbital of the O–F bond. The elementary act of electron transfer is accompanied by the formation of a cationoid centre on the carbon atom, as represented by the following scheme.

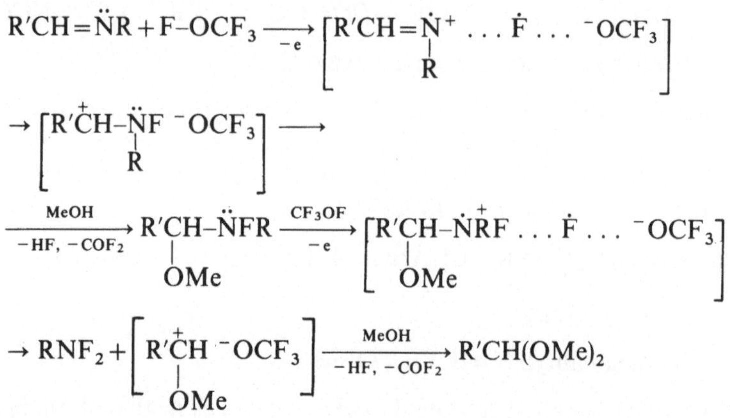

$$R'CH = \ddot{N}R + F\text{-}OCF_3 \xrightarrow{-e} \left[R'CH = \overset{+}{\underset{R}{N}} \ldots \dot{F} \ldots {}^-OCF_3 \right]$$

$$\rightarrow \left[\overset{+}{R'CH}\text{-}\underset{R}{\ddot{N}F} \ {}^-OCF_3 \right] \longrightarrow$$

$$\xrightarrow[-HF, \ -COF_2]{MeOH} R'CH\text{-}\underset{OMe}{\ddot{N}FR} \xrightarrow[-e]{CF_3OF} \left[R'CH\text{-}\overset{+}{\underset{OMe}{N}}RF \ldots \dot{F} \ldots {}^-OCF_3 \right]$$

$$\rightarrow RNF_2 + \left[\overset{+}{R'CH} \ {}^-OCF_3 \right] \xrightarrow[-HF, \ -COF_2]{MeOH} R'CH(OMe)_2$$

The above scheme explains how fluorination of the double bond proceeds when the reaction is conducted in the absence of MeOH: in the non-solvating medium, the trifluoromethoxide is stabilized by the fluoride ion elimination, the latter being consumed for stabilisation of the cationoid centre.

The reactions with ionic intermediates change to the reactions with radical intermediates when the starting reagents are perfluorinated imines. Thus the radical mechanism is achieved in the reaction of CF_3OF with perfluoroethyl-methyleneimine [135], forming perfluoroethylmethyl-N-fluoroamine. The reaction occurs only on heating the reaction mixture to 523 to 573 K, and, in view of the fact that another product is COF_2, it may be represented by the scheme.

$$CF_3OF \rightarrow CF_3O^{\cdot} + F$$

$$CF_3O^{\cdot} \rightarrow COF_2 + F$$

$$C_2F_5N = CF_2 + 2F \rightarrow C_2F_5NFCF_3$$

The reactions of $CF_2(OF)_2$ and $CF_3CF(OF)_2$ with perfluoroguanidine, described in the patent literature [136,137], obviously belong to the same type. The radical nature of these processes is indicated by a great number of products.

$$CF_2(OF)_2 + (F_2N)_2C = NF \rightarrow CF_3NF_2 + CF_2(NF_2)_2 + CF(NF_2)_3$$

$$+ (NF_2)_3CN + CF_3OOC(O)F$$

$$+ CF_3OOCF_2OF + CF_3OOF$$

$$+ (CF_3OO)_2CO + CF_3OOOCF_3$$

Another example of radical-type reactions is the interaction of trifluoromethyl hypofluorite with fluorosulfonyl isocyanate, which shows the feasibility of the direct synthesis of the N-OCF_3 derivative [138].

$$FSO_2N = C = O + CF_3OF \xrightarrow{h\nu} FSO_2N(OCF_3)C(O)F$$

3.4.5 Reactions with Compounds Containing The C=C Bond

The interaction of hypofluorites with compounds containing the C=C bonds has been most widely covered in the literature, and synthetically it is the most interesting reaction. Before discussing separate papers, let us consider in short some general issues.

First of all it should be noted that to choose the reaction type and conditions, it is practical to make a preliminary estimation of the electron density of the double bond and its distribution between the carbon atoms. Generally, the higher the electron density of the C=C bond, the more reactive is the unsaturated reagent and, consequently, the milder are the reaction conditions.

In predicting the reaction results it is necessary to take into account that the attack on the double bond is almost always directed at the point of the highest electron density. This attack is, however, frequently hindered by steric factors. The effect of polar and steric factors shows itself in the fact that some classes of unsaturated compounds or hypofluorites with bulky substituents (such as perfluoroalkyl) react in a different way compared to the unsubstituted analogues.

This led to the ambiguous interpretation of the reactivity of hypofluorites with unsaturated compounds: in terms of the free-radical and electrophilic mechanisms.

One of the first attempts to clear up this question was made in a review [9], where the author treated the electrophilic mechanism as unlikely. Here we give a more detailed treatment, discussing first the reactions that are undoubtfully radical, and then—the ones which admit an alternative interpretation to the electrophilic mechanism.

The available experimental data on the reactions of hypofluorites with unsaturated compounds containing the C=C bond allow one to divide all the transformations into four classes: fluorination of multiple bonds, fluoroalkoxylation (addition of the CF_3O and F fragments), polymerisation (or telomerisation), and substitution of hydrogen or complex groups involved in a multiple bond. The latter type of transformation has only a formal relation to substitution; as a matter of fact, these reactions proceed via the addition–elimination stage.

One of the specific indications to the radical nature of the reactions of hypofluorites with unsaturated compounds is their self-explosive character, as shown by the authors of [139]. They also found that it is possible to avoid explosions by diluting the reagents with an inert gas. However too high dilution suppresses self-initiation and requires UV irradiation, as in the case of ethylene.

$$CH_2=CH_2 + CF_3OF \xrightarrow{h\nu} CF_3OCH_2CH_2F$$

Safe conditions are provided by combining dilution of the reagents with an inert gas with deep (to 90 K) cooling. When these conditions are applied to the reaction of tetrafluoroethylene with $CF_2(OF)_2$, yield of the target product reaches 80–90% [140].

$$CF_2=CF_2 + CF_2(OF)_2 \rightarrow CF_2(OCF_2CF_3)_2$$

But this is not always the case. Thus in the reaction of $CF_2(OF)_2$ with 1,2-dichloroethylene, yield of the diether is only 9%.

$$CHCl=CHCl + CF_2(OF)_2 \rightarrow CF_2(OCHClCHFCl)_2$$

Addition of hypofluorites to chlorine-containing olefins always proceeds in a more complex way than the addition to perfluoroalkenes, and its results call for consideration of this question on analogy with elemental fluorine.

Addition of fluorine to chloroolefins is known to proceed, as a rule, with formation of higher-chlorinated alkanes than may be expected from the reaction route. For example, 1,2-dichlorodifluoroethylene, in addition to 1,2-dichloro-tetrafluoroethane, gives 1,1,2-trichloro- and 1,1,2,2-tetrachloroethanes [141].

The reason of for the formation of by-products is elimination of atomic chlorine from the β-position relative to the radical centre in the intermediate fluoroalkyl radicals. This phenomenon was well studied in the case of the reversible reaction of bromination of olefins [142], and has been strictly proved by recording the ESR spectrum of atomic chlorine in the diffusion flame of fluorine with tetrachloroethylene [143].

$$\underset{\underset{Cl}{|}}{\overset{\overset{Cl}{|}}{F\dot{C}}}{-}\dot{C}Cl_2 \rightarrow \underset{\underset{Cl}{|}}{F}C{=}CCl_2 + Cl$$

Naturally, the removed halogen atom further participates in the halogenation of the double bond of the starting olefin, or is transformed by recombination to molecular halogen which competes with fluorine. In the case of 1,2-dichlorodifluoroethylene, this chain process involves the following elementary stages.

$$CFCl=CFCl + F_2 \rightarrow CF_2Cl\dot{C}FCl + F$$

$$CF_2Cl\dot{C}FCl \rightarrow CF_2=CFCl + Cl$$

$$CFCl=CFCl + Cl \rightarrow CFCl_2\dot{C}FCl$$

$$CFCl_2\dot{C}FCl + F_2 \rightarrow CFCl_2CF_2Cl + F$$

$$Cl + Cl \rightarrow Cl_2$$

$$CFCl_2\dot{C}FCl + Cl_2 \rightarrow CFCl_2CFCl_2 + Cl$$

The same is expected for the reaction of 1,2-dichlorodifluoroethylene with trifluoromethyl hypofluorite. Indirect evidence for this is offered by the results of [144], where the authors indicate that dechlorination of the products of this reaction by zinc in DMSO leads to a mixture of perfluoromethylvinyl ether, tetrafluoroethylene, and chlorotrifluoroethylene. As formation of the latter olefin may not be explained by the peculiarities of the dechlorination reaction, one can only suggest the presence of the $CFCl_2CF_2Cl$ admixture in the main product $CF_3OCFClCF_2Cl$.

$$CFCl=CFCl + CF_3OF \rightarrow CF_3OCFClCF_2Cl + ClCF_2CF_2Cl$$
$$+ CFCl_2CF_2Cl$$
$$\xrightarrow[-ZnCl_2]{Zn} CF_3OCF=CF_2 + CF_2=CF_2 + CF_2=CFCl$$

Thus the first and rather essential conclusion about the nature of the reactivity of hypofluorites is the fact that they behave in a similar way as elemental fluorine. This conclusion is preparatively important in view of the researchers' orientation to search for the by-products of various transformations of olefins. When the reactivity of multiple bond is low and the activity of hypofluorite is small, formation of by-products proceeds at the expense of hypofluorite. In such a way perfluoro-*tert*-butyl hypofluorite reacts with perfluorocycloolefins under the UV irradiation [145].

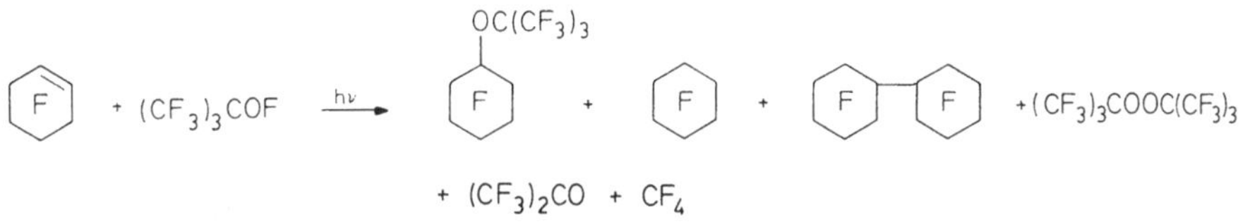

In the absence of the UV irradiation, the reaction does not proceed.

With increased reactivity of the multiple bond, the UV irradiation is no longer required. Thus, $(CF_3)_3COF$ and hexafluoropropylene, preliminarily mixed at a low temperature in perfluoropentane, react with slow heating to form the products of addition chiefly at the difluoromethylene group [146].

$$(CF_3)_3COF + CF_2=CFCF_3 \rightarrow (CF_3)_3COCF_2CF_2CF_3$$
$$(CF_3)_3COF + CF_2=CFOR_F \rightarrow (CF_3)_3COCF_2CF_2OR_F$$

The products of addition of the alkoxy group at the 2-position do not exceed 5%.

Under similar conditions, perfluoro-*tert*-butyl hypofluorite can be added to hexafluorobenzene. Orientation of the addition is such that the reaction leads to one of the possible isomers—perfluoro-4-*tert*-butoxy-1,4-cyclohexadiene [147].

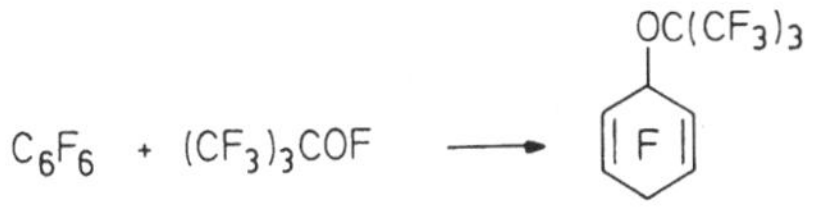

As opposed to this, the reaction of hexafluorobenzene with CF_3OF gives two isomers, with a two-fold predominance of perfluoro-5-trifluoromethoxy-1,3-cyclohexadiene.

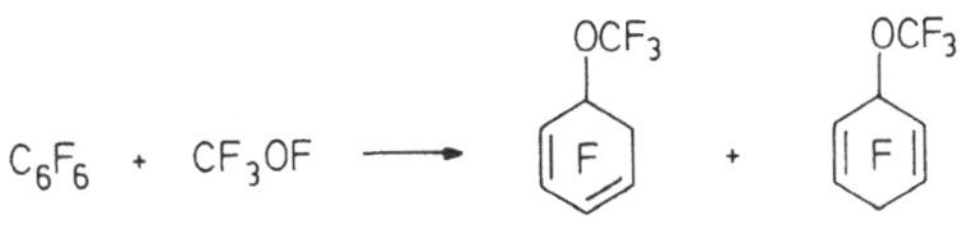

Trifluoromethyl hypofluorite strongly differs in its behavior from perfluoro-*tert*-butyl hypofluorite. Thus, with perfluoropropylene it also gives two isomers [148,149].

$$CF_2=CFCF_3 + CF_3OF \rightarrow CF_3OCF_2CF_2CF_3 + (CF_3)_2CFOCF_3$$
$$66\% 33\%$$

Perfluoro-*tert*-butyl hypofluorite does not react with octafluorotoluene, whereas CF_3OF adds to it forming cyclohexadiene and cyclohexene, though in more rigid conditions than in the case of C_6F_6.

$$C_6F_5CF_3 + CF_3OF \rightarrow CF_3C_6F_6OCF_3 + CF_3C_6F_8OCF_3$$

Unfortunately, the authors of [147] do not indicate position of CF_3O groups in the ring, which might be important for clearing up some mechanistic details of the reaction. Nevertheless, on the basis of the low reactivity of octafluorotoluene with CF_3OF, they regard the process to be "electrophilic addition involving fluoronium ions".

The authors of [147] are right if they have in view the decreased electron density on the O–F fluorine (the "pseudoelectrophilic" fluorine). But this is doubtful if they suggest participation of the fluorine cation in the reaction (cf. [150]). To clear up this question, let us consider other works.

Recent studies [149] show the hypothesis of electrophilic addition of hypofluorites to olefins to be erroneous. Enormous factual material indicates that hypofluorites, such as CF_3OF, show low stereospecificity of addition and no regiospecificity. Thus interaction of CF_3OF with 1,1-difluoroethylene and 1,1-dichloroethylene proceeds against the Markovnikov rule.

$$\overset{\delta-}{CH_2}=\overset{\delta+}{CF_2} + CF_3OF \rightarrow CF_3OCH_2CF_3$$
$$\overset{\delta-}{CH_2}=\overset{\delta+}{CCl_2} + CF_3OF \rightarrow CF_3OCH_2CFCl_2$$

1,1-Dichlorodifluoroethylene reacts with CF_3OF, forming 1,1-dichlorotetra-fluoroethane and trifluoromethoxylation products (a mixture of regioisomers), and *cis*-1,2-dichloroethylene is transformed to the equimolar mixture of *threo*- and *erythro*-dichlorotetrafluoromethylethyl ethers instead of one *erythro*-isomer expected in the case of the electrophilic addition mechanism.

$$\overset{\delta+}{CF_2}=\overset{\delta-}{CCl_2} + CF_3OF \rightarrow CF_3CFCl_2 + CF_3OCF_2CFCl_2 + CF_3OCCl_2CF_3$$
$$25\% 63\% 12\%$$

$$CHCl=CHCl + CF_3OF \rightarrow CF_3OCHClCHFCl$$

erythro : *threo* = 1 : 1

The authors of [149] explain these results from the viewpoint of the predominance of steric factors (over the polar ones), as it should be in the polar radical processes [142].

The predominant formation of trifluoromethyl-β,β-dichlorotrifluoroethyl ether (63%) from 1,1-dichlorodifluoroethylene seemingly indicate the electrophilic addition, as it corresponds to the above-shown electron density distribution in the olefins. The same is suggested by the results of addition of CF_3OF to perfluoropropylene and bromotrifluoroethylene, where the content of "normal" (from the point of view of electrophilic addition) regioisomers is 66% and 80% respectively [149].

$$\overset{\delta+}{CF_2}=\overset{\delta-}{CF}CF_3 + CF_3OF \rightarrow CF_3OCF_2CF_2CF_3$$
$$66\%$$

$$\overset{\delta+}{CF_2}=\overset{\delta-}{CF}Br + CF_3OF \rightarrow CF_3OCF_2CF_2Br + CF_3CFBrOCF_3$$
$$80\% \qquad\qquad 20\%$$

However these facts find simple and convincing explanation just in terms of the radical mechanism involving the predominant attack of olefin by the trifluoromethoxyl radical formed in the chain process, but not by the fluorine atom of hypofluorite.

$$CF_2=CCl_2 + CF_3OF \rightarrow CF_3\dot{C}Cl_2 + CF_3O^{\cdot}$$

$$CF_2=CCl_2 + CF_3O^{\cdot} \rightarrow CF_3OCF_2\dot{C}Cl_2$$

$$CF_3OCF_2\dot{C}Cl_2 + CF_3OF \rightarrow CF_3OCF_2CFCl_2 + CF_3O^{\cdot} \quad etc$$

The observed reaction route results from the high reactivity of the CF_3O radical. The latter circumstance allows this radical to attack the electron-deficient (but less screened) carbon atom of the double bond. Hence yield of the "electrophilic" addition product increases with the increased screening effect of substituent, though the reaction follows the radical mechanism.

The above-considered peculiarity of addition of CF_3OF to olefins is not a single one. For example, 1,1-dihaloethylenes (see above) and fluoroethylene add the trifluoromethoxyl radical chiefly at the methylene group [149].

$$\overset{\delta-}{CH_2}=\overset{\delta+}{CH}F + CF_3OF \rightarrow CF_3OCHFCH_2F + CF_3OCH_2CHF_2$$
$$13\% \qquad\qquad 87\%$$

Here the polar and steric factors act in concord, and the orientation of the $CF_3O^{\cdot}$ radical corresponds to its expected electrophilic behaviour. The observed orientation of the addition has been experimentally supported by the photolysis of bis-trifluoromethyl peroxide in the presence of the above-mentioned olefins [151].

$$CF_3OOCF_3 \xrightarrow{h\nu} 2CF_3O^{\cdot}$$

$$CH_2=CF_2 + CF_3O^{\cdot} \rightarrow CF_3OCH_2\dot{C}F_2$$

$$CH_2=CHF + CF_3O^{\cdot} \rightarrow CF_3OCH_2\dot{C}HF$$

The electrophilic addition here would suggest bonding of the methylene group with the fluorine atom, which does not take place in reality.

$$CH_2=CF_2 + CF_3OF \rightarrow FCH_2CF_2OCF_3$$

It is extremely important for understanding the nature of hypofluorites that their reactions with most of known olefins are self-initiated. The only exceptions are perfluorocyclohexene and perfluorocyclopentene, which react under the UV irradiation. The equimolar mixtures of each of the two latter compounds with CF_3OF form the respective ethers with yields close to quantitative at a temperature as low as 213 K [145].

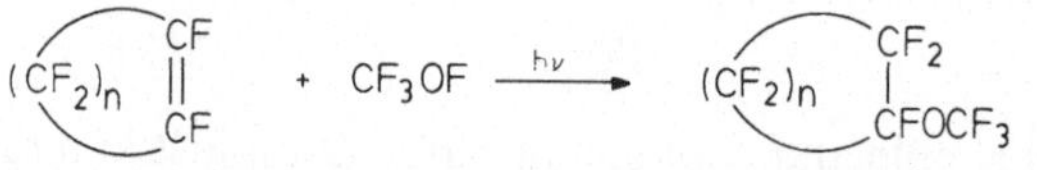

The energy of UV irradiation in this case is certainly consumed for the initial dissociation of CF_3OF to radicals, whereupon the chain process occurs.

Stating the fact of self-initiation of the reaction of hypofluorites with olefins and their change to the radical process, it is reasonable to ask: what are the reasons for the formation of radicals in the system? To answer this question, it is again necessary to refer to the one-electron transfer theory.

The initiation stage of the process may formally be represented as a bimolecular act reflecting the homolysis of the π-bond of olefin and the O–F bond of hypofluorite.

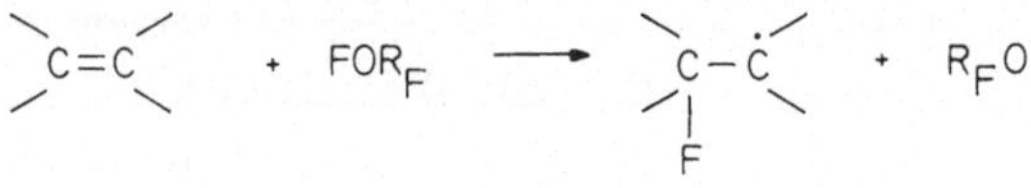

However, closer consideration shows that the elementary act of chain initiation must be represented by the radical-ion reaction, in which the transition state involves formation of an exciplex (radical-ion pair). Its further transformations are produced by the electrostatic interaction between different-sign charges, as shown in the scheme below.

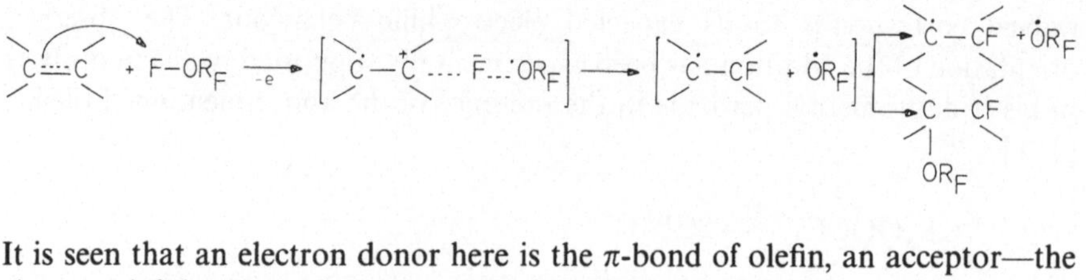

It is seen that an electron donor here is the π-bond of olefin, an acceptor—the electron-deficient (pseudoelectrophilic) fluorine atom of hypofluorite. In terms of quantum chemistry this would imply the electron transfer from HOMO of olefin's π-bond to LUMO of the hypofluorite's O–F bond. Consequently, the higher the energy difference between the interacting orbitals, the more reactive will be the olefin.

The ability of hypofluorites to cause polymerisation possibly extends to all highly reactive olefins. In 1959 Allison and Cady [139] pointed out that the polymer formation in the reaction of tetrafluoroethylene with CF_3OF. Chloro-trifluoroethylene [152,153] easily undergoes the same reaction as well as the chloro-, fluoro-, and hydrogen-containing olefins [154]. The reactions lead to liquid or solid polymeric materials possessing valuable properties. As polymeris-ation initiators, it is also recommended to use difluoromethane-bis-hypofluorite and higher hypofluorites.

CF_3OF and $CF_2(OF)_2$ have a rather high initiating ability. Thus CF_3OF promotes easy polymerisation of perfluoro-2-butyne [155] and the aromatic perfluorocarbons where the multiple bond is known to be less active than in aliphatic olefins.

At the initial stage of the reaction of CF_3OF with fluorinated aromatics, there occurs addition to multiple bonds. Thus, octafluoronaphthalene forms mono- and bis-adducts *1–4* [156–159].

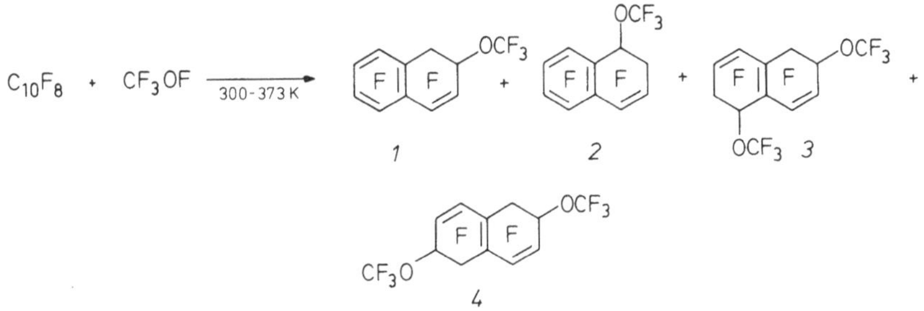

Compounds *3* and *4* react further with the trifluoromethoxyl radicals, forming biradicals *5* and *6*, the principal structural units of the polymer.

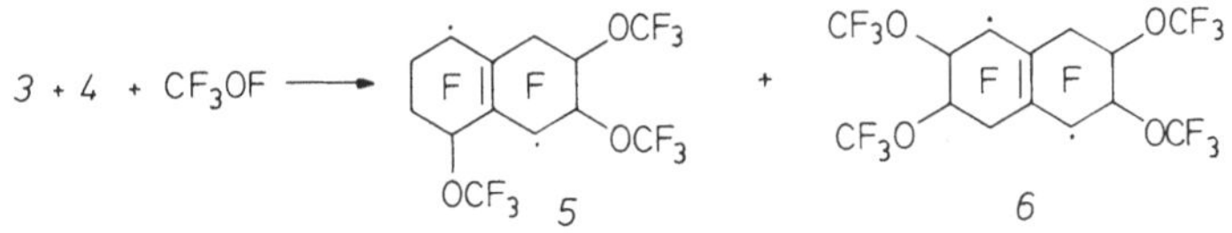

The structure of the products of the reaction of octafluoronaphthalene with $CF_2(OF)_2$ indicates that the key intermediate here is the fluorooxydifluoro-methoxyl radical [158,159].

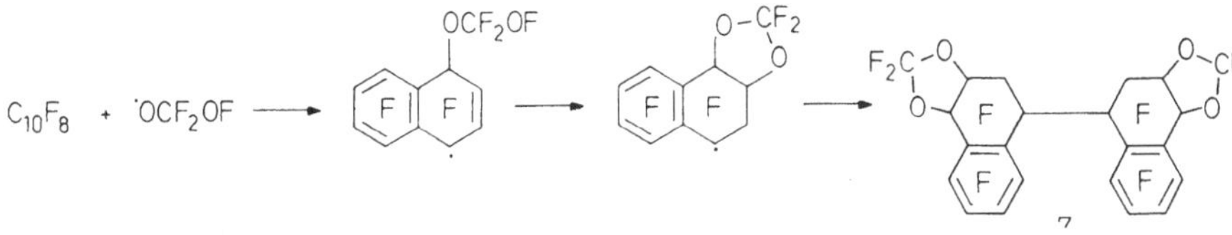

Dioxolanes are the primary products of the reaction of $CF_2(OF)_2$ and hexa-fluorobenzene [160].

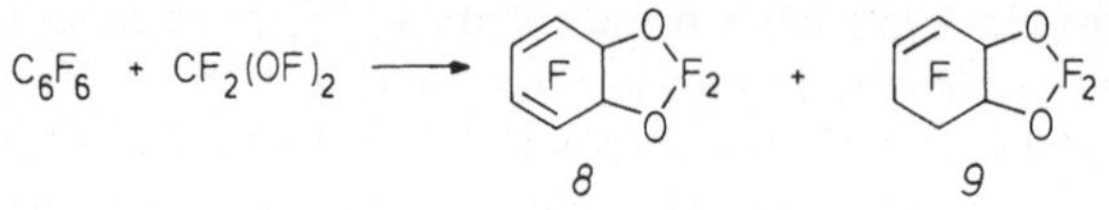

Under conditions of photochemical initiation, the yield of cyclohexene (9) is much higher than that of diene (8), as the process involves partial isomerisation of hexafluorobenzene to hexafluorobicyclo[2,2,0]hexadiene-2,5 (10). The latter is transformed by $CF_2(OF)_2$ to a mixture of mono- and dioxolanes (11–13) containing an insignificant amount of oxirane (14) [162,162].

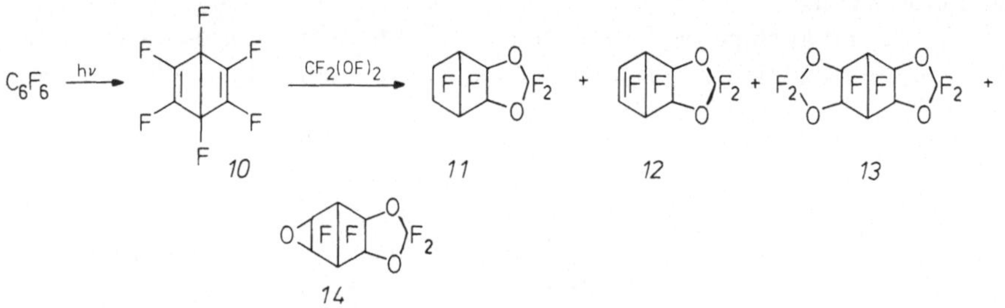

In the thermal reaction of $CF_2(OF)_2$ with diene (10), compounds 11–14 are not formed because of the instability of the $FOCF_2O^.$ radical in these conditions.

Owing to the high reactivity and the presence of two fluorooxy groups, polymerisation of hexafluorobenzene and octafluoronaphthalene by $CF_2(OF)_2$ proceeds rather effectively. The structures of the polymer products are very complex and their representation requires much space. For example, polymer formed from C_6F_6 and $CF_2(OF)_2$ involves cyclohexadiene and cyclohexene fragments 15 and 16 [163].

The molecular mass of these polymers reaches 2500. These are low-melting point substances.

The example of perfluoropyridine shows that polymerisation by hypofluori-tes extends to practically the whole series of perfluoroaromatic carbons. With $CF_2(OF)_2$ pentafluoropyridine reacts in milder conditions than hexafluoro-benzene, giving a viscous liquid with molecular mass of 1000 to 2000, the yield of which is about 70% at the mole ratio of hypofluorite to C_5F_5N 1:3 [164].

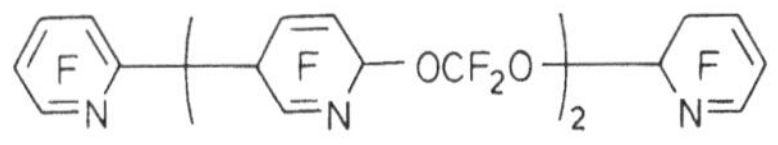

17

Further treatment of compound 17 by hypofluorite leads to its partial fluorination and dioxomethylation with formation of product 18 containing about 18% of azacyclohexene (19) and azacyclohexadiene (20) as impurities, in the 1:3 mole ratio.

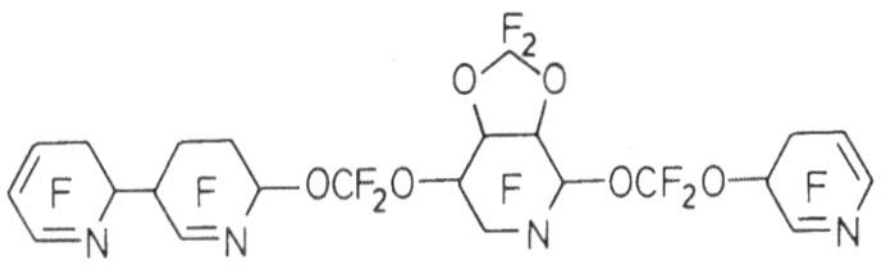

18 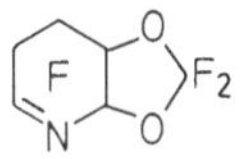   19 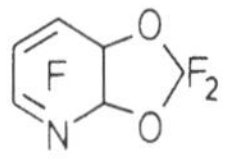   20

When the mole ratio of pentafluoropyridine to hypofluorite is raised to 1:1, the yield of products 19 and 20 increases to 75%.

The presence of a reactive functional group in the ring of perfluoroaromatic compounds imparts some specific features to the process of their interaction with hypofluorites. Thus the products of the reaction of CF_3OF with pentafluorophenol are perfluorinated cyclohexadienones and phenoxycyclohexadienone [165].

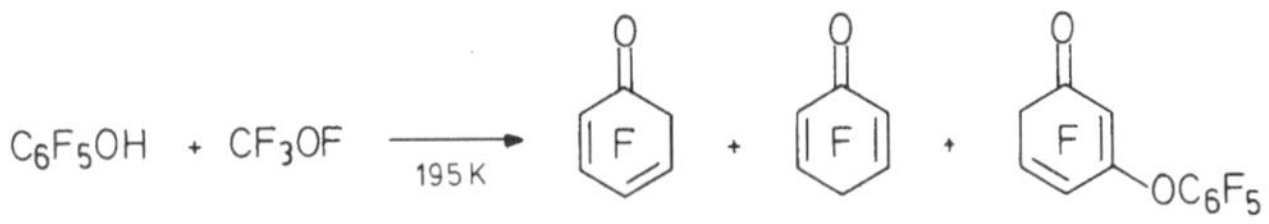

As in the reaction of C_6F_5OH with hydrogen peroxide, the key intermediate here seems to be the pentafluorophenoxyl radical [166].

A rather interesting case unambiguously supporting the radical nature of the reaction of perfluorinated unsaturated compounds with hypofluorites was described in patent [167]. Trifluoromethyl hypofluorite was suggested for use as an initiator of oxidation of perfluoroisopropylethylene by oxygen to polyoxydifluoromethylenecarbonyl fluorides.

$$(CF_3)_2CFCF=CF_2 + CF_3OF + O_2 \rightarrow CF_3O(CF_2O)_nCF_2COF$$

The by-product is perfluoroisobutyryl fluoride, which suggests formation of the respective alkyl radicals at the stage of initiation; these react with oxygen, forming radical peroxides.

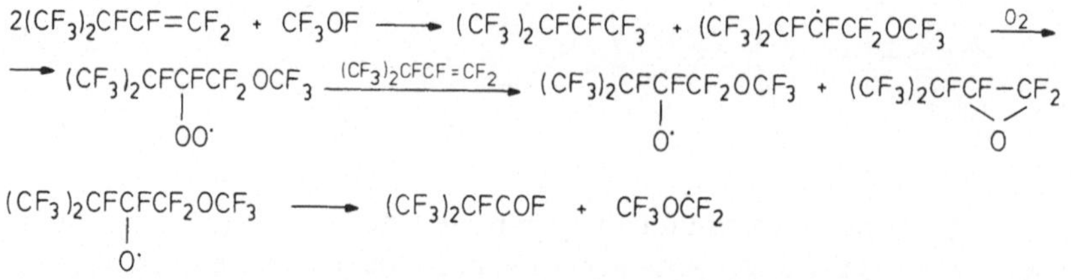

$$2(CF_3)_2CFCF=CF_2 + CF_3OF \longrightarrow (CF_3)_2CF\dot{C}FCF_3 + (CF_3)_2CF\dot{C}FCF_2OCF_3 \xrightarrow{O_2}$$

$$\longrightarrow (CF_3)_2CFCFCF_2OCF_3 \xrightarrow{(CF_3)_2CFCF=CF_2} (CF_3)_2CFCFCF_2OCF_3 + (CF_3)_2CFCF-CF_2$$
$$\underset{OO^\cdot}{|} \qquad\qquad \underset{O^\cdot}{|} \qquad \underset{O}{\diagdown\diagup}$$

$$(CF_3)_2CFCFCF_2OCF_3 \longrightarrow (CF_3)_2CFCOF + CF_3O\dot{C}F_2$$
$$\underset{O^\cdot}{|}$$

The above literature data seems to be quite enough to doubt the explanation of the reactivity of hypofluorites in terms of the electrophilic mechanism. At the same time, these data justify consideration of other types of the reactions of hypofluorites with unsaturated compounds in terms of the one-electron transfer mechanism. In considering this question, we shall first of all deal with the reactions whose result may be defined as formal fluorination proceeding with substitution of hydrogen at a multiple bond by fluorine.

The fluorinating ability of hypofluorites with respect to unsaturated hydrogen-containing compounds was found in the attempts to substitute them for a more rigid and less selective reagent—elemental fluorine. This ability has been confirmed with reference to many compounds: steroids [168–173], anhydrosugars [174–180], styrene [181], stilbene [182], diphenylacetylene [169,173,183], uracyl derivatives [184–190], and a variety of aromatic compounds [191–194].

Treatment of benzene with CF_3OF in dichloromethane (273 K, 2 h) affords monofluoro-, difluoro-, and trifluoromethoxybenzenes with a general yield of more than 40% [194].

$$C_6H_6 + CF_3OF \rightarrow C_6H_5F + 1,4\text{-}C_6H_4F_2 + C_6H_5OCF_3$$

Toluene and xylene react in a similar way, but apart from the nuclear fluorination, the reaction involves partial substitution of hydrogen in the methyl group.

$$C_6H_5CH_3 + CF_3OF \rightarrow 2\text{-}FC_6H_4CH_3 + 4\text{-}FC_6H_4CH_3 + C_6H_5CH_2F$$

As seen from the scheme, the methyl group produces the *ortho-* and *para-* orienting effect. The same effect is shown by other electron-donating groups. Thus, phenol [195], and the naphthalene and anthracene derivatives [192,193] react in a similar way.

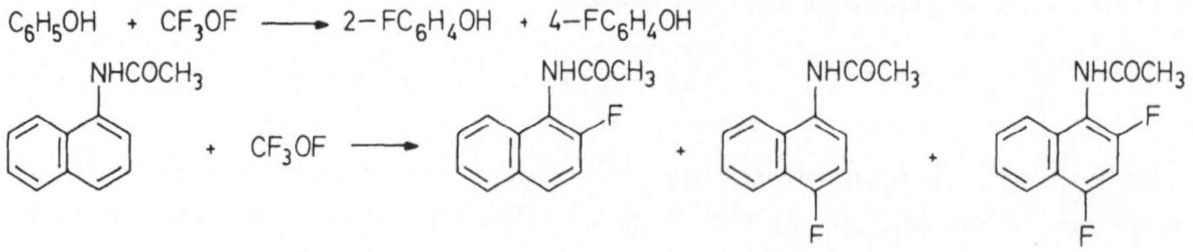

$$C_6H_5OH + CF_3OF \longrightarrow 2\text{-}FC_6H_4OH + 4\text{-}FC_6H_4OH$$

The observed reaction route would seem to correspond to the classical idea of electrophilic substitution. But this mechanism cannot account for the formation of trifluoromethoxybenzene and benzyl fluoride from benzene and toluene respectively. Furthermore, the electrophilic mechanism cannot explain ketonisation which frequently occurs upon treatment of aromatic hydroxy-, alkoxy-, acetoxy, and amino-derivatives.

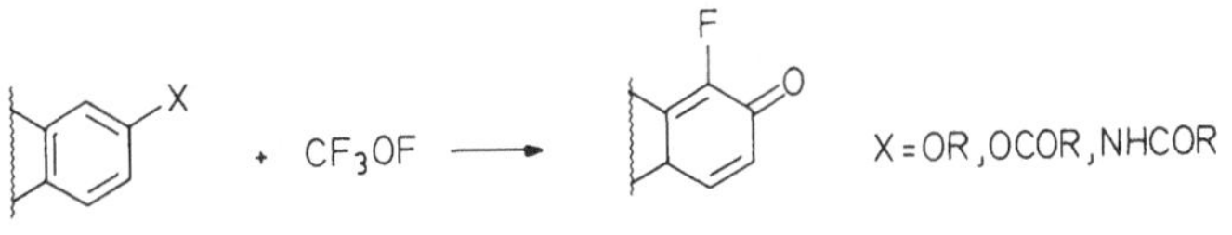

$X = OR, OCOR, NHCOR$

The latter type of transformations seems to be of the general character, as the substituted heterocyclic compounds, such as uracyl derivatives [171,172], and the silyl ethers of enols [196] react in a similar way.

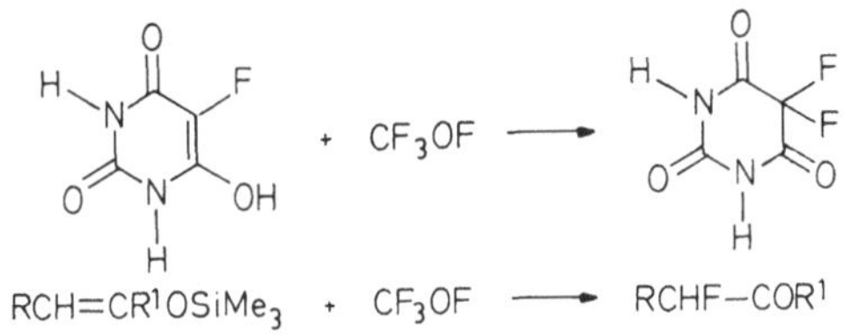

$RCH=CR^1OSiMe_3 \; + \; CF_3OF \longrightarrow RCHF-COR^1$

When the C=C bond of uracyl does not bear the above substituents, fluorination proceeds in the ordinary way [185–187].

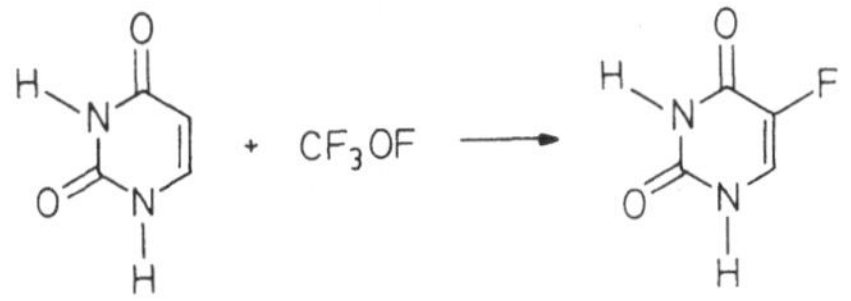

The same refers to pyrazole derivatives. Though here again the reaction proceeds with partial destruction of the starting substance (dealkylation).

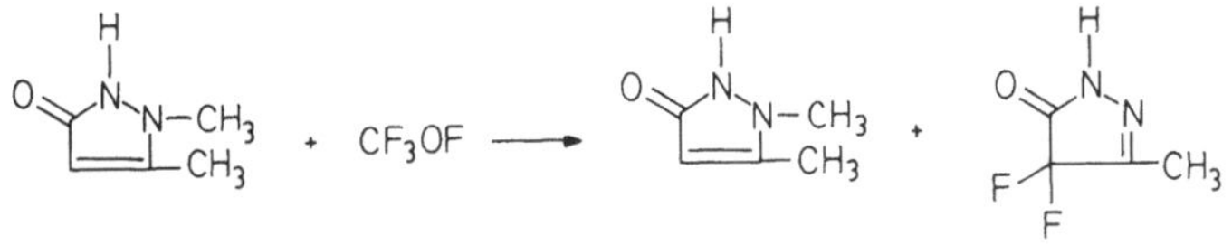

In the case of fluorination of uracyl and stilbene by trifluoromethyl hypofluorite, there is a genetic relation with fluorination of imines considered earlier: both unsaturated compounds in the presence of MeOH give the respective methoxy-derivatives [182,186].

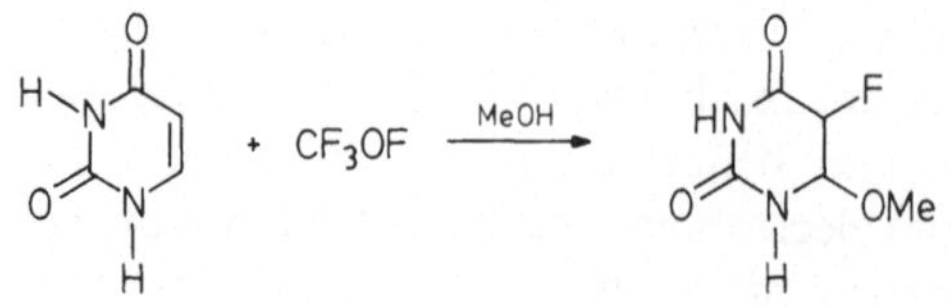

PhCH=CHPh + CF$_3$OF $\xrightarrow{\text{MeOH}}$ PhCHF—CH(OMe)Ph

The most essential feature of these reactions is their occurrence in rather mild conditions (195 K). In this case *trans*-stilbene is transformed to a mixture of *threo*- and *erythro*-isomers in the ratio of 2:1, and *cis*-stilbene gives the same (but reverse) ratio of stereoisomers [182]. When *trans*-stilbene is treated with CF$_3$OF in the presence of Et$_2$O, the stereospecificity of addition increases to the ratio of *threo* to *erythro* 4:1, in the case of *cis*-stilbene the *erythro* to *threo* ratio becomes 3:1 [182].

$$PhCH=CHPh + CF_3OF \xrightarrow{\text{Et}_2\text{O}} PhCHFCH(OCF_3)Ph$$

The total yield of products of addition to *cis*-stilbene approaches 60%; to *trans*-stilbene, slightly exceeds 40%. In addition, the products of fluorination of double bond are formed in yields of 40 and 56% respectively.

$$PhCH=CHPh + CF_3OF \rightarrow PhCHF\text{–}CHFPh$$

It is interesting to note that the stilbene *trans*-isomer is in general more liable to fluorination. In the absence of polar solvents (in the CFCl$_3$ solution), the ratio of the addition to fluorination products for *trans*-stilbene is about 60:40; for *cis*-stilbene, 87:12.

The analysis of reasons, which brought about the hypothesis about the electrophilic nature of hypofluorites, shows that it has also been substantiated by stereospecificity of addition, apart from the above-mentioned *ortho*- and *para*-orientation of substitution of hydrogen. As suggested in [182, 193], the first stage of the reaction involves formation of the carbonium ion, whose further transformations determine the experimental conditions and the character of substituents at the double bond. In the inert solvents, the carbonium ion is stabilised by capturing the counterion CF$_3$O$^-$. The electron-donating solvents, such as Et$_2$O, stabilise the carbonium centre, giving difluoride. A similar effect is produced by the electron-donating substituents bonded with the carbonium centre. When the reaction is carried out in the presence of MeOH, the carbonium centre adds methoxide.

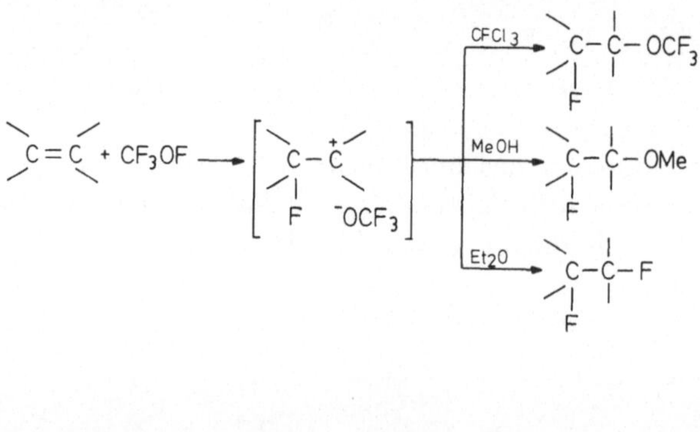

However many facts contradict these ideas. Thus, apart from the above-mentioned examples of formation of benzyl fluoride and tri-fluoromethoxy-benzene, the study of the reaction of stilbene with pentafluoroethyl hypofluorite is contradictory [197]. The results of the reaction of CF_3OF with 1,1-diphenyl-ethylene producing at least five different products cannot be explained in terms of the electrophilic mechanism [198].

$$Ph_2C{=}CH_2 + CF_3OF \xrightarrow[195\ K]{CH_2Cl_2} Ph_2C(OCF_3)CH_2F + Ph_2CFCH_2F$$

$$+\ Ph_2C{=}CHF + Ph_2C(OCF_3)CHF_2 + Ph_2\underset{\underset{CH_2F}{|}}{C}CH{=}CPh_2$$

Explanation by this mechanism of the synthesis of α-fluoroketones from the respective vinyl acetates is very objectionable [199–201].

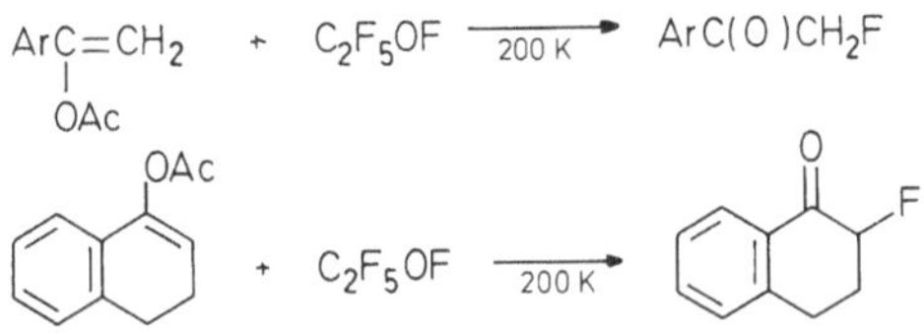

It is noteworthy that the results of the reactions of some of these compounds with CF_3OF and elemental fluorine are very similar. Thus, the reaction of fluorine with 1,1-difluoroethylene gives three out of five products—analogues of the reaction with CF_3OF [202].

$$Ph_2C{=}CH_2 + F_2 \rightarrow Ph_2CFCH_2F + Ph_2C{=}CHF + Ph_2CFCHF_2$$

The reaction of fluorine and diphenylacetylene in MeOH leads to the formation of the respective mono- and dimethoxy-derivatives [202].

$$PhC{\equiv}CPh + 2F_2 \xrightarrow{MeOH} PhCF_2CF_2Ph + PhCF(OMe)CF_2Ph$$

$$+\ PhC(OMe)_2CF_2Ph$$

In the absence of methanol, only tetrafluorodiphenylethane is formed, which is the analogue of the product of the reaction between diphenylacetylene and CF_3OF [170,203].

$$PhC{\equiv}CPh + 2CF_3OF \rightarrow PhCF_2CF(OCF_3)Ph$$

The impossibility in principle to generate the fluorine cation by the heterolysis of the F–F bond under the action of unsaturated C=C compounds makes us reject the electrophilic mechanism of the reaction involving fluorine. However, since they are self-initiated at rather low temperatures, an alternative mechanism to account for the observed results seems to be the one-electron transfer. In this case it would be better to substitute the widely used term "electrophilic" by the term "electron affinity". The electron affinity of the fluorine molecule is known to be maximal among all known substances, and therefore it may be regarded as a powerful electron acceptor. On the other hand, the similarity of chemical behaviour and reaction results between fluorine and hypofluorites allow to regard hypofluorites from the same viewpoint.

Application of the one-electron transfer concept to the reaction of hypofluorites with unsaturated CH-compounds, in contrast to the electrophilic mechanism, accounts for practically all the above-mentioned "anomalous" results. Their full analysis would have taken too much space, if it is required at all. It will be enough to note some general items taken from the fundamental works dealing with this question (see, e.g. [132, 133, 142]) and completed by our data:

a) the aromatic and related compounds are oxidated to radical cations;
b) radical cations are stabilised by proton elimination, being transformed to radicals;
c) the regioselectivity of hydrogen substitution in the aromatic ring is determined by the electron acceptor attack at a position of the highest electron density of the ring;
d) in the radical cations formed from the unsaturated aliphatic compounds, rotation round the carbon-carbon bond is hindered, therefore addition of hypofluorite fragments (as well as fluorine) occurs predominantly at the *cis*-position, determining the stereospecificity;
e) due to the formation of a radical ion pair in the electron transfer, the reaction system is considerably influenced by the solvation and cell effects.

Therefore formation of benzyl fluoride from toluene may be represented as a series of transformations, in which the key intermediate is the benzyl radical.

$$C_6H_5CH_3 + CF_3OF \xrightarrow[-e]{} C_6H_5CH_3^{+\cdot} \xrightarrow[-H^+]{} C_6H_5CH_2^{\cdot} \xrightarrow[-CF_3O^\cdot]{CF_3OF} C_6H_5CH_2F$$

To explain ketonisation of the hydroxy-, alkoxy-, and acetoxy-derivatives, radical decay of the intermediate seems to be more preferable.

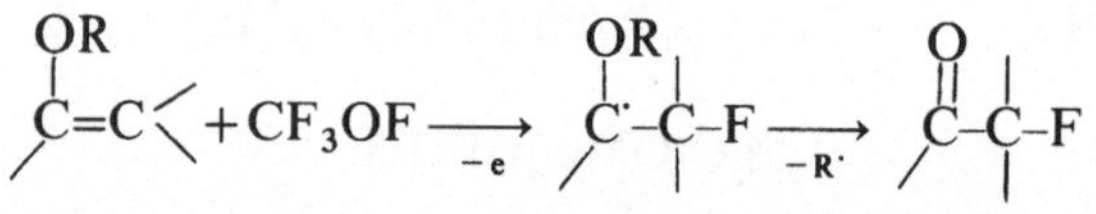

However dealkylation as an alternative reaction route cannot be excluded.

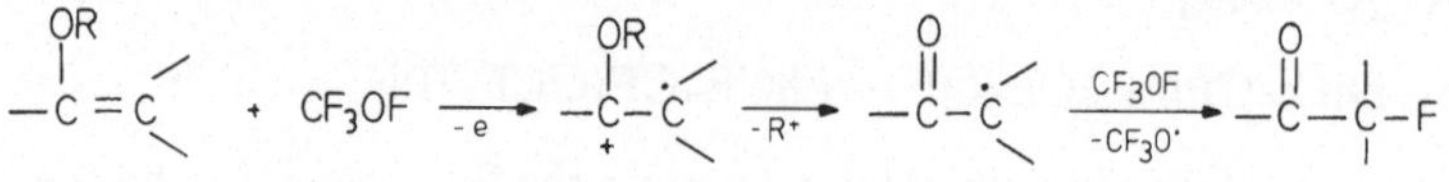

Finally, formation of trifluoromethoxybenzene may be explained by the radical substitution of hydrogen by the trifluoromethoxyl radical.

$$C_6H_6 + CF_3O^\cdot \xrightarrow[-H^\cdot]{} C_6H_5OCF_3$$

For the general model of one-electron transfer, the transition state should obviously be regarded as a superposition of structures.

$$\left[\overset{+}{\underset{\diagdown}{C}} - \overset{\cdot}{\underset{\diagup}{C}} \cdots \dot{F} - \bar{O}CF_3 \right] \longleftrightarrow \left[\overset{\cdot}{\underset{\diagdown}{C}} - \overset{+}{\underset{\diagup}{C}} \cdots \bar{F} - \dot{O}CF_3 \right]$$

The left-hand structure is realised when the carbonium centre may be stabilised by the electron-donating substituents, and just this structure is specific to non-fluorinated compounds. In the radical part of this structure, the electrons are in the field of opposite signs and, consequently, should have antiparallel spins (singlet) allowing their instant recombination. As a consequence, the system is transformed into the ion pair, thus simulating the electrophilic character of the reaction.

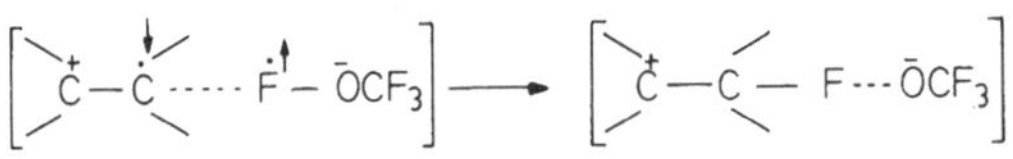

$$\left[\overset{+}{\underset{\diagdown}{C}} - \overset{\downarrow}{\underset{\diagup}{C}} \cdots \overset{\uparrow}{F} - \bar{O}CF_3 \right] \longrightarrow \left[\overset{+}{\underset{\diagdown}{C}} - \overset{}{\underset{\diagup}{C}} - F \cdots \bar{O}CF_3 \right]$$

As to the right-hand structure, its realisation should be provided by the electron-accepting substituents. In this case, the process follows the radical mechanism, whose features were discussed above.

3.4.6 Substitution Reactions

This section considers the substitution of hydrogen atoms and other groups at a saturated carbon and nitrogen. Besides, the section also deals with the reactions which it was senseless to consider in a separate chapter, as they are very few (S- and P-fluorination).

Detailed studies of hydrogen substitution with CF_3OF in the methane series have shown that this process requires heating to between 393 and 473 K. The reaction of CF_3OF with methane, fluoromethane, and chloroform at a low pressure and under dilution with nitrogen is sluggish. Difluoromethane, di-chloromethane, and chlorofluoromethane are much more reactive [204–206].

Fluorination of alkanes by trifluoromethyl hypofluorite is a free-radical chain process, as in the case of elemental fluorine. The rate ratio of decay and deactivation of species in collisions has been estimated in work [206].

A convincing proof of the energy-induced chain branching is the reaction of CF_3OF with CH_2FCl, resulting in the formation of perfluoroethylmethyl ether and tetrafluoromethane according to the following scheme [205].

$$CF_3OF + CH_2CFCl \overset{\Delta}{\longrightarrow} CF_3O^{\cdot} + {}^{\cdot}CHFCl + HF$$

$$CF_3O^{\cdot} + CH_2CFCl \rightarrow {}^{\cdot}CHFCl + [CF_3OH]$$

$$CF_3OF + {}^{\cdot}CHFCl \rightarrow CF_3O^{\cdot} + CHF_2Cl^*$$

$$CHF_2Cl^* \rightarrow \ddot{C}F_2 + HCl$$

$$\ddot{C}F_2 + CF_3OF \rightarrow \dot{}CF_3 + CF_3O\dot{}$$

$$\dot{}CF_3 + CF_3OF \rightarrow CF_4 + CF_3O\dot{}$$

$$2\ddot{C}F_2 \rightarrow CF_2=CF_2$$

$$CF_2=CF_2 + CF_3O\dot{} \rightarrow CF_3OCF_2\dot{C}F_2 \xrightarrow{CF_3OF} CF_3OCF_2CF_3$$

$$[CF_3OH] \rightarrow COF_2 + HF$$

The rate constant for the reaction $\dot{}CF_3 + CF_3OF \rightarrow CF_4 + CF_3O\dot{}$ at room temperature is equal to $2 \cdot 10^8$ cm^3 mol^{-1} sec^{-1} [207]. In the system $CH_2F_2 + CF_3OF$, chain branching is provided by the reactions.

$$\dot{}CHF_2 + CF_3OF \rightarrow CHF_3^* + CF_3O\dot{}$$

$$CHF_3^* \rightarrow \ddot{C}F_2 + HF$$

The $CF_3O\dot{}$ radical is much more reactive than $CH_3O\dot{}$. The rate constant of hydrogen abstraction from methane by $CF_3O\dot{}$ was reported to be $\sim 1.5 \cdot 10^8$ cm^3 mol^{-1} sec^{-1} [208]. The rate constant of the elementary reaction of fluorine abstraction from CF_3OF by alkyl radicals also has a high value of $\sim 10^8$ cm^3 mol^{-1} sec^{-1} [209]. Easy hydrogen abstraction by $CF_3O\dot{}$ and easy reaction of alkyl radicals with CF_3OF may possibly underlie fluorination of steroids [171,210], adamantane derivatives [211], lactoses [180], etc.

The generation of atomic fluorine and highly active trifluoromethoxyl radical by the UV irradiation is widely used in the synthesis of various fluorine-containing compounds. This method extends to fluorination of aromatic compounds, for example, benzene and toluene [212].

$$C_6H_6 + CF_3OF \xrightarrow{h\nu} C_6H_5F$$

$$C_6H_5CH_3 + CF_3OF \xrightarrow{h\nu} 2\text{-}FC_6H_4CH_3 + C_6H_5CH_2F$$

The photochemical fluorination method is characterised by a remarkable selectivity.

$$C_6H_5CH_2CH(NH_2)COOH + CF_3OF \xrightarrow{h\nu}$$

$$3\text{-}FC_6H_4CH_2CH(NH_2)COOH \quad [213]$$

$$4\text{-}RC_6H_4NHCOCH_3 + CF_3OF \xrightarrow[\mu F]{h\nu} 4\text{-}R\text{-}2\text{-}FC_6H_3NHCOCH_3 \quad [214]$$

$$PhCH_2CH_2Ph + CF_3OF \xrightarrow{h\nu} PhCF_2CF_2Ph \quad [215]$$

A great role, though not clear as yet, is played in these reactions by HF. Upon photochemical fluorination in hydrogen fluoride there occurs C-fluorination of alkylamines [216], carboxylic acids, cycloalkanes, and polymer materials, such as polyethylene, polystyrene, and polycaprolactame [9]. The selectivity of photochemical fluorination by trifluoromethyl hypofluorite in HF is of special interest, as this method affords fluorinated amino acids.

$$CH_3CX(NH_2)COOH + CF_3OF$$

$$\xrightarrow[HF]{hv} CH_2FCX(NH_2)COOH \quad [217,218]$$

X = H, D

$$CH_3CH(NH_2)(CH_2)_2COOH + CF_3OF$$

$$\xrightarrow[HF]{hv} CH_2FCH(NH_2)(CH_2)_2COOH \quad [219]$$

$$NH_2(CH_2)_4CH(NH_2)COOH + CF_3OF$$

$$\xrightarrow[HF]{hv} NH_2(CH_2)_2CHFCH_2CH(NH_2)COOH \quad [220]$$

Activation of the CH bonds of the reaction centres here possibly results from the fact that they are adjacent to the protonated amino group. This follows from ease of hydrogen substitution by fluorine in the CH acids [221].

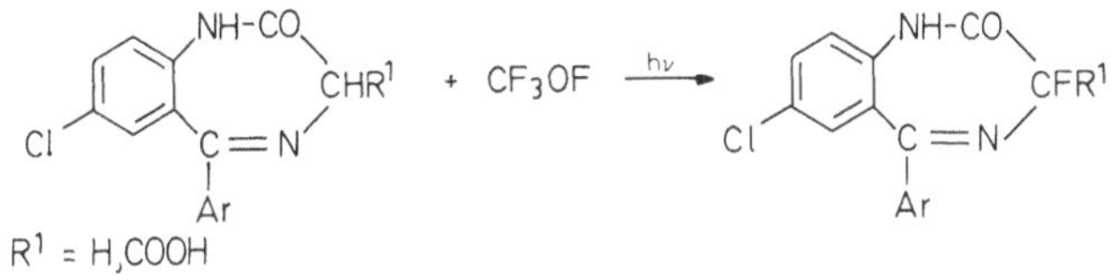

$R^1 = H, COOH$

N-Fluorination reactions do not require the external initiation sources as opposed to the C-fluorination reactions. Thus N-acyl and N-sulfonyl amides smoothly react with CF_3OF in inert solvents at temperatures as low as 273 K, forming initially the respective N–F amides [222–224].

$$RR'NH + CF_3OF \xrightarrow[273 K]{CFCl_3} RR'NF$$

R = alkyl, cyclo-alkyl, aryl; R' = COR, SO_2R

Upon further treatment of N-fluoroamides with CF_3OF, there occurs the nitrogen–carbon bond cleavage to give N,N-difluoroamines.

The reaction of CF_3OF with secondary amines also proceeds with formation of N-F derivatives. There have been attempts to compare the activity of CF_3OF and F_2 with aziridines, which have the same results, except for the N-formyl fluoride formation in the case of trifluoromethyl hypofluorite [225–227].

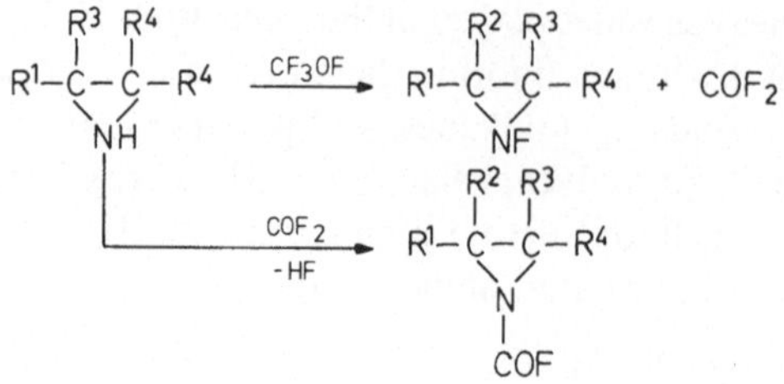

R_1, R_2, R_3, R_4 = H, alkyl, aryl

As opposed to this, the reactions of *N*-acyl and *N*-sulfonyl aziridines with CF_3OF proceed with cycle opening to produce the respective *N*-fluoroamides or -amines [226].

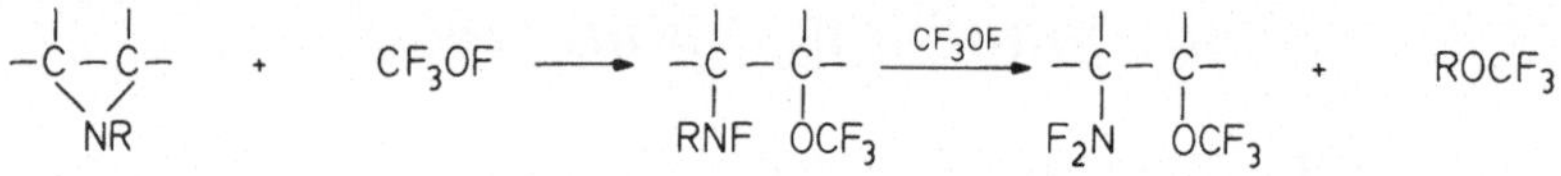

The liability to destruction is shown by the sulfur-containing aliphatic compounds: thiols undergo dehydrosulfination [228], and dialkyl disulfides—S–S bond cleavage [229].

$$(CH_3)_2\underset{\displaystyle SH}{C}CH_2COOH + CF_3OF \rightarrow (CH_3)_2C{=}CHCOOH$$

$$CF_3SSCF_3 + CF_3OF \rightarrow CF_3SF_3 + (CF_3)_2S$$

By contrast with this, cyclic sulfides react with CF_3OF without cycle cleavage, giving difluorosulfuranes [230,231].

$$\underbrace{CH_2CH_2CH_2CH_2}_{S} + CF_3OF \xrightarrow[195\ K]{} \underbrace{CH_2CH_2CH_2CH_2}_{SF_2}$$

A similar behaviour is shown by the cyclic compounds containing trivalent phosphorus [232].

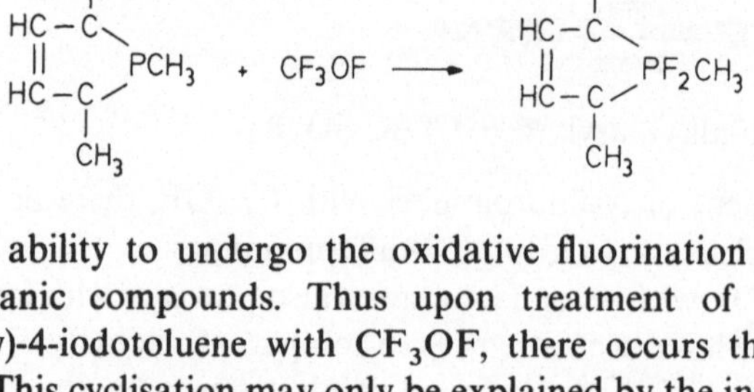

The ability to undergo the oxidative fluorination is also inherent in the iodoorganic compounds. Thus upon treatment of 3-(hexafluoroisopropylhydroxy)-4-iodotoluene with CF_3OF, there occurs the cyclisation involving iodine. This cyclisation may only be explained by the intermediate formation of the difluoroiodo-derivative [233].

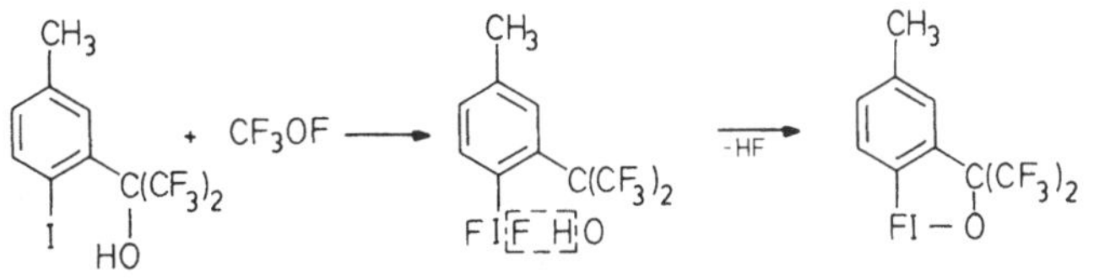

The three latter reactions again illustrate common features in the chemical behaviour of CF_3OF and elemental fluorine. In particular, fluorine and tri-fluoromethyl hypofluorite give qualitatively similar results in the reactions with aromatic mercury- and tin-organic compounds, where CF_3OF is especially effective [234,235].

$$(C_6H_5)_4Sn + CF_3OF \rightarrow C_6H_5F$$
$$22\%$$

$$(C_6H_5)_2Hg + CF_3OF \rightarrow C_6H_5F$$
$$83\%$$

Fluorination of organometal compounds by CF_3OF is suggested to follow the electrophilic mechanism. But participation of elemental fluorine in this reaction makes this interpretation doubtful.

In terms of the electrophilic mechanism, one more interesting class of the reactions of CF_3OF is considered in literature—those with diazoketones [235–237].

$$RCOCHN_2 + CF_3OF \rightarrow RCOCHF_2 + RCOCH(OCF_3)F + RXC\!\!-\!\!CHF$$
$$\diagdown O \diagup$$

$$X=F, OCF_3$$

But this viewpoint also may not be considered as final, as it does not account for the formation of α-oxides. The one-electron transfer approach, which admits the reaction route via the resonance-stabilised radical with two reaction centres, seems to be more suitable.

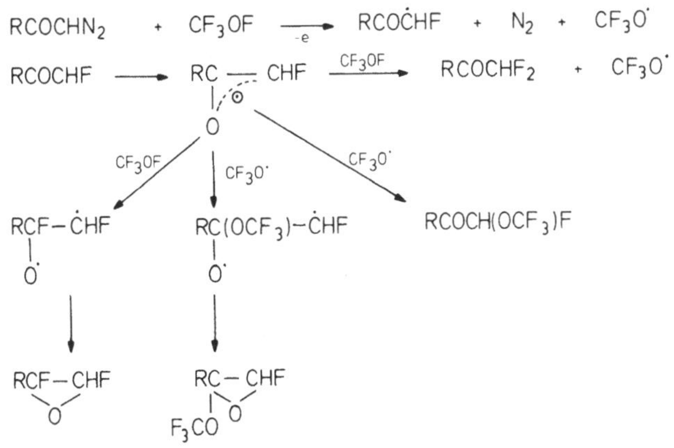

3.5   Conclusion

In conclusion, it is necessary to underline the great practical and theoretic importance of organic hypofluorites.

The use of hypofluorites as fluorinating agents is mostly dictated by the milder reaction conditions, the higher selectivity and increased yields of target products as compared with elemental fluorine. This is extremely important in the synthesis of biologically active substances, such as 5-fluorouracyl, fluorobarbituric acids, fluoroalanine, fluorosteroids, etc.

An interesting feature of hypofluorites is their ability to cause polymerisation and telomerisation of different fluorinated unsaturated compounds. Though the advances here are small, in future this direction in the chemistry of hypofluorites seems to be promising. This also refers to the reaction of addition of hypofluorites at the multiple bond of unsaturated compounds as a method of direct introduction of perfluoroalkoxy groups into organic molecules.

The most widely spread and accessible of organic hypofluorites—trifluoromethyl hypofluorite—is a unique object of studies presenting certain interest for the chemical theory. Being a source of easily identifiable CF_3O fragment, CF_3OF is extremely valuable for the studies and quantitative estimation of steric, inductive, and mesomeric substituent effects in various reactions, as kind of a label for the most possible site of attack by a reagent.

Of equal theoretic importance are the questions connected with the problem of the reactivity of hypofluorites with substances that are potential electron donors (anions, multiple bonds, covalent compounds with fixed centres of excessive electron density). This field is expected to have significant advances in the nearest future. Solution of this problem is stimulated, first, by its contradictory nature, and, second, by its promise to bridge the gap in our understanding of the fluorine phenomenon.

3.6   Preparations

1. Trifluoromethyl hypofluorite [26]
A dry 150 ml Monel Hoke cylinder was charged with 10 g of CsF dried in vacuum at 450 to 475 K and 30 stainless steel 10 mm d. balls. The cylinder was then shaken for several hours to ensure fine grinding of the catalyst. A 1.72 mmole of COF_2 was condensed into the pre-evacuated reactor at 77 K, and then fluorine (1.73 mmoles) was introduced into the reactor at this temperature. The reactor was then placed in a 195 K bath and allowed to stand for 4 h. After cooling the reactor to 77 K, the unreacted fluorine was pumped out through a trap containing KOH. The product was then removed for further purification. Practically pure CF_3OF was obtained in a 97% yield.

The procedure is of a general character and is suitable for the preparation of any perfluoroalkyl hypofluorites.

2. Difluoromethane-bis-hypofluorite [239]

Into a nickel pipe reactor containing 330 g of CsF at 300 K, CO_2 and F_2 (diluted with He at $1:1$) were fed at the rate of 30 ml min^{-1} and 60 ml min^{-1} respectively. The reaction products were condensed in a trap (195 K). The product contained 95% $CF_2(OF)_2$, 4% CO_2, and 1% CF_4.

3. Reactions of CF_3OF with olefins [149]

All reactions were carried out in 100-ml glass bulbs fitted with glass-Teflon valves. Separation of volatile products was done by trap to trap distillation.

The olefin (2 to 3 mmol) was first condensed into the reactor, and approximately 20 mmol of $CFCl_3$ or CF_2Cl_2 was condensed onto it at 77 K. The contents were then warmed to 295 K to form a homogeneous solution. The reactor was then cooled to 77 K, and a stoichiometric amount of CF_3OF was condensed into the reactor. The reactor was placed in a cold bath and allowed to warm up slowly (16–24 h) to 295 K. The products were separated by distilling through traps cooled to 243 K, 210 K (the addition products were collected in 30–80% yields) and 77 K.

4. 5-Fluorouracyl [188]

A mixture of 0.336 g (3 mmol) of uracyl, 6 ml of trifluoroacetic acid and 20 ml of water was slowly added to a cooled (to 195 K) solution of 0.47 g (4.5 mmol) of CF_3OF in 50 ml of $CFCl_3$ placed in autoclave. The mixture was vigorously stirred for 15 h at room temperature, then the excess CF_3OF was removed with nitrogen and the solvent was removed under reduced pressure. The solid residue was sublimed at 483 to 503 K (0.5 mm Hg), and after recrystallization from the methanol-ether mixture, 5-fluorouracyl was obtained in a 85% yield.

5. N-Methyl-N-fluoro-4-fluorobenzamide [224]

CF_3OF (23 mmol) diluted with nitrogen ($1:1$) was slowly bubbled through a solution of 2.5 g (16 mmol) of N-methyl-4-fluorobenzamide in 25 ml of $CHCl_3$ at room temperature. After usual aqueous work-up the residual oil was chromatographed on silica gel. Elution with hexane gave 0.12 g of trifluoromethyl 4-fluorobenzoate and 0.15 g of 4-fluorobenzoyl fluoride. Elution with a hexane-dichloromethane ($1:1$) mixture gave N-methyl-N-fluoro-4-fluorobenzamide.

3.7   References

1. Shellhamer DF, Curtis CM, Hollingsworth DR (1982) Tetrahedron Lett. 31:2157
2. Kemmit RDW, Sharp DWA (1965) Adv. Fluor. Chem. 4:217
3. Banks RE (1964) Fluorocarbon and their derivatives. Oldbourne, London
4. Hoffmann CJ (1964) Chem. Revs 64:91
5. Lustig M, Shreeve JM (1973) Adv. Fluor. Chem. 7:175
6. Sharts CM, Sheppard WA (1974) Org. Reactions 21:243
7. Shreeve JM (1983) Adv. Inorg. Chem. Radiochem. 26:119
8. Nikitin IV (1986) Khimiya kislorodnyh soedineniy galogenov. Nauka, Moscow
9. Mukhametshin FM (1980) Usp. Khim. 49:1260

10. Kellog KB, Cady GH (1948) J. Amer. Chem. Soc. 70:3983
11. Cady GH, Krayat J, Wielson KS (1966) Inorg. Synthesis 8:165
12. Brauer G (ed) (1975) Handbuch der Präparativen Anorganischen Chemie, 3rd edn., vol 1, Enke, Stuttgart
13. Kapralova GA, Buben SM, Chaikin AM (1975) Kinetika i Kataliz 16:591
14. Talbott RL (1968) J. Org. Chem. 33:2095
15. US Pat 3179702 (1962)
16. Schumacher HJ (1968) Photochem. Photobiol. 7:755
17. Jubert AH, Siecre JE, Schumacher HJ (1969) Z. phys. Chem. (BRD) 67:138
18. Czarnowski J, Schumacher HJ (1969) Z. phys. Chem. (BRD) 68:149
19. Veyre R (1969) Pour obtenir le grande docteur de L'Universite de Lyon sciences physiques, Lyon
20. Aymonino PJ (1964) Proc. Chem. Soc. 10:341
21. Lopez MJ, Castellano E, Schumacher HJ (1974) J. Photochem. 3:97
22. Czarnowski J, Schumacher HJ (1974) Z. phys. Chem. (BRD) 92:329
23. Brit. Pat 1484823 (1974)
24. Kennedy RC, Cady GH (1973) J. Fluor. Chem. 3:341
25. Young DE, Anderson LR, Fox WB (1970) Inorg. Nucl. Chem. Lett. 6:341
26. Lustig M, Pitochelli AR, Ruff JK (1967) J. Amer. Chem. Soc. 89:2841
27. Tompson PJ (1967) J. Amer. Chem. Soc. 89:4316
28. Ruff JK, Pitochelli AR, Lustig M (1966) J. Amer. Chem. Soc. 88:4531
29. Schack CJ, Christe KO (1979) J. Fluor. Chem. 14:519
30. Meshri DT, Shreeve JM (1968) J. Amer. Chem. Soc. 90:1711
31. Lustig M, Ruff JK (1967) J. Chem. Soc. Chem. Commun. 17:870
32. Bernstein PA, Hohorst FA, DesMarteau DD (1971) J. Amer. Chem. Soc. 93:3382
33. US Pat 4499024 (1983)
34. Cauble RL, Cady GH (1967) J. Amer. Chem. Soc. 87:5161
35. Hohorst FA, Shreeve JM (1967) J. Amer. Chem. Soc. 87:1809
36. Hohorst FA (1968) Inorg. Synth. 11:143
37. Redwood ME, Willis CJ (1965) Can. J. Chem. 43:1893
38. Ponomarenko VA, Krukovskii SP, Alybina AJu (1973) Ftorsodershchashchie geterozepnye polimery. Nauka, Moscow
39. Liebman JF, Jarvis BB (1975) J. Fluor. Chem. 5:41
40. Mukhametshin FM, Povrosnik SV (1986) Zh. Org. Khim. 23:945
41. Martineau E, Milne JB (1971) J. Chem. Soc. Chem. Commun. 21:1327
42. Cady GH (1971) Ann. asoc. quim. Argent. 59:125
43. Wechsberg M, Cady GH (1969) J. Amer. Chem. Soc. 91:4432
44. Hudlicky M (1976) Comprehensive chemistry of organic fluorine compounds. Wiley, New York
45. Nikonorov YuI (1978) In: V Vses. Symp. Khim. Neorg. Ftoridov. Nauka, Moscow
46. Nikolaev NS, Sukhoverkhov VF, Shishkov YuD, Alenchikova IF (1968) Khimiya galoidnyh soedinenii ftora. Nauka, Moscow
47. Czarnowski J, Schumacher HJ (1970) Z. phys. Chem. (BRD) 73:68
48. Neirinski RD, Lambrecht RM, Walf AP (1978) Int. J. Appl. Rad. Isotopes 29:323
49. Prager JH, Thompson PG (1965) J. Amer. Chem. Soc. 87:230
50. US Pat 3415865 (1968)
51. Thompson PG, Prager JH (1967) J. Amer. Chem. Soc. 89:2263
52. Prager JH (1966) J. Org. Chem. 31:392
53. US Pat 3442927 (1966)

54. Thompson PG (1967) J. Amer. Chem. Soc. 89:1811
55. US Pat 3420866 (1966)
56. US Pat 3692815 (1964)
57. Schack CJ, Christe KO (1979) J. Fluor. Chem. 14:519
58. Yakubovich AYa, Englin MA, Makarov SP (1960) Zh. Obshch. Khim. 30:2374
59. Talbott RL (1965) J. Org. Chem. 30:1429
60. US Pat 3344194 (1964)
61. Anderson LR, Fox WB (1967) J. Amer. Chem. Soc. 89:4313
62. Solomon IJ, Kacmarek AJ, Sumida WK, Raney JK (1972) Inorg. Chem. 11:195
63. Marantz LB (1965) J. Org. Chem. 30:4380
64. Solomon IJ, Kacmarek AJ, Keith NJ, Raney JK (1968) J. Amer. Chem. Soc. 90:6557
65. Ginsburg VA, Tumanov AA (1968) Zh. Obshch. Khim. 38:1410
66. Sekija A, DesMarteau DD (1980) Inorg. Chem. 19:1328
67. Czarnowski J, Castellano E, Schumacher HJ (1969) Z. phys. Chem. (BRD) 65:225
68. Wilt PM, Jones EA (1968) J. Inorg. Nucl. Chem. 30:2933
69. Wilt PM, Jones EA (1967) J. Inorg. Nucl. Chem. 29:2108
70. Smardzewski RR, Fox WB (1975) J. Fluor. Chem. 6:417
71. Buckley P, Weber JP (1974) Can. J. Chem. 52:942
72. Bartell LS, Diodati FP (1971) J. Molec. Struct. 8:395
73. Cady GH, Merrit CJ (1962) J. Amer. Chem. Soc. 84:2260
74. Phillips L, Wray V: J. Chem. Soc. (B) 1971:2068
75. Phillips L, Wray V: J. Chem. Soc. Perkin Trans. II 1972:220
76. Thynne JCJ, Macheil KAG (1970) Int. J. Mass. Spectr. Ion Phys. 5:95
77. Foss GD, Pitt DA (1968) J. Phys. Chem. 72:3512
78. Sieiro C, Conzalez-Diaz P, Smeyers YG (1975) J. Molec. Struct. 25:345
79. Walker N, Fox WB (1979) J. Magn. Res. 34:295
80. Mitchell RW, Merrit JA (1967) J. Molec. Spectrosc. 24:128
81. Aymonino PJ (1965) J. Inorg. Nucl. Chem. 27:2675
82. Olsen JF (1978) J. Molec. Struct. 49:361
83. Olsen JF (1977) J. Fluor. Chem. 9:471
84. Kiss AJ, Hagittai J (1982) Z. Naturforsch. 37A:134
85. Ebrahem KA, Webb GA (1976) Org. Magn. Res. 8:317
86. Jug K, Nande DN (1980) Theor. Chim. Acta 57:131
87. Marsden CJ, DesMarteau DD (1977) Inorg. Chem. 16:2354
88. DesMarteau DD (1980) Inorg. Chem. 19:1690
89. Glidewill C (1980) J. Molec. Struct. 67:35
90. Wahi PK, Patel ND (1980) Can. J. Spectrosc. 25:70
91. Ghibandi E, Colussi AJ (1984) Inorg. Chem. 23:635
92. Huston JL, Studier MN (1979) J. Fluor. Chem. 13:235
93. DesMarteau DD (1972) Inorg. Chem. 11:193
94. Porter RS, Cady GH (1957) J. Amer. Chem. Soc. 79:5628
95. Ghibandi E, Colussi AJ (1981) Int. J. Chem. Kinet. 13:591
96. Cobos CJ, Jubert AH (1979) React. Kinet. Catal. Lett. 11:313
97. Croce AE, Castellano E (1982) J. Photochem. 19:303
98. Vanderkooi N, Fox WB (1967) J. Chem. Phys. 47:3634
99. Aymonino PJ, Blesa MA (1972) Z. phys. Chem. (BRD) 80:129
100. Vedeneev VI, Teitelboim MA, Shoikhet AA: Izv. Akad. Nauk SSSR. Ser. Khim. 1977:535
101. US Pat 3100083 (1963)

102. Roberts HL: J. Chem. Soc. 1964:4538
103. US Pat 3230264 (1962)
104. US Pat 3230263 (1962)
105. Kennedy RC, Levy JB (1972) J. Phys. Chem. 76:3480
106. Levy JB, Kennedy RC (1972) J. Amer. Chem. Soc. 94:3302
107. Descamps B, Forst W (1976) J. Phys. Chem. 80:933
108. Descamps B, Forst W (1975) Can. J. Chem. 53:1442
109. US Pat 3202718 (1960)
110. Christe KO, Pilipovich D (1971) J. Amer. Chem. Soc. 93:5111
111. Aymonino PJ (1968) Photochem. Photobiol. 7:761
112. Aymonino PJ: Chem. Commun. 1965:241
113. Blesa MA, Aymonino PJ (1968) Ann. asoc. quim. Argent. 56:101
114. Blesa MA, Aymonino PJ (1968) Ann. asoc. quim. Argent. 56:113
115. Varreti E, Aymonino PJ: Chem. Commun. 1967:680
116. Croce AE, Schumacher HJ (1982) Int. J. Chem. Kinet. 14:647
117. DesMarteau DD (1970) Inorg. Chem. 9:2179
118. Anderson LR, Fox WB (1970) Inorg. Chem. 9:2182
119. Berstein P, Hohorst FA, DesMarteau DD (1971) Amer. Chem. Soc. Polym. Prep. 12:378
120. Walker NS, DesMarteau DD (1975) J. Fluor. Chem. 5:127
121. Fraser GW, Shreeve JM (1967) Inorg. Chem. 6:1711
122. US Pat 3679709 (1970)
123. Bailey RE, Cady GH (1970) Inorg. Chem. 9:1930
124. Morton JR, Preston KF (1973) J. Phys. Chem. 77:2645
125. Brit. Pat 1439923 (1972)
126. Leroy J, Dudragne F, Adenis JC (1973) Tetrahedron Lett. 29:2771
127. US Pat 4008217 (1975)
128. Barton DHR, Hesse RH, Klose TR, Pechet MM: J. Chem. Soc. Chem. Commun. 1975:97
129. US Pat 3737533 (1971)
130. Mak-Kleland VI (1966) Usp. Khim. 35:509
131. Bilevich KA, Ohlobystin OYu (1968) Usp. Khim. 37:2162
132. Bagdasariyan KS (1984) Usp. Khim. 53:1073
133. Todres ZV (1986) Ion-radikaly v organicheskoi khimii. Khimiya, Moscow
134. Chanon M: Bull. soc. chim. France 1985:209
135. Moldavskii DD, Temchenko VG, Antipenko GL (1971) Zh. Org. Khim. 7:44
136. US Pat 3585218 (1967)
137. US Pat 3541128 (1966)
138. Noftle RE, Shreeve JM (1961) Inorg. Chem. 7:687
139. Allison JAC, Cady GH (1959) J. Amer. Chem. Soc. 81:1089
140. Hohorst FA, Shreeve JM (1968) Inorg. Chem. 7:624
141. Miller W, Koch S (1957) J. Amer. Chem. Soc. 79:3084
142. Nonhebel DC, Walton JC (1974) Free radical chemistry. University Press, Cambridge
143. Rusin LYu, Chaikin AM, Shilov AE (1964) Kinetika i kataliz, 5:1121
144. Durrell WS, Stump EC, Westmoreland G, Padgett CD (1965) J. Polym. Sci. 3A:4065
145. Toy MS, Stringham RS (1975) J. Fluor. Chem. 5:481
146. Toy MS, Stringham RS (1975) J. Fluor. Chem. 5:25
147. Toy MS, Stringham RS (1975) J. Fluor. Chem. 5:31
148. PosSatos AM, Schumacher HJ (1984) J. Int. Chem. Kinet. 16:103

149. Johri KK, DesMarteau DD (1983) J. Org. Chem. 48:242
150. Barton DHR, Hesse RH, Pechet MM, Ganguly AK: J. Chem. Soc. Chem. Commun. 1972:122
151. Elson IH, Mao SW, Kochi JK (1975) J. Amer. Chem. Soc. 97:335
152. US Pat 4533762 (1983)
153. US Pat 4577044 (1985)
154. US Pat 4588796 (1985)
155. US Pat 3931132 (1970)
156. US Pat 4393198 (1981)
157. US Pat 4321359 (1980)
158. Toy MS, Stringham RS (1979) Amer. Chem. Soc. Polym. Prep. 20:328
159. Toy MS, Stringham RS (1980) J. Polym. Sci. Polym. Lett. Ed. 18:229
160. Toy MS, Stringham RS (1978) J. Polym. Sci. Polym. Lett. Ed. 16:2781
161. Toy MS, Stringham RS (1978) J. Fluor. Chem. 13:23
162. Toy MS, Stringham RS (1978) Amer. Chem. Soc. Polym. Prep. 19:534
163. Toy MS, Stringham RS (1977) Amer. Chem. Soc. Polym. Prep. 18:438
164. Toy MS, Stringham RS (1979) J. Polym. Sci. Polym. Lett. Ed. 17:561
165. Soelch RR, Maller GW, Lemal DM (1985) J. Org. Chem. 50:5845
166. Bogachev AA, Kobrina LS, Yakobson GG (1986) Zh. Org. Khim. 22:2578
167. US Pat 4460514 (1984)
168. Barton DHR, Hesse RH, Tazzia G, Pechet MM: J. Chem. Soc. Chem. Commun. 1969:1497
169. Barton DHR, Danks LJ, Ganguly AK, Pechet MM: J. Chem. Soc. Chem. Commun. 1969:227
170. US Pat 3687943 (1969)
171. Barton DHR, Hesse RH, Markwell RE, Pechet MM, Rozen S (1976) J. Amer. Chem. Soc. 98:3036
172. Chavis C, Mousseron-Canet M: Bull. soc. chim. France 1971:632
173. Barton DHR (1969) J. Loughborough Uni. Technol. Chem. Soc. 7:37
174. Adamson J, Foster AB, Westwood JH (1971) Carbohydr. Res. 18:345
175. Podesva J, Pacak J (1973) Chem. Listy 67:785
176. Foster AB, Westwood JH (1973) Pure Appl. Chem. 35:147
177. Bischofberger K, Abraham BJ: J. Chem. Soc. Perkin Trans. I 1975:2457
178. Butchard CG, Kent PW (1971) Tetrahedron 27:3457
179. Adamson J, Foster AB, Hall LD (1970) Carbohydr. Res. 15:351
180. Adamson J, Marcus DM (1970) Carbohydr. Res. 13:314
181. Levy JB, Sterling DM (1985) J. Org. Chem. 50:5615
182. Barton DHR, Hesse RH, Jackman GP, Ogunkoya L, Pechet MM: J. Chem. Soc. Perkin Trans. I 1974:739
183. Barton DHR, Danks LJ, Ganguly AK, Hesse RH, Tazzia G, Pechet MM: J. Chem. Soc. Perkin Trans. I 1976:101
184. BDR Pat 2149504 (1970)
185. Robins JM, MacCoss M, Naik SR (1976) J. Amer. Chem. Soc. 98:7381
186. Robins JM, Ramani G, MacCoss M (1975) Can. J. Chem. 53:1302
187. Earl RA, Tabsent LB (1972) J. Heterocycl. Chem. 9:1441
188. Barton DHR, Hesse RH, Toh HT, Pechet MM (1972) J. Org. Chem. 37:329
189. Lin TS, Gao YS, Mancini WR (1983) J. Med. Chem. 26:1691
190. Miyashita O, Matsumura K, Shimadzu H, Hashimoto N (1981) Chem. Pharm. Bull. 29:3181
191. Patrick TB, Cantrell GL, Chang CI (1979) J. Amer. Chem. Soc. 101:7434

192. Patrick TB, Hayward EC (1974) J. Org. Chem. 39:2120
193. Airey J, Barton DHR (1974) Ann. quim. Real. Soc. exp. y quim. 70:871
194. Fifalt MJ, Olezak RT, Mundhenke RF, Bieron JF (1985) J. Org. Chem. 50:4576
195. BDR Pat 2136008 (1970)
196. Middleton WJ, Birgham EM (1980) J. Amer. Chem. Soc. 102:4840
197. Lerman O, Rozen S (1980) J. Org. Chem. 45:4122
198. Patrick TB, Cantrell GL, Sandra MJ (1980) J. Org. Chem. 45:1409
199. Rozen S, Menahem Y (1979) Tetrahedron Lett. 8:725
200. Rozen S, Menahem Y (1980) J. Fluor. Chem. 16:19
201. Rozen S, Menahem Y: J. Chem. Soc. Chem. Commun. 1979:479
202. Merrit RF (1966) J. Org. Chem. 31:3871
203. Hesse RH (1978) Isr. J. Chem. 17:60
204. Teitelboim MA, Shoikhet AA, Kaplunov MF (1981) Kinetika i kataliz 22:298
205. Teitelboim MA, Shoikhet AA, Vedeneev VI (1981) Kinetika i kataliz 22:564
206. Teitelboim MA, Shoikhet AA, Vedeneev VI (1981) Kinetika i kataliz 22:852
207. Shoikhet AA, Teitelboim MA, Vedeneev VI (1983) Kinetika i kataliz 24:225
208. Shoikhet AA, Teitelboim MA (1982) Dep. VINITI 08.09.82, N 4800–82
209. Wang NY, Rowland FS (1985) J. Phys. Chem. 89:5159
210. Barton DHR (1970) Pure Appl. Chem. 21:285
211. Barton DHR, Hesse RH, Markwell RE, Pechet MM, Rozen S (1976) J. Amer. Chem. Soc. 98:3034
212. Kollonitsch J, Barash J (1970) J. Amer. Chem. Soc. 92:7494
213. Austrian Pat 3225524 (1971)
214. US Pat 3775444 (1971)
215. France Pat 2126545 (1971)
216. US Pat 3839170 (1972); US Pat 4030994 (1973)
217. Sweden Pat 380255 (1972)
218. Kollonitsch J, Barash J (1976) J. Amer. Chem. Soc. 98:5591
219. US Pat 4431817 (1984)
220. US Pat 4427661 (1984)
221. BDR Pat 2460360 (1974)
222. US Pat 3917688 (1973)
223. Brit. Pat 1437074 (1972)
224. Barton DHR, Hesse RH, Pechet MM, Toh HT: J. Chem. Soc. Perkin Trans. I 1974:732
225. Sequin M, Adenis JC, Michaud C, Basselier JJ (1980) J. Fluor. Chem. 16:506
226. Sequin M, Adenis JC, Michaud C, Basselier JJ (1980) J. Fluor. Chem. 16:37
227. Sequin M, Adenis JC, Michaud C, Basselier JJ (1980) J. Fluor. Chem. 16:201
228. Kollonitsch J, Marburg S, Perkins LM (1975) J. Org. Chem. 41:3107
229. Ratcliffe CT, Shreeve JM (1968) J. Amer. Chem. Soc. 90:5403
230. Denney DB, Denney DZ (1973) J. Amer. Chem. Soc. 95:4064
231. Denney DB, Denney DZ (1973) J. Amer. Chem. Soc. 95:8191
232. Death NJ, Denney DZ (1974) Phosphorus 3:205
233. Amey R, Martin JC (1979) J. Org. Chem. 44:1779
234. Bryce MR, Chambers RD, Millins ST (1984) J. Fluor. Chem. 26:533
235. Millins ST, Bryce MR, Chambers RD (1987) J. Fluor. Chem. 35:64
236. Leroy J (1981) J. Org. Chem. 46:206
237. Leroy J, Wakselman C: J. Chem. Soc. Perkin Trans. I 1978:1224
238. Wakselman C, Leroy J: J. Chem. Soc. Chem. Commun. 1976:611
239. US Pat 4499024 (1985)

4 Higher Fluorides of Group V and VI Elements as Fluorinating Agents in Organic Synthesis

Georgii Georgievich Furin and Vadim Viktorovich Bardin

Institute of Organic Chemistry, Pr. Lavrent'eva 9, 630090 Novosibirsk, USSR

Contents

4.1 Introduction

Recent advances have necessitated a survey of the data on the utility of fluorides MF_5 and MF_6 in the fluorination of organic molecules, since many of them have become commercially available and it has been found that they have synthetically useful properties. Some reactions of these fluorides were discussed in reviews [1–4], among which the most exhaustive was [2]. At the same time, new data have appeared revealing new potentialities of these reagents, which make these interesting but exotic substances practically useful and important.

As the data on the basic properties of MF_n are scattered in many references—monographs, reviews and papers, some of them being contradictory, it is reasonable to start every section with brief physico-chemical properties of the fluorides on the basis of the most precise (in our view) data.

4.2   Group V Pentafluorides and Their Derivatives in Fluoroorganic Synthesis

Phosphorus and arsenic pentafluorides are gases, antimony and vanadium pentafluorides are viscous associated liquids, whereas niobium, tantalum and bismuth pentafluorides are solids (Table 1). All of them are vigorously hydrolysed, and are converted by the reducing agents to stable trifluorides, excluding vanadium which forms the stable paramagnetic tetrafluoride apart from trifluoride.

4.2.1   Antimony Pentafluoride and Fluoroantimonates

Antimony pentafluoride is a strong fluorinating agent. It should be borne in mind that it is also one of the most powerful Lewis acids, whose presence may give rise to other processes apart from fluorination (see, e.g. [7,8]). It acts as a strong fluorinating agent in the reactions of chlorine (bromine, iodine) exchange for fluorine at a saturated carbon atom, and in the oxidative fluorinations of multiple bonds in polyhalogenated alkenes.

The processes of halogen exchange for fluorine at a saturated carbon atom in the reactions with SbF_5 or $SbCl_5$–HF, and the application of these reactions in organic synthesis are described in [1]. We shall give some examples to illustrate these transformations. It should be noted that the reactions of SbF_5 afford fluorine-containing alkanes, ketones, and ethers from polyhalogenated derivatives, but, as a rule, it is impossible to have all halogens replaced by fluorine.

Heating of 1-H-4-chloroperfluorobutane with SbF_5 at 175 °C for 3 h leads to the formation of 1-H-perfluorobutane with a 50% yield [9]. Boiling of methyl pentachloroethyl ether with antimony pentafluoride yields methyl (α,α-difluorotrichloroethyl) ether (84%), with residual chlorine atoms unsubstituted

Table 1. Properties of group V pentafluorides [6]

Fluoride	Mol. wt.	M.p. (°C)	B.p. (°C)	Density (g cm^{-3}) (°C)	Heat of formation (kJ mol^{-1})
PF_5	126	−93.7	−88.6	5.805(g l^{-1})	−1593
AsF_5	170	−79.8	−52.8	2.33(−52.8)	−1237
SbF_5	217	8.3	142.7	2.99(23)	−1379 [5a]
BiF_5	304	151.4 [5b]	230 [5b]	5.52(25)	
VF_5	146	19.5	48.0	2.18(19)	−1481
NbF_5	188	79.5	234.5	3.29(25)	−1814
TaF_5	276	96	229.2	4.98(15)	−1828

by fluorine [10]. At the same time, Dear [11] reported fluorination of β,β,β-trichlorohexafluoro-*tert*-butanol by antimony pentafluoride, giving perfluoro-*tert*-butanol with a 92% yield.

$$CHF_2CF_2CF_2CF_2Cl + SbF_5 \xrightarrow[175\,°C,3h]{} CHF_2CF_2CF_2CF_3$$
$$50\%$$

$$CCl_3CCl_2OCH_3 + SbF_5 \xrightarrow[140\ to\ 150\,°C]{} CCl_3CF_2OCH_3$$
$$84\%$$

$$CCl_3C(CF_3)_2OH + SbF_5 \xrightarrow[35\ to\ 65\,°C]{} (CF_3)_3COH$$
$$92\%$$

Under the action of SbF_5 iodine is replaced by fluorine in iodo-perfluoroalkanes; here the efficiency of fluorination by SbF_5 is close to that in the reactions with chlorine and bromine trifluorides [12].

$$CF_3(CF_2)_nCF_2I + SbF_5 \rightarrow CF_3(CF_2)_nCF_3 \quad (n=1–8)$$
$$(60–70\%)$$

It is worthwhile mentioning the use of antimony pentafluoride as a fluorooxidant. First it was found in the reactions of SbF_5 with polychlorinated alkenes which are transformed to polyfluoroalkanes. Thus on heating with SbF_5, hexachlorobutadiene forms 2,3-dichlorohexafluoro-2-butene and a small amount of 2-chloroheptafluoro-2-butene [13].

$$CCl_2=CCl-CCl=CCl_2 + SbF_5 \rightarrow CF_3CCl=CClCF_3 + CF_3CF=CClCF_3$$

In a similar way proceeds fluorination of hexachlorocyclopentadiene by antimony pentafluoride, leading to 1,2-dichlorohexafluorocyclopentene in a 50% yield [14,15]. In contrast to this, octachloro-2,5-diaza-1,5-hexadiene reacts with SbF_5 at 20 °C with the substitutive fluorination and cyclization. Heating of the resulting product with antimony pentafluoride leads to the substitution of the residual chlorine atom and formation of the immonium salt [16].

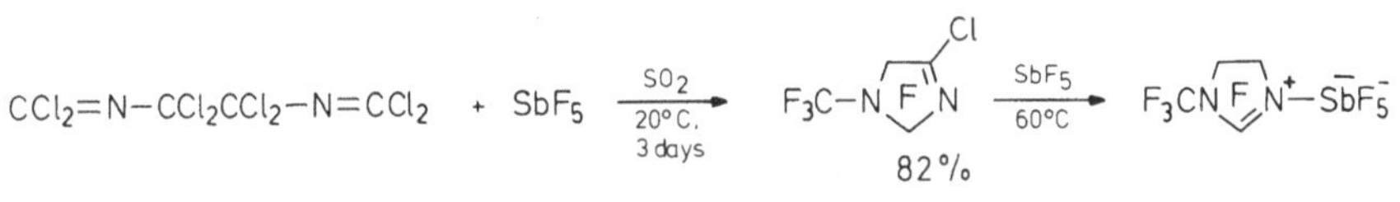

The polyfluorinated ethylene derivatives C_2F_3X react with SbF_5 at the atmospheric pressure with addition of two fluorine atoms, forming the respective polyfluoroethanes [17].

$$CF_2=CFX + SbF_5 \xrightarrow[50\,°C]{} CF_3CF_2X \quad (X=H,\ F,\ Cl)$$

Perfluoropropylene is fluorinated by SbF_5 at 70 to 80 °C (pressure) giving perfluoropropane (44% yield) but at the same time the dimer $(CF_3)_2CFCF=CFCF_3$ is formed (26% yield) [18]. The reaction of 1,1-difluoro-ethylene with antimony pentafluoride (50 °C) unexpectedly gave tris (β,β,β-trifluoroethyl)difluorostibine [17]. In the presence of catalytic amounts of SbF_5, the higher terminal polyfluoroalkanes are isomerised to the respective internal alkenes without fluorination [19].

Perhalo derivatives of cyclohexene are much more stable towards SbF_5 than fluorine-containing ethylenes and do not react with it at temperatures below 100 °C. At 140 to 150 °C 4-bromo- and 1-iodononafluorocyclohexenes add two fluorine atoms to form bromo(iodo)undecafluorocyclohexanes [20]. It is interesting that 1-chlorononafluorocyclohexene disproportionates in the same conditions to 1,2-dichlorooctafluorocyclohexene and perfluorocyclohexene.

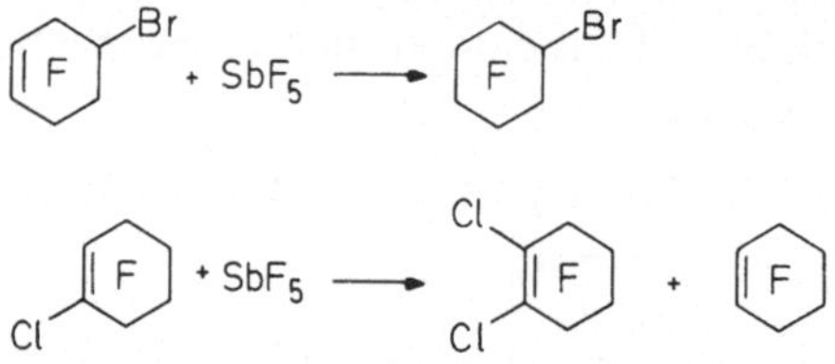

On heating with antimony pentafluoride at $\geqslant 100$ °C, polyhaloaromatic compounds are fluorinated to the respective derivatives of cyclohexene. This was first found by McBee with co-authors [14], when they treated hexachlorobenzene with SbF_5 at 125 °C, which resulted in 1,2-dichlorooctafluorocyclohexene. Later it appeared that, apart from this compound, the reaction gives 1,2,4-trichloro-heptafluorocyclohexene, 1,2,4,4-tetrachlorohexafluorocyclohexene, and traces of 1,2-dichlorohexafluorocyclopentene [21].

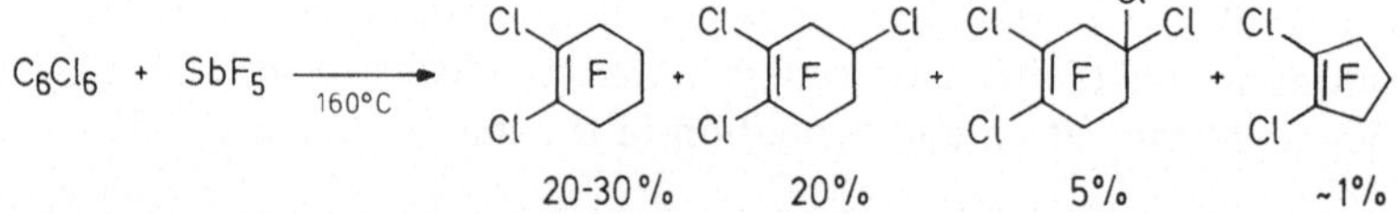

According to [22], antimony pentafluoride reacts in a similar way with perchloronaphthalene and perchlorodiphenyl, but the structure of the products of fluorination was not determined.

Upon heating with SbF_5 in a sealed tube, the pentafluorobenzene derivatives C_6F_5R (R = F, Cl, Br, I) are fluorinated to form mainly 1-R-nonafluorocyclo-hexenes [23]. Apart from them, the reaction products contain perfluorocyclo-hexene, R-undecafluorocyclohexane (R = Br, I) and 1,2-dichlorooctafluorocyclo-hexene (R = Cl). Pentafluorophenol reacts with SbF_5 at 100 °C in an open system to give perfluoro-2-cyclohexenone, whereas pentafluoroanisole and pentafluor-oaniline are completely resinified.

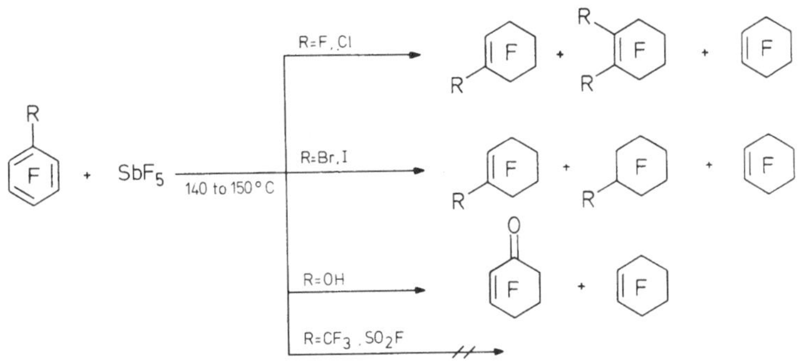

Octafluorotoluene and pentafluorobenzenesulfonyl fluoride are not fluorinated by antimony pentafluoride even after prolonged heating at 140 to 160 °C in a sealed tube.

Perfluorinated diphenyl [24] and naphthalene [24,25] react with SbF$_5$ upon heating, with addition of only four fluorine atoms.

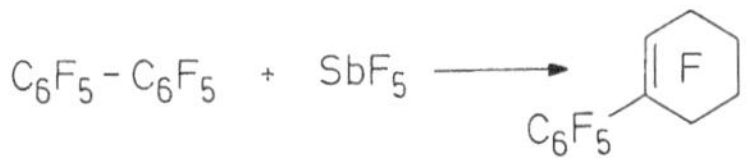

2-R-Heptafluoronaphthalenes (R = H, Cl, CF$_3$) treated with antimony pentafluoride are converted to perfluorotetralin and 6-R-perfluorotetralins (R = H, Cl) [24]. But heating of 2-bromoheptafluoronaphthalene and SbF$_5$ unexpectedly led mainly to perfluoro-1-methylindan [26]. As this is also the product of the reaction 2-bromoundecafluorotetralin with SbF$_5$ [27], the transformation of polyfluoronaphthalenes to the respective tetralins under the action of SbF$_5$ may be regarded as the common property of these compounds.

For the reaction of SbF$_5$ with perfluoromethylindans, there has been found an interesting dependence of fluorination route on the position of the CF$_3$ group. It appeared that fluorination of perfluoro-2-methylindan occurs only in the aromatic ring, whereas fluorination of other isomeric methylindans (as well as perfluoroindan) leads to cleavage of the aliphatic ring [28].

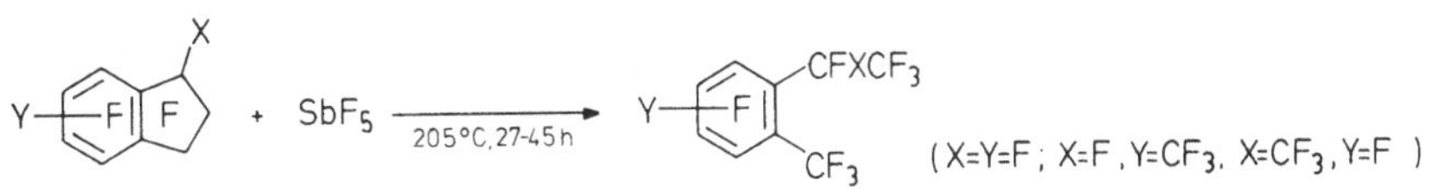

The same authors reported that in the presence of the catalytic amounts of bromine or iodine, perfluorinated indan, 4- and 5-methylindans react with SbF_5 to form perfluoro[4.3.0]bicyclo-1-nonene. From perfluorotetralin they obtained perfluoro[4.4.0]bicyclo-1-decene. The reactions were proved to proceed via the initial bromofluorination of the aromatic ring with subsequent fluorodebromination.

In the reaction of perfluorobenzocyclobutene with Br_2–SbF_5 the only product is 2-bromoperfluoroethylbenzene.

The aromatic hydrocarbons are not fluorinated by antimony pentafluoride, but are resinified for the most part. In the case of the reactions of SbF_5 with benzene derivatives, it has been shown that, apart from resinified products, the reaction mixture contains arylstibines [29].

Antimony pentafluoride may be used for fluorinations not only at the carbon atom. For example, pentafluorobenzenesulfonyl chloride treated with SbF_5 (20 to 25 °C) is quantitatively converted to pentafluorobenzenesulfonyl fluoride [30]. Pentafluorobenzenesulfenyl chloride and pentafluorophenyldifluoro-phosphine are oxidised by SbF_5 to form the salts of the pentafluorophenyl-difluorosulfonium and pentafluorophenyltrifluorophosphonium cations respectively [31,32]. On heating with SbF_5 phenyldichlorophosphine yields phenyltetrafluorophosphorane [33]. An example of the use of antimony penta-fluoride to obtain fluorosilanes from chloro derivatives is the transformation of β-chloro-α,α-difluoroethyltrichlorosilane to β-chloro-α,α-difluoroethyltrifluoro-silane [34]. It is interesting that interaction of tris-tert-butylsilane with SbF_5 also leads to the substituted fluorosilane [35].

Milder fluorinating agents are phenyltetrafluorostibine and diphenyl tri-fluorostibine. They provide smooth substitution of chlorine by fluorine in benzotrichloride and phenylpentachloroethane without involving the aromatic ring [36].

4.2.2 Vanadium Pentafluoride

Despite the fact that vanadium pentafluoride has been known for more than 30 years, its chemical properties have been little studied. In contrast to other group V element pentafluorides, this fluoride was found to be a very weak Lewis acid, but a substantially stronger fluorooxidant. Vanadium pentafluoride is easily soluble in HF, SO_2FCl, $CFCl_3$, ClF_3, BrF_3, BrF_5, SbF_5, Cl_2, Br_2 but has poor solubility in cyclo-C_4F_8, perfluorohexane, SF_6, AsF_5 [37,38]. With tetrahalom-ethanes CX_4 vanadium pentafluoride reacts at 20 °C, forming CF_4, CF_3X (X = Cl, Br, I), and CF_2Cl_2 (X = Cl) [39]. Carbon disulfide is transformed to bis(trifluoromethyl) di- and trisulfide and sulfur tetrafluoride, with VF_5 reduced to the trifluoride [40]. Heptane and cyclohexane reduce vanadium pentafluoride to the tetrafluoride, but the structure of the organic products formed is unknown [41]. With pyridine VF_5 reacts vigorously even at $-78\,°C$, forming a dark-brown precipitate insoluble in pyridine excess [42]. Under the action of ammonia, ethylenediamine, and tetramethylethylenediamine, VF_5 is reduced without fluorination of organic compounds [42].

Recently there have started the systematic studies of fluorination of poly-halogenated unsaturated and aromatic compounds by vanadium pentafluoride. Fluorination of terminal polyfluoroalkenes readily proceeds at -20 to $-30\,°C$ in $CFCl_3$ to give fluorine-containing alkanes [43].

$$CF_2{=}CXR_F + VF_5 \rightarrow CF_3CFXR_F$$

$$(X = Cl,\ R_F = C_2F_5,\ X = F,\ R_F = (CF_2)_4CFClCF_2Cl)$$

In a similar process tetrachloroethylene is transformed to 1,2-difluorotetra-chloroethane (72%).

Internal perfluorinated alkenes react with VF_5 under more severe conditions. Thus the transformation of isomeric perfluoromethylpentenes to perfluoro-2-methylpentane requires heating them with a 3-fold excess of VF_5 at 100 to 150 °C, but even under these conditions conversion of perfluoroalkenes is low. At the same time, replacement of vinyl fluorines by chlorine apparently facilitates fluorination of the C=C bond by VF_5, as 2,3-dichlorohexafluoro-2-butene is transformed to the respective alkane in the reaction with VF_5 at 20 to 30 °C. The effect of the nature of substituent on the rate of fluorination is more pronounced in the case of perfluoromethyl vinyl ether: unlike other terminal alkenes, this ether is fluorinated slowly even at 40 to 50 °C.

Vanadium pentafluoride easily fluorinates polyhalogenated 1,3-dienes [43]. At -20 to $-30\,°C$ 2-chloroperfluoro-1,3-butadiene is converted to cis-, trans-2-chloroperfluoro-2-butene and 2-chloroperfluorobutane (yields 56 and 8% respectively). As the alkene formed does not further react with VF_5 in these conditions, the polyfluorinated alkane is presumably the product of fluorine 1,2-addition to the starting fluorine-containing diene and subsequent fluorination of terminal butene.

$$CF_2{=}CCl{-}CF{=}CF_2 + VF_5 \xrightarrow{\ CFCl_3\ } CF_3CCl{=}CFCF_3 + CF_3CFClCF_2CF_3$$

Perchlorinated 1,3-butadiene in the same conditions adds 4 fluorine atoms, forming nearly equal amounts of hexachloro-1,2,3,4-tetrafluorobutane and hexachloro-1,1,2,4-tetrafluorobutane. Prolongation of the reaction from 0.5 to 3 h does not alter the isomer ratio.

As for tetrachloroethylene, there is no exchange of chlorine atoms for fluorine. It is not clear as yet, whether hexachloro-1,1,2,4-tetrafluorobutane is the primary product or is obtained from another isomer.

Polyfluorinated cycloalkenes are fluorinated by VF_5 in more rigid conditions than the terminal polyfluoroolefins, and their reactivity approaches that of the internal polyfluoroalkenes. The addition of two fluorine atoms to the 1,2-difluorovinyl fragment of polyfluorocycloalkene proceeds at 50 to 60 °C and the fluoro-containing cycloalkanes are obtained in yields of 60 to 70% [20].

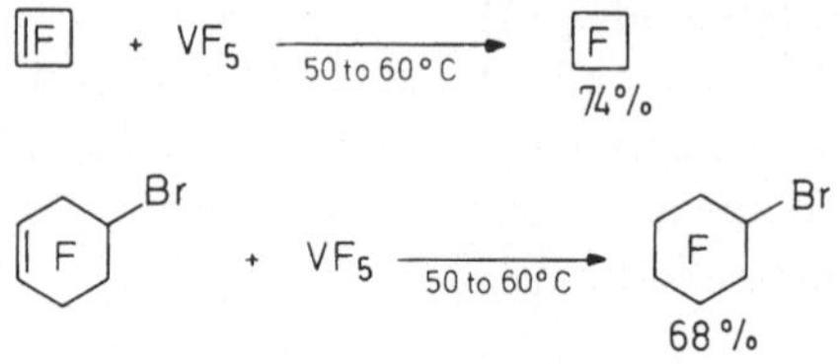

Replacement of the vinyl fluorines by chlorine leads to the increased reactivity of polyfluorocycloalkene, whereas the perfluoroalkyl groups strongly hinder the C=C bond fluorination. For example, 4-bromoperfluoro-1-methylcyclohexene does not react with VF_5 at 150 °C, and is slowly fluorinated only at 250 °C. At the same time, 1,2-dichlorooctafluorocyclohexene and 1,2-dichlorohexafluorocyclopentene are transformed to 1,2-dichloroperfluorocycloalkanes at 25 °C.

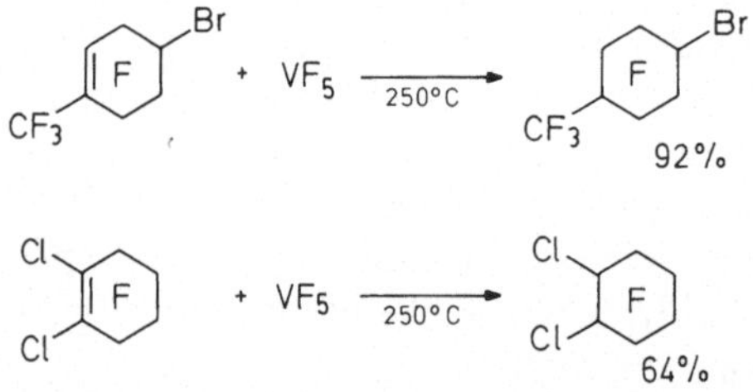

The polyfluorinated derivatives of 1,4-cyclohexadiene have two formally independent C=C bonds, therefore the relative rate of fluorine addition in the reaction with VF_5 is determined by the electronic nature of substituents at the vinyl carbon atoms [44].

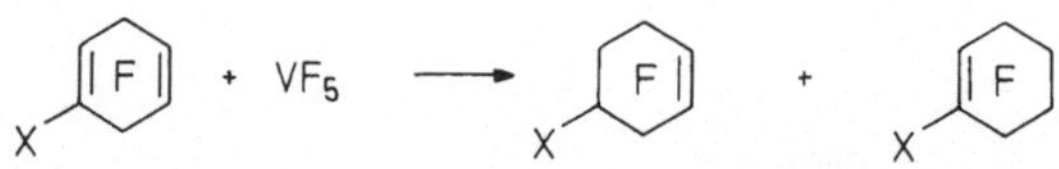

X	Yield (%)		Temperature (°C)
F		91	25
H	75	4	−25
Cl	59	27	−25
CF$_3$	—	92	25

It should be noted that 1-methylheptafluoro-1,4-cyclohexadiene is fluorinated by vanadium pentafluoride at 20 to 25 °C not only via the fluorine addition at the C=C bond but also via substitution of hydrogen by fluorine in the alkyl group.

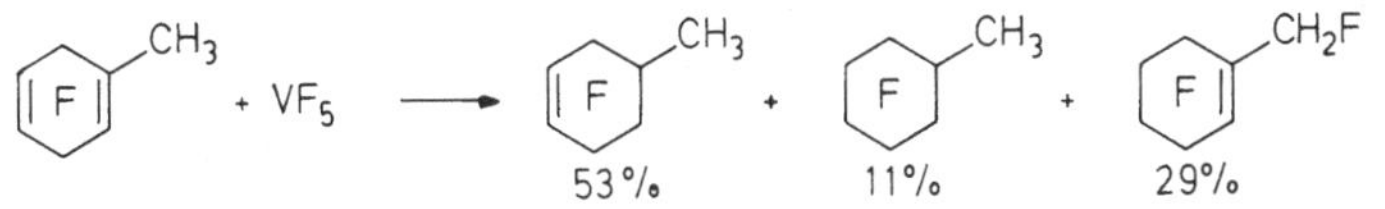

53%   11%   29%

The reactions of vanadium pentafluoride with polyfluorinated aromatic compounds have been comprehensively studied. In these reactions, the products of fluorination of perfluoroaromatics at -30 to $-20°C$ in SO_2FCl or $CFCl_3$ are the derivatives of 1,4-cyclohexadiene and cyclohexene, whereas the products of fluorine 1,2-addition to the aromatic ring have not been found [45,46]. The fluorination degree of decafluorodiphenyl, octafluoronaphthalene, and decafluoroanthracene mainly depends on the amount of vanadium pentafluoride. However even in a large excess of VF_5, the end products of fluorination at -20 to $20°C$ are again unsaturated compounds [45,46] (cf. [43]).

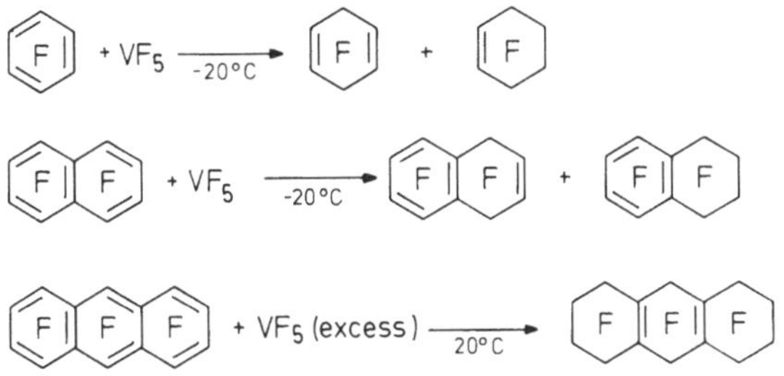

Treatment of pentafluorobenzenes C_6F_5X with vanadium pentafluoride leads to formation of a mixture of polyfluorinated 1,4-cyclohexadienes and cyclohexenes [45,46].

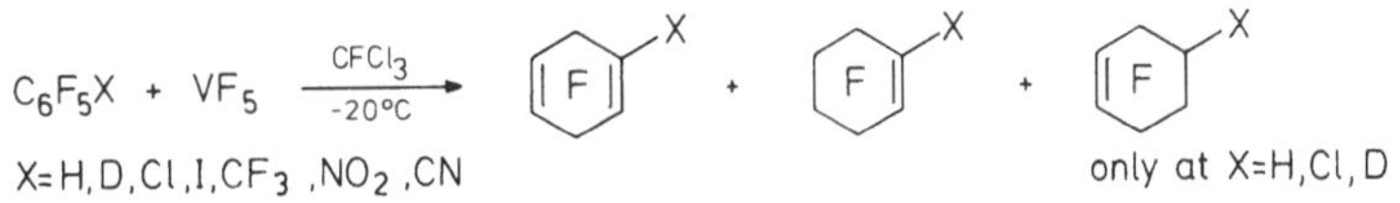

$X = H, D, Cl, I, CF_3, NO_2, CN$ only at $X = H, Cl, D$

Fluorination of methoxy- and difluoromethoxypentafluorobenzene also results in alkoxyundecafluorocyclohexane and 3-alkoxyheptafluoro-1,4-cyclohexadiene. The low-stable 3-methoxycyclohexadiene is converted in these conditions to hexafluoro-2,5-cyclohexadien-1-one, and 3-difluoromethoxyheptafluoro-1,4-cyclohexadiene was isolated [47].

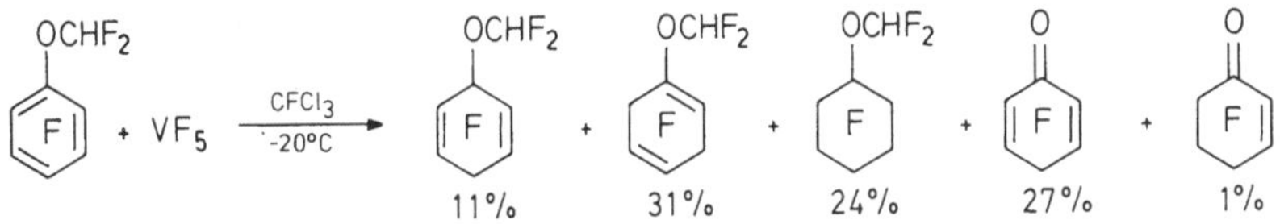

11%   31%   24%   27%   1%

It should be noted that pentafluorophenol is not fluorinated by VF_5 but oxidised to the pentafluorophenoxyl radicals, which are transformed to perfluorinated phenoxycyclohexadienones and hexafluoro-2,5-cyclohexadien-1-one [48].

The interaction of VF_5 with alkylpentafluorobenzenes proceeds in a more complex manner. Not only the benzene ring is fluorinated, but also the α-hydrogen atom is substituted by fluorine even in heptafluorotoluene [47].

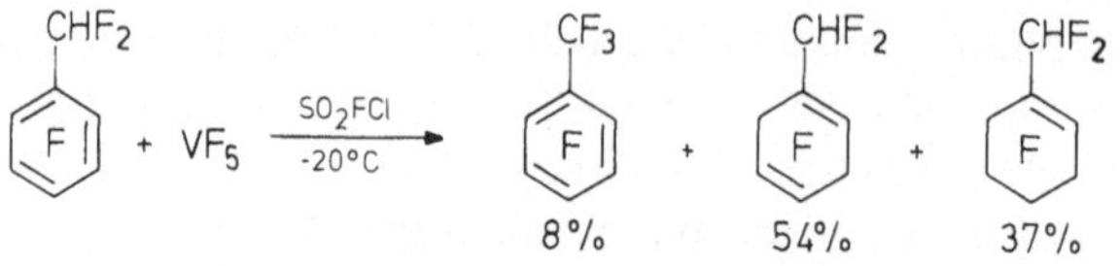

Thus vanadium pentafluoride is a promising fluorinating agent characterised by the fluorinating and oxidative properties. The data available suggest that the relative reactivity of polyfluorinated compounds decreases in the series:

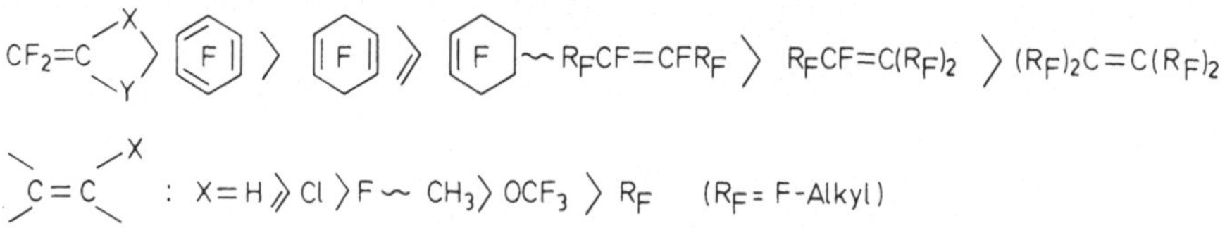

In this case the chlorine, bromine and iodine atoms, and the OAlk, NO_2, and CN groups do not hinder fluorination of the polyfluorinated unsaturated and aromatic compounds and are not changed by VF_5. There is no cleavage of the carbon–deuterium bond in the fluorination of deuteropentafluorobenzene by VF_5 [45], nor of the carbon–hydrogen bond in alkoxypentafluorobenzenes [47], but the alkyl hydrogen atoms in the allyl or benzyl position may be substituted by the fluorine atoms.

The fluorinations are carried out by adding VF_5 in a solvent or without it to the solution of a substrate in SO_2FCl, $CFCl_3$ or CH_3CN. Hydrogen fluoride readily dissolves VF_5, but is less suitable because of the low solubility of polyhalogenated organic compounds in it. Suitable reactors are of quartz, polytetrafluoroethylene, polychlorotrifluoroethylene (Kel-F), stainless steel, nickel, or copper. When necessary (at a temperature above 50 °C), the process is conducted in a steel or nickel tube with a copper or Teflon seal, but in this case the reaction mixture should be shaken or stirred because of the formation of insoluble vanadium tetrafluoride. The fluorination products are easily separated either by pouring them onto ice after hydrolysis of the reaction mixture, or by distilling them off from the reactor. In the latter case the residual VF_4 may be converted to VF_5 by the reaction with fluorine at 180 to 220 °C.

4.2.3 Phosphorus Pentafluoride and Fluorophosphoranes

Literature contains few data on the use of phosphorus pentafluoride for fluorination of organic compounds. Phosgene was reported [49] to be converted by PF_5 under rigid conditions (temperature, pressure) to dichlorodifluoro-methane. Tris (*tert*-butyl)silane treated with PF_5 in 1,1,2-trichlorotrifluoro-ethane is transformed to tris(*tert*-butyl)fluorosilane [50]. Interaction of tetraphenyltin with PF_5 in an autoclave leads to the formation of phenyltetra-fluorophosphorane and triphenyltin hexafluorophosphate. Heating of the latter yields fluorotriphenyltin [51].

$$(C_6H_5)_4Sn + PF_5 \xrightarrow[135\,°C,\ 20\,h]{} C_6H_5PF_4 + (C_6H_5)_3SnPF_6$$

$$(C_6H_5)_3SnPF_6 \xrightarrow{\Delta} (C_6H_5)_3SnF + PF_5$$

With phenol phosphorus pentafluoride forms phenoxyphosphoranes $(PhO)_nPF_{5-n}$ ($n = 1-3$) and $(PhO)_4P^+PF_6^-$, but no fluorination of the aromatic ring occurs [52].

There are reports of the fluorination of the benzaldehyde carbonyl group with diphenyltrifluorophosphorane, leading to benzal fluoride (moderate yield) [52]. Phenyltetrafluorophosphorane vigorously reacts with aldehydes and ketones, forming fluorine-free polymer products. Hexamethyldisiloxane reacts with $PhPF_4$ to give fluorotrimethylsilane (yield (93%), and propionic anhydride forms propionyl fluoride (yield 91%) [53]. Succinic anhydride is transformed to succinyl difluoride in more rigid conditions (120 to 140 °C, yield 80%), but maleic anhydride polymerises in the reaction with $PhPF_4$.

Alcohols react with phenyltetrafluorophosphorane, forming organic fluori-des and olefins in low yields. However, if, instead of alcohols, $PhPF_4$ is made to react with their trimethylsilyl ethers, the reaction smoothly leads to phenyl-alkoxytrifluorophosphoranes, which decompose upon heating with liberation of alkyl fluorides [54–56].

$$ROH + Me_3SiCl \rightarrow ROSiMe_3 \xrightarrow{PhPF_4} ROPF_3Ph \rightarrow RF + PhPOF_2$$

$$R = Me,\ Et,\ i\text{-}Pr,\ MePrCH-,\ cyclo\text{-}C_6H_{11},\ t\text{-}Bu$$

Phenol and 2,2,2-trichloroethanol form very stable phosphoranes $ROPF_3Ph$ ($R = Ph,\ CCl_3CH_2$); their attempted transformation to the respective fluorides by heating to 150 to 200 °C failed.

For the direct transformation of alcohols to fluorides, the use of triphenyldi-fluorophosphorane [57] and diphenyltrifluorophosphorane [58] has been sug-gested. The syntheses are carried out at 130 to 180 °C in acetonitrile for 6 to 10 h. Yield of the alkyl fluoride depends on the structure of the alkyl group: primary aliphatic alcohols are transformed to alkyl fluorides in 50–80% yields; secondary alkyl fluorides are formed with lower yields, and cyclohexanol gives only cyclohexene.

$$ROH + Ph_2PF_3 \xrightarrow{MeCN} RF + Ph_2POF + HF$$

$$R = C_5H_{11}(62\%),\ C_8H_{17}(76\%),\ PhCH_2(32\%),\ PhCH_2CH_2(52\%),$$
$$MePrCH(54\%),\ CH_2ClCH_2CH_2(64\%)$$

As a fluorinating agent, tetraphenylphosphonium hydrogen difluoride may be used [59]. Unlike mono-, di, and triphenyl fluorophosphoranes, this substance exists in the ionic form $Ph_4P^+ HF_2^-$, so it may be easily prepared [60] and dried. It enjoys good thermal stability and excellent solubility in polar solvents, such as CH_3CN, and will also dissolve in many less polar solvents on warming. This substance acts as a nucleophilic agent. Thus $PhCH_2Br$ may be completely converted to $PhCH_2F$ after 2.5 h at 52 °C using two equivalents of Ph_4PHF_2 in MeCN. 1-Iodoheptane and 1-bromodecane are converted to the corresponding fluorides at 80 to 130 °C after 2 h (46–70% yields). The substitution of chlorine atoms by fluorine in the activated aromatic compounds also proceeds smoothly.

$$ArCl + Ph_4PHF_2 \rightarrow ArF$$

$$Ar = 2, 4\text{-}(NO_2)_2C_6H_3, 2, 6\text{-}Cl_2\text{-}4\text{-}CF_3C_6H_2$$

At the same time, 1,2-dinitrobenzene reacts with Ph_4PHF_2, giving 2-fluoronitrobenzene (70% yield). The reaction of 2-chloro-6-nitrobenzonitrile with the tetraphenylphosphonium hydrogen difluoride in DMSO at room temperature gives 2-chloro-6-fluorobenzonitrile (100% yield), exclusively as a result of fluorodenitration. On the other hand, the reaction with RbF (150 °C, DMSO) was reported to give both 2-fluoro-6-nitrobenzonitrile and 2-chloro-6-fluorobenzonitrile (approx. 1 : 3) [61]. The reason of this difference is not clear.

4.2.4 Arsenic, Niobium, Tantalum and Bismuth Pentafluorides

There is no published evidence of the use of these fluorides for fluorination of organic compounds, though AsF_5, NbF_5, and TaF_5 have found use as strong Lewis acids. With the exception of BiF_5, they are not characterised by the oxidative fluorination reactions, and, as far as one can judge by publications, they show little reaction in halogen or oxygen exchange for fluorine. Niobium and tantalum pentafluorides dissolve in aromatic hydrocarbons, pyridine, DMF, DMSO, acetonitrile, propylene carbonate, forming complexes with them, but no fluorination occurs [2,62]. Bismuth pentafluoride is a strong fluorooxidant, but it is slightly soluble in most known organic and inorganic solvents, including HF, and its reactions with organic substances have not been investigated as yet.

4.3 Group VI Hexafluorides in Fluoroorganic Synthesis

Chromium hexafluoride is thermally unstable and decomposes at −80 °C to chromium pentafluoride and fluorine [63]. The rest of group VI hexafluorides are quite stable, being volatile substances (Table 2). Molybdenum, tungsten and uranium hexafluorides are easily hydrolysed. In contrast to this, sulfur hexafluoride is chemically inert and reacts with organic compounds only under very severe conditions [64]. Selenium and tellurium hexafluorides are more reactive than SF_6, but have not found use as fluorinating agents for organic synthesis.

Table 2. Properties of group VI hexafluorides [6]

Fluoride	Mol. wt.	M.p. (°C)(MPa)	B.p. (°C)	Density (g cm^{-3})(°C)	Heat of formation (kJ mol^{-1})
MoF$_6$	210	17.6	33.9	2.55(17.6)	−1557
WF$_6$	298	2.0	17.3	3.44	−1721
UF$_6$	352	64.0(0.14)	56.3(subl.)	5.06	−2148
SF$_6$	146	−50.7(0.227)	−63.6(subl.)	1.88(−50.5)	−1221
SeF$_6$	193	−34.6(0.2)	−46.6(subl.)	3.25(−28) (g l^{-1})	−1029
TeF$_6$	242	−37.7(0.1088)	−38.6(subl.)	2.56(−38)	−1318

Therefore further treatment will involve examples of fluorinations of organic compounds by molybdenum, tungsten and uranium hexafluorides.

4.3.1   Molybdenum Hexafluoride

Molybdenum hexafluoride does not react with perfluorinated alkanes and cycloalkanes, whereas with hexene and cyclohexane it forms coloured solutions, which slowly decompose upon standing. Benzene and xylene reduce MoF$_6$, but the structure of the organic products of the reaction is unknown [2]. With CCl$_4$ molybdenum hexafluoride at 150 to 220 °C undergoes the substitution reaction leading to CFCl$_3$, CF$_2$Cl$_2$, and CF$_3$Cl. As a result of a vigorous reaction of CS$_2$ with MoF$_6$, one obtains MoF$_5$, sulfur, and (CF$_3$)$_2$S$_2$ [65]. At the same time, MoF$_6$ has unlimited solubility in hexachlorobutadiene at 5 to 35 °C, and does not react with it, nor with tetrachloroethylene [66,67].

The oxidative properties of MoF$_6$ have been used to transform phosphines to fluorophosphoranes [68]. For that purpose the dichloromethane solutions of substituted phosphine and MoF$_6$ were mixed at −60 °C in the dry argon atmosphere, and then heated to room temperature to give fluorophosphoranes with 30–40% yields.

$$R_1R_2R_3P + MoF_6 \xrightarrow[-60 \text{ to } 20\,°C]{CH_2Cl_2} R_1R_2R_3PF_2$$

(R$_i$ = Et, Bu, Ph, i-Pr, i-Bu)

An important aspect of the chemistry of molybdenum hexafluoride is its reactions with oxygen-containing compounds, especially the carbonyl compounds. The aliphatic and aromatic aldehydes quickly react with MoF$_6$ at 20 °C in the presence of BF$_3$ to afford the geminal difluorides (yields 20–55%). In a similar way proceed the reactions of MoF$_6$ with the aliphatic, alicyclic, and aromatic ketones [69,70].

$$PhC(O)R + MoF_6 \xrightarrow[-15 \text{ to } 20\,°C]{BF_3/CH_2Cl_2} PhCF_2R$$

R = Ph(55%), H(41%), CF$_3$(38%)

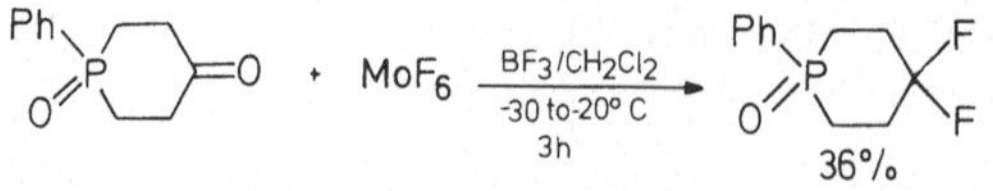

Alcohols and acids do not react with MoF_6 in these conditions.

It should be noted that in the absence of a catalyst, molybdenum hexafluoride fluorinates ketones only at temperatures above 100 °C [71].

Molybdenum hexafluoride has been used to obtain aryl trifluoromethyl ethers from aryl chlorothioformiates [72]. For that purpose the reagents are mixed at −25 °C in a metal reactor and heated. Yields of the target products are 70–90%.

$$RC_6H_4OH + Na \rightarrow RC_6H_4ONa \xrightarrow{CSCl_2} RC_6H_4OCSCl$$

$$\xrightarrow{MoF_6} RC_6H_4OCF_3$$

$$R = H(40\%), 2\text{-Me}(70\%), 3\text{-Me}(61\%), 4\text{-Me}(87\%), 4\text{-Cl}(90\%),$$
$$3\text{-F}(66\%), 4\text{-F}(95\%), 3\text{-CF}_3(70\%), 4\text{-Br}(89\%)$$

Chloroformiates are much more difficult to fluorinate with molybdenum hexafluoride. For example, phenyl chloroformiate was only transformed to phenyl fluoroformiate in a 30% yield, and there is no substitution of the carbonyl oxygen by fluorine. It means that the C=S group is more reactive toward MoF_6 than the C=O group.

Under the action of MoF_6 in mild conditions, the acyl and aroyl chlorides and the respective carboxylic acids are only transformed to the acyl (or aroyl) fluorides [71–77]. Dissolution of MoF_6 in acetic acid proceeds with evolution of heat, the temperature of the solution rising to 40 to 60 °C, to form acetyl fluoride (quantitative yield) and a colourless solution of $MoOF_4$ in acid excess [75]. If the fluorination is conducted at 130 to 140 °C in an autoclave, the products are the respective 1,1,1-trifluoroalkanes with 60–80% yields. The chlorine and bromine atoms in α-position to the carboxyl group remain intact in this process, but the dichloro- and difluoroacetic acids undergo decarboxylation (or decarbonylation) leading to the methane derivatives [73].

$$CH_2RCOOH + MoF_6 \rightarrow CH_2RCF_3$$

$$R = H(63\%), Cl(88\%), Br(89\%)$$

$$CF_3COOH + MoF_6 \xrightarrow{130\,°C,19\,h} CF_3COF$$

$$CHCl_2COOH + MoF_6 \rightarrow CHCl_2CF_3 + CHClF_2 + CHF_3 + CO + CO_2$$

The latter result is in disagreement with the data of [75], where the difluoroacetic acid and MoF_6 are reported to react for 20 h at 190 °C to form pentafluoroethane with a 60% yield.

Aliphatic esters react with molybdenum hexafluoride under the same conditions as acids. For example, ethyl acetate is transformed to 1,1,1-trifluoroethane with a 46% yield upon heating with two equivalent of MoF_6 at 130 °C for 16.5 h [73].

Fluorination of the aromatic carboxylic acids is illustrated by the reaction of the 3- and 4-pyridinecarboxylic, and 2,6-pyridinedicarboxylic acids [75]. Substitution of oxygen by fluorine proceeds in rigid conditions, requiring the use of 3.5 to 5 mole of MoF_6 per mole of the acid.

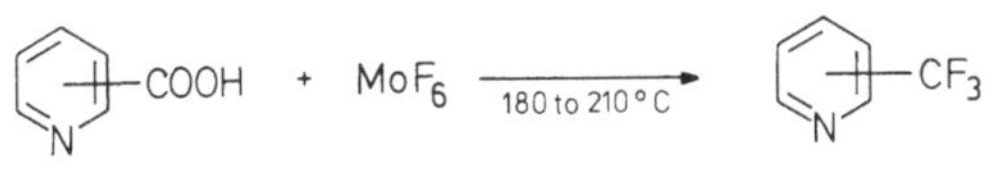

The reaction is performed in an autoclave, the yield of the trifluoromethylpyridines being 60–80%.

The transformations of aroyl chlorides to the substituted benzotrifluorides under the action of molybdenum hexafluoride proceed more readily than of the acids themselves [73]. Aroyl chloride and MoF_6 are mixed at -20 °C in the ratio of 3:1 (mol), and the mixture is heated to 130 °C. The yields of the target products strongly depend on the substituent position in the ring, the ortho-substituted benzotrifluorides always obtained in low yields.

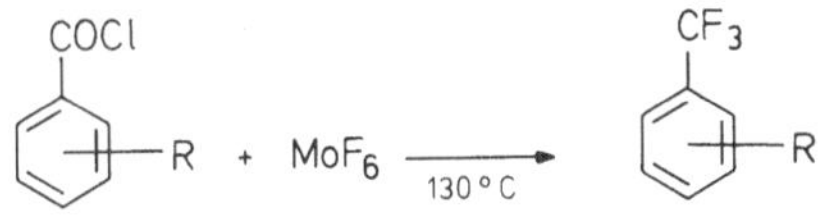

R = 2-F (1%), 4-F (55%), 2-Cl (1%), 4-Cl (17%), H (42%)

The presence of electron-accepting substituents in the aromatic ring facilitates fluorination of the chlorocarbonyl group, whereas the electron-donating substituents considerably slow down the process. For example, 4-methylbenzoyl chloride was converted to 4-methylbenzotrifluoride with an 8% yield.

Thiobenzoyl chloride reacts with MoF_6 in the same manner as benzoyl chloride; the yield of benzotrifluoride in both cases is close to 40% [73].

Puy [71] has shown that molybdenum hexafluoride at 130 °C transforms 2,2,2-trifluoroethanol to 1,1,1,2-tetrafluoroethane (85%), and ethylene oxide gives 1,1-difluoroethane with a 42% yield. The reaction of MoF_6 with dimethyl sulfite has been reported, where the products are methyl fluoride and methyl fluorosulfinate [78]. There are no other examples of fluorination of these functional groups in literature, therefore it is early to judge how general these transformations are. It should be noted that, unlike carbonyl compounds, carbonyl hydrazones react with molybdenum hexafluoride without formation of fluoroorganic products [79].

4.3.2 Tungsten Hexafluoride

Tungsten is less reactive than MoF_6. It dissolves in hexane, cyclohexane, CCl_4, perfluoroalkanes, tetrachloroethylene, benzene, xylene, toluene, hexafluorobenzene, and acetonitrile with formation of the donor-acceptor type complexes, but there is no reaction deeper than that [2].

According to [80], upon defreezing of a mixture of dimethyl sulfoxide and tungsten hexafluoride (2:1) from -196 to $0\,°C$, there takes place a vigorous reaction with the liberation of *symm*-difluorodimethyl ether. The reaction of DMSO with sulfur tetrafluoride proceeds in a similar way [80]. Dialkyl sulfites readily react with WF_6, forming alkyl fluorosulfinates and alkoxytungsten pentafluoride [81]. Compounds of both types decompose upon heating with the liberation of alkyl fluorides.

$$(MeO)_2SO + WF_6 \xrightarrow[-40\,°C]{} MeOSOF + MeOWF_5$$

$$MeOSOF \rightarrow MeF + SO_2$$

$$MeOWF_5 \xrightarrow[135\,°C]{} MeF + WOF_4$$

Diphenyl sulfite is also converted by WF_6 to $PhOWF_5$, but, unlike alkoxytungsten pentafluoride, this compound does not decompose below $180\,°C$.

Formation of CH_3OWF_5 is observed in the reaction of tungsten hexafluoride with trimethyl phosphite [81], tetramethoxysilane, trimethoxyboron, and pentamethoxyniobium [82]. The latter two reactions give also methyl fluoride.

Tungsten hexafluoride can react with carbonyl compounds, forming the geminal difluorides, in the same way as MoF_6 [83]. The reaction proceeds with dialkyl ketones, aliphatic aldehydes, and dialkyl carbonates. The fluorination is carried out in anhydrous dichloromethane at 50 to $150\,°C$, in the presence of BF_3 or BF_3OEt_2, but even in these conditions the yield of target products is $\sim 20\%$.

$$R_1COR_2 + WF_6 \xrightarrow{BF_3/CH_2Cl_2} R_1CF_2R_2$$

$R_1 = R_2 = Pr(22\%)$, $R_1 = H$, $R_2 = C_5H_{11}(20\%)$, $R_1 = H$,
$R_2 = C_6H_{13}(21\%)$, $R_1R_2 = \text{cyclo-}C_6H_{10}(19\%)$, $\text{cyclo-}C_5H_8(17\%)$

The carboxylic acids, their anhydrides and acyl chlorides form only acyl fluorides [73,83]. Thioacetic-S-acid reacts with WF_6 to give acetyl fluoride [84].

$$RCOCl + WF_6 \xrightarrow[120\ to\ 150\,°C,\ 6\ h]{BF_3/CH_2Cl_2} RCOF$$

$$R = CH_3(18\%),\ C_5H_{11}(23\%)$$

$$CH_3C(S)OH + WF_6 \rightarrow CH_3COF$$

On heating with tungsten hexafluoride, alcohols are dehydrated, and the alkenes formed are polymerised. Aldehyde and ketone hydrazones on treatment with

WF_6 (0 °C, 1 h, $C_2F_3Cl_3$) decompose to the carbonyl compounds, which do not undergo further reaction with WF_6 under these conditions [85].

4.3.3 Uranium Hexafluoride

In contrast to MoF_6 and WF_6, uranium hexafluoride vigorously reacts at room temperature with saturated hydrocarbons, causing their carbonization and being itself reduced to UF_4. At the same time, UF_6 dissolves in CH_2Cl_2, $CHCl_3$, CCl_4, pentachloroethane, perfluoroalkanes, acetonitrile, SO_2 and SO_2FCl [2,3,86,87]. With carbon tetrachloride, uranium hexafluoride starts to react only at 150 °C, giving chlorofluoromethanes. Ethylene and trichloroethylene react with uranium hexafluoride at a low temperature. Thus the interaction of UF_6 with $CHCl=CCl_2$ at -50 °C leads to the formation of chlorofluoroethanes and UF_4. The aromatic hydrocarbons react with UF_6 at room temperature with carbonisation. As shown in [80,88], UF_6 quickly reacts at 20 to 25 °C with the perfluorinated benzene, toluene, and p-xylene, but the reaction products have not been characterized. Uranium hexafluoride reacts at -90 to -100 °C with methanol in $CFCl_3$, giving $MeOUF_5$ and HF. Heating of the solution obtained to room temperature results in the formation of a green precipitate, but no details have been reported for the process [89].

With acetic acid, uranium hexafluoride reacts at 40 to 60 °C, yielding acetyl fluoride, UO_2F_2, and HF [77]. But with trifluoroacetic acid the reaction proceeds as follows [77,90].

$$2CF_3COOH + 2UF_6 \rightarrow CF_3COF + COF_2 + CO_2 + 2HUF_7$$

A typical distinction of the reactions of uranium hexafluoride with aldehydes from the respective reactions of MoF_6 and WF_6 is the fact that in the aldehyde group fluorine is substituted for hydrogen but not oxygen [91,92]. The reaction is carried out at 0 °C in the solution of 1,1,2-trichlorotrifluoroethane. The carbonyl component may be represented by the aliphatic and aromatic aldehyde.

$$RCHO + UF_6 \xrightarrow[\text{0 °C, 2 h}]{C_2F_3Cl_3} RCOF + HF + UF_4$$

$$R = C_4H_9(29\%),\ C_6H_{13}(47\%),\ C_6H_5(40\%),\ 2\text{-}BrC_6H_4(35\%),$$
$$4\text{-}CH_3OC_6H_4(5\%)$$

Formation of $C_6H_{13}COOMe$ and $PhCOOMe$ in the reactions of $C_6H_{13}CHO$ and PhCHO with uranium hexafluoride (0 °C, 2 h), with subsequent treatment of the reaction mixture with MeONa, provides an indirect support of these results [86].

The non-enolisable ketones (fluorenone, benzophenone) do not react in these conditions with uranium hexafluoride, though 2-adamantanone is transformed to 2,2-difluoroadamantane with a 41% yield [92].

Alcohols are dehydrated by UF_6 to the aldehydes, which are subsequently fluorinated to acyl fluorides [92].

$$ArCRHOH + UF_6 \rightarrow ArCOR + ArCOF$$

Ar	R			
C_6H_5	H	1	:	2(40%)
$4\text{-}CH_3C_6H_4$	H	1	:	2(38%)
C_6H_5	C_6H_5			(45%)
C_6H_5	CH_3			(64%)

Oximes, hydrazones, tertiary amines R_2CHNMe_2 and ethers R_2CHOMe are converted by uranium hexafluoride to the carbonyl compounds. In a similar way occurs cleavage of carboxylic acid hydrazones [91,92].

$$R_2C=N-NXY + UF_6 \xrightarrow[0\,°C,\ 2\,h]{C_2F_3Cl_3} (R_2C=N-\overset{+}{N}XY)F^- \xrightarrow{H_2O} R_2C=O$$

$$\underset{UF_5}{} \qquad\qquad 50\text{–}96\%$$

$$R_2CHOCH_3 + UF_6 \xrightarrow[25\,°C,\ 1\,h]{C_2F_3Cl_3} R_2C=O$$

$$RCONHNH_2 + UF_6 \xrightarrow{45\,°C} RCOOH \quad (R\text{—alkyl, aryl})$$

Trimethylchlorosilane has been shown [93] to react with uranium hexafluoride at a temperature as low as $-78\,°C$ in $CFCl_3$ to give trimethylfluorosilane. This observation is interesting in that it demonstrates the dependence of the Si–Cl bond reactivity on other substituents at the silicon atom, since UF_6 reacts with $SiCl_4$ at approx. $25\,°C$ [94]:

4.4 Practical Recommendations and Preparations

In conclusion we would like to give some practical recommendations for the use of the penta- and hexafluorides considered in this chapter in the synthesis of fluoroorganic compounds to facilitate the choice of a suitable fluorinating agent.

Antimony pentafluoride is a convenient reagent for the synthesis of poly-fluoroalkanes from haloperfluoroalkanes R_FX ($X = Cl$, Br, I) by the exchange of X for F at 60 to $170\,°C$. The aliphatic radical R_F may contain one or two hydrogen atoms, and the functional groups C–O–C, and $>C=O$, but the hydrocarbon analogues are resinified in the reaction with SbF_5.

Antimony and vanadium pentafluorides are strong fluorooxidants, VF_5 being more reactive than SbF_5. Using these fluorides, it is possible to transform polyhaloaromatic compounds to the polyfluorinated cyclohexadienes, cyclohex-enes, and cyclohexanes, whereas polyhaloalkenes and -cycloalkenes may be converted to the respective alkanes (phosphorus fluorides, as well as molybdenum, tungsten, and uranium hexafluorides are not suitable for that). It is important to know that in the processes of oxidative fluorination by antimony pentafluoride, the exchange of the Cl (Br, I) atoms for fluorine is possible, but this does not occur in the reactions involving VF_5. Fluorination by VF_5 proceeds in far milder conditions than fluorination by SbF_5.

The phenyl-substituted phosphorus (V) derivatives (except for Ph_4PHF_2) and hexafluorides MoF_6 and WF_6 may be used for substituting oxygen by fluorine in the functional groups COOR, C=O, and C–OH. The most suitable reagent for this is molybdenum hexafluoride, which is a rival to SF_4 and fluorosulfuranes. Currently MoF_6 stands second (after SF_4) in importance as an agent for the transformation of the carbonyl and carboxyl groups to the difluoromethylene and trifluoromethyl groups respectively. The aromatic aldehydes and ketones react with MoF_6 in the presence of BF_3 to form geminal difluorides with an approx. 50% yield. The presence of substituents F, Cl, Br, NO_2 in the ring does not prevent fluorination. Molybdenum hexafluoride is more selective than SF_4, therefore arylketones may be fluorinated by it without involvement of groups COOR, $CONR_2$, CN, and POR_2. The *ortho*-substituted arylketones do not react with MoF_6. Molybdenum hexafluoride is a convenient reagent for the synthesis of aryl trifluoromethyl ethers from aryl chlorothioformiates.

Primary and secondary alcohols may be easily converted to alkyl fluorides using Ph_2PF_3 or Ph_3PF_2, but for the fluorination of carbonyl compounds they are unsuitable.

Fluorination of oxygen-containing compounds by WF_6 proceeds in relatively severe conditions and gives small yields. Hence this fluoride may hardly be recommended for the synthesis. At the same time, UF_6 allows to obtain acyl fluorides from alcohols and aldehydes in one step.

1. Fluorination with SbF_5

a) *Perfluoroheptane* [12]. 1-Iodoperfluoroheptane (2 g) was placed with an excess of SbF_5 in a small autoclave, which was heated at 250 °C for 8 h, then gradually to 320 °C for 24 h. Distillation gave perfluoroheptane (66% yield).

A similar procedure yields perfluoropropane (67%), perfluorobutane (70%), perfluoropentane (65%), perfluorohexane (68%), and perfluorodecane (66%).

b) *Perfluorotetralin* [25]. Octafluoronaphthalene (1.60 g, 5.9 mmol) and SbF_5 (5.60 g, 25.8 mmol) were heated for 15 min at 100 °C with intermittent shaking of the flask. Then the mixture was heated to approx. 250 °C (the temperature of a bath) for 0.5 h, with simultaneous distilling off of the product into a trap containing cold water. The organic layer was washed with water and dried over $MgSO_4$ to give 1.45 g (71%) of perfluorotetralin (b.p. 163 to 165 °C).

2. Fluorination with VF_5

a) *1,2-Dichlorodecafluorocyclohexane* [20]. In a 10 ml nickel or steel tube was placed VF_5 (7.0 g, 48 mmol), the tube was cooled to −10 to −20 °C, and 1,2-dichlorooctafluorocyclohexene (3.0 g, 10.1 mmol) was added. The tube was stoppered and shaken at 50 to 60 °C for 3 h. The reaction mixture was cooled to −10 °C, then poured onto ice, the organic layer was washed with water, dried over $MgSO_4$ and distilled. The yield of 1,2-dichlorodecafluorocyclohexane was 95%.

3. Fluorination with Fluorophosphoranes

a) *Synthesis of diphenyltrifluorophosphorane* [58]. Antimony trifluoride (56.0 g, 0.31 mol) was added to Ph_2PCl (49.0 g, 0.22 mol) in five portions, with vigorous stirring. The reaction was slightly exothermic, but cooling was not necessary. After being warmed at 60 °C for 5 h, the reaction mixture was distilled in vacuum in a N_2 current. Antimony trichloride was distilled off first at 120 °C (18 mm Hg), then Ph_2PF_3 (b.p. 135 °C at 4 mm Hg) (22.8 g) was obtained.

b) *Alkyl fluorides* [58]. Freshly distilled Ph_2PF_3 (10.2 g, 42 mmol), alcohols (19 mmol) and acetonitrile (30 ml) were heated in a stainless steel autoclave at 150 °C for 10 h with stirring. After being cooled to room temperature the layer was washed with 30% NaOH and water, and distilled after being dried with Na_2SO_4. Yields of the alkyl fluorides were as follows: pentyl fluoride (62%), *sec*-pentyl fluoride (54%), benzyl fluoride (32%), phenethyl fluoride (52%), 3-chloropropyl fluoride (64%), octyl fluoride (76%, the product was obtained at 170 °C).

c) *Benzyl fluoride* [59]. Tetraphenylphosphonium hydrogen difluoride (1.88 g, 50 mmol) was dissolved in dry acetonitrile (20 g), and the solution was heated with stirring to 52 °C. The substrate, benzyl bromide (0.43 g, 25 mmol), was added to the warm solution, and the reaction mixture was kept for 2.5 h at 52 °C. Addition of dry ether (100 g) resulted in precipitation of all inorganics. Filtration of this solution followed by evaporation to dryness gave pure benzyl fluoride (0.25 g, 44 mmol, 88%).

4. Fluorination with MoF_6

a) *1,1-Difluorocyclohexane* [69]. In a 500 ml flask equipped with a stirrer, a reflux condenser and a gas inlet, 200 ml of anhydrous dichloromethane was stirred with 14 ml (35.6 g, 169 mmol) of molybdenum hexafluoride. The initially blue solution became green, then yellow-brown. The solution was cooled to 0 °C, and a slight current of boron trifluoride was passed through it. Then the solution was cooled to -15 to -10 °C, and a solution of 16.6 g (170 mmol) of the distilled cyclohexanone in 90 ml of dichloromethane was added to it dropwise. The resulting dark-red solution was kept at room temperature for 4 h, then anhydrous NaF was added to it, and the mixture was passed through the alumina column. The solvent was distilled off to give 10.7 g (53%) of the product, b.p. 80 °C.

4.5 References

1. Barbour AK, Belf LJ, Buxton MW (1963) Adv. Fluor. Chem. 3:181
2. Orekhov VT (1977) Usp. Khim. 46:799; (1977) Chem. Abs. 87:61793
3. Gubkina NI, Sokolov SV, Krylov EI (1966) Usp. Khim. 35:2219; (1967) Chem. Abs. 66:54654
4. Buslaev YuA, Kokunov YuV (1983) Koord. Khim. 9:723; (1983) Chem. Abs. 99:132622
5. a) Richards GW, Woolf AA (1971) J. Fluor. Chem. 1:129; b) O' Donnell TA (1976) in: Comprehensive Inorganic Chemistry, vol. 2. Pergamon, Oxford, p 1009

6. Efimov AI (1983) Svoystva neorganicheskikh soedinenii. Khimiya, Leningrad
7. Yakobson GG, Furin GG: Synthesis 1980:345
8. Olah GA (ed) (1963–1965) Friedel–Crafts and related reactions, vols 1–4, Interscience, New York
9. US Pat 2490764 (1949); (1950) Chem. Abs. 44:303
10. US Pat 2803665 (1957); (1958) Chem. Abs. 52:2047
11. Dear RFA: Synthesis 1970:361
12. Haszeldine RN: J. Chem. Soc. 1953:3761
13. US Pat 2436357 (1948); (1948) Chem. Abs. 42:5465
14. McBee ET, Wiseman PA, Bachman GR (1947) Ind. Eng. Chem. 39:415
15. US Pat 2449233 (1948); (1949) Chem. Abs. 43:678
16. Pawelke G, Bürger H, Brauer DJ, Wilke J (1987) J. Fluor. Chem. 36:185
17. Belen'kii GG, Kopaevich YuL, German LS, Knunyants IL: Izv. Akad. Nauk SSSR. Ser. Khim. 1972:983; (1972) Chem. Abs. 77:75296
18. Kopaevich YuL, Belen'kii GG, Mysov EI, German LS, Knunyants IL (1972) Zh. Vses. Khim. Obshchest. 17:236; (1972) Chem. Abs. 77:33864
19. Belen'kii GG, Savicheva GI, Lur'e EP, German LS: Izv. Akad. Nauk SSSR. Ser. Khim. 1978:1640; (1978) Chem. Abs. 89:162986
20. Bardin VV, Avramenko AA, Petrov VA, Krasil'nikov VA, Karelin AI, Tushin PP, Furin GG, Yakobson GG (1987) Zh. Org. Khim. 23:593
21. Leffler AJ (1959) J. Org. Chem. 24:1132
22. US Pat 2553217 (1951); (1951) Chem. Abs. 45:9079
23. Bardin VV, Furin GG, Yakobson GG (1978) Izv. Sib. Otd. Akad. Nauk SSSR. Ser. Khim. Nauk. Vyp 6:142; (1979) Chem. Abs. 90:121057
24. Pozdnyakovich YuV, Steingarts VD (1978) Zh. Org. Khim. 14:2237; (1979) Chem. Abs. 90:54714
25. USSR Pat 491605 (1974); (1976) Chem. Abs. 84:30536
26. Pozdnyakovich YuV, Bardin VV, Shtark AA, Steingarts VD (1979) Zh. Org. Khim. 15:656; (1979) Chem. Abs. 91:20153
27. Bardin VV, Furin GG, Yakobson GG (1979) J. Fluor. Chem. 14:455
28. Karpov VM, Mezhenkova TV, Platonov VE, Yakobson GG: Bull. Soc. Chim. France 1986:980
29. Olah GA, Schilling P, Gross I (1974) J. Amer. Chem. Soc. 96:876
30. Furin GG, Terent'eva TV, Rezvukhin AI, Yakobson GG (1974) Izv. Sib. Otd. Akad. Nauk SSSR. Ser. Khim. Nauk. Vyp. 6:135; (1975) Chem. Abs. 82:72600
31. Yakobson GG, Furin GG, Terent'eva TV (1974) Zh. Org. Khim. 10:799; (1974) Chem. Abs. 81:25234
32. Furin GG, Terent'eva TV, Rezvukhin AI (1978) Zh. Obshch. Khim. 45:473
33. US Pat 2904588 (1958); (1960) Chem. Abs. 54:2254
34. Haszeldine RN, Rogers DI, Tipping M: J. Chem. Soc. Dalton Trans. 1976:1056
35. Dexheimer EM, Spialter L, Smithson LD (1975) J. Organomet. Chem. 102:21
36. Yagupol'skii LM, Kondratenko NV, Popov VI (1977) Zh. Org. Khim. 13:613; (1977) Chem. Abs. 87:22599
37. Fowler BR, Moss KC (1979) J. Fluor. Chem. 14:485
38. Bardin VV, Furin GG, Avramenko AA, Krasil'nikov VA, Tushin PP, Karelin AI, Yakobson GG (1984) Zh. Org. Khim. 20:343; (1984) Chem. Abs. 101:72338
39. Clark H, Emeleus H: J. Chem. Soc. 1957:2119
40. Canterford J, O'Donnell TA (1967) Inorg. Chem. 6:541
41. Hammond PR, Lake PR: J. Chem. Soc. (A) 1971:3800

42. Cavell RG, Clark HC (1961) J. Inorg. Nucl. Chem. 17:257
43. Petrov VA, Bardin VV, Furin GG, Avramenko AA, Galakhov MV, Krasil'nikov VA, Karelin AI, Tushin PP, Yakobson GG (1987) Zh. Org. Khim. 23:43
44. Avramenko AA, Bardin VV, Furin GG, Karelin AI, Krasil'nikov VA, Tushin PP (1988) Zh. Org. Khim. 24:1443
45. Bardin VV, Avramenko AA, Furin GG, Yakobson GG, Krasil'nikov VA, Tushin PP, Karelin AI (1985) J. Fluor. Chem. 28:37
46. Furin GG, Bardin VV, Avramenko AA (1988) In: 5th Regular Meeting of Soviet-Japanese Fluorine Chemists, 25–26 Jan 1988, Tokyo, p IV-1
47. Avramenko AA, Bardin VV, Karelin AI, Krasil'nikov VA, Tushin PP, Furin GG, Yakobson GG (1986) Zh. Org. Khim. 22:2584
48. Avramenko AA, Bardin VV, Karelin AI, Krasil'nikov VA, Tushin PP, Furin GG, Yakobson GG (1985) Zh. Org. Khim. 21:822; (1985) Chem. Abs. 103:141551
49. Brit. Pat 848561 (1960); (1961) Chem. Abs. 54:2164
50. Weidenbruch M, Pesel H, Peter H, Steichen R (1977) J. Organomet. Chem. 141:9
51. Sharp DWA, Winfield JM: J. Chem. Soc. 1965:2278
52. Il'in EG, Kalov U, Kolditz L, Buslaev YuA (1982) Dokl. Akad. Nauk SSSR 266:123; (1983) Chem. Abs. 98:10690
53. Schmutzler R (1964) Inorg. Chem. 3:410
54. Robert DU, Riess JG: Tetrahedron Lett. 1972:847
55. Koop H, Schmutzler R (1971/1972) J. Fluor. Chem. 1:252
56. Charlon C, Luu-Duc C (1986) Ann. Pharm. France 44:123
57. Kobayashi Y, Akashi C (1968) Chem. Pharm. Bull. 17:1009
58. Kobayashi Y, Akashi C, Morinaga K (1968) Chem. Pharm. Bull. 17:1784
59. Brown SJ, Clark JH (1985) J. Fluor. Chem. 30:251
60. Brown SJ, Clark JH: J. Chem. Soc. Chem. Commun. 1983:1256
61. Attina M, Cacace F, Wolf AP (1983) J. Label. Compounds Radiopharm. 20:501
62. Winfield JM (1976) In: Inorg. Chem. Ser. One, vol 5, Transition Metals 1. Butterworth, London, p 280
63. Glemser O, Roesky H, Hellberg KH (1963) Angew. Chem. 75:346
64. Opalovskii AA, Lobkov EU (1975) Usp. Khim. 44:193; (1975) Chem. Abs. 82:148830
65. O'Donnell TA, Stewart DF (1966) Inorg. Chem. 5:1434
66. Galkin NP, Bertina LE, Orekhov VT, Paklenkov EA (1975) Zh. Fiz. Khim. 49:2454; (1976) Chem. Abs. 84:22734
67. Galkin NR, Bogdanov GV, Fedorov VD, Orekhov VT (1971) Zh. Neorg. Khim. 16:496; (1971) Chem. Abs. 74:80345
68. Mathey F, Bensoam J (1972) C.r. Acad. Sci. Paris 274:1095
69. Mathey F, Bensoam J (1971) Tetrahedron 27:3965
70. Mathey F, Miller G (1973) C.r. Acad. Sci. Paris 277:45
71. Puy MV (1975) J. Fluor. Chem. 13:375
72. Mathey F, Bensoam J: Tetrahedron Lett. 1973:2253
73. Mathey F, Bensoam J (1973) C.r. Acad. Sci. Paris 276C:1569
74. Mathey F (1978) Inform. Chim. 174:233
75. Shustov LD, Nikolenko LN, Senchenkova TM (1983) Zh. Obshch. Khim. 53:103; (1983) Chem. Abs. 98:14332
76. Mathey F, Bensoam J (1975) Tetrahedron 31:391
77. Nikolaev NS, Kharitonov YuYa, Sadikova AT, Rasskazova TA, Kozorezov AZ: Izv. Akad. Nauk SSSR. Ser. Khim. 1972:757; (1972) Chem. Abs. 77:66823
78. Walker DW, Winfield JM (1972) J. Fluor. Chem. 1:376

79. Olah GA, Welch J, Prakash GKS, Ho TL: Synthesis 1976:808
80. Darragh JJ, Noble AM, Sharp DW, Winfield JM (1970) J. Inorg. Nucl. Chem. 32:1745
81. Noble AM, Winfield JM: J. Chem. Soc. (A) 1970:501
82. Walker DW, Winfield JM (1972) J. Inorg. Nucl. Chem. 34:759
83. Haas A, Maciej T (1982) J. Fluor. Chem. 20:581
84. Buslaev YuA, Kokunov YuV, Chubar YuD (1974) Dokl. Akad. Nauk SSSR 217:93; (1974) Chem. Abs. 81:98811
85. Olah GA, Welch J: Synthesis 1976:809
86. Goosen A, McCleland CW, Venter PJ, Venter MW (1987) S. Afr. J. Chem. 40:30
87. Brownstein S (1987) J. Fluor. Chem. 37:21
88. Hammond PR, McEwan WS: J. Chem. Soc. (A) 1971:3812
89. Vergamini PJ: J. Chem. Soc. Chem. Commun. 1979:54
90. Sadikova LT, Nikolaev NS, Rassakazova TA (1970) Zh. Neorg. Khim. 15:2012; (1970) Chem. Abs. 73:83353
91. Olah GA, Welch J (1976) J. Amer. Chem. Soc. 98:6717
92. Olah GA, Welch J (1978) J. Amer. Chem. Soc. 100:5396
93. Downs AJ, Gardner CJ: J. Chem. Soc. Dalton Trans. 1984:2127
94. O'Donnell TA, Stewart DF, Wilson P (1966) Inorg. Chem. 5:1438

5 Halogen Fluorides in Organic Synthesis

Ludmila Solomonovna Boguslavskaya and Nikolai Nikolaevich Chuvatkin

Institute of Polymers, 606006 Dzerzhinsk, USSR

Contents

5.1   Introduction

The pioneering works on the synthesis of halogen fluorides were undertaken in the 1930s. Studies of the properties of these compounds were carried out in the 1940s and 1950s in connection with the development of nuclear energetics and rocket techniques [1–3]. After a short period of time, methods have been worked out for the synthesis of all stable or recordable halogen fluorides: ClF, ClF$_3$, ClF$_5$, BrF, BrF$_3$, BrF$_5$, IF, IF$_3$, IF$_5$, and IF$_7$, and their physico-chemical properties have been studied [1–4].

Despite well-developed methods of synthesis and commercial availability, some of these reagents had been little used, if at all, in organic synthesis [1–7]. Halogen fluorides had been thought to be just as aggressive as elemental fluorine, and their use had been therefore restricted to the reactions with polyhalogenated compounds [5–7]. Recently there has been a sharp growth of interest in halogen fluorides as the fluorinating agents, as conditions have been found for the controlled selective introduction of fluorine into organic molecules. Parallel with the preparative studies of halogen fluorides, the reactions of stoichiometric equivalents of halogen monofluorides "HlgF in situ" were studied, the source of which are the solutions of N-haloamides (or N-haloamines) in anhydrous HF [5,8,10]. Comparison of the reactivity of halogen monofluorides and their stoichiometric equivalents afforded useful information on the mechanism of fluorination by halogen fluorides. This chapter demonstrates not only synthetic utility of halogen fluorides as the reagents for selective fluorination, but also contribution of studies on the reactions of this reagents to the organic chemical theory. Fluoroorganic synthesis from halogen fluorides is being actively developed and the most interesting applications of these reagents seem to be in future.

5.2 Synthesis and Properties of Halogen Fluorides

All halogen fluorides were first synthesized by the reaction of elemental fluorine with other halogens or their compounds [1,2,6]. Synthesis directly from their elements still remains the most convenient method for the preparation of halogen fluorides both in industry and laboratories.

Chlorine monofluoride is obtained by passing the equimolar amounts of fluorine and chlorine at 200 to 250 °C through a copper or nickel tube reactor [1,6].

$$Cl_2 + F_2 \rightarrow 2ClF$$

Formation of chlorine monofluoride proceeds by the radical chain mechanism [11] and is accompanied by the evolution of a relatively small amount of heat. The kinetic and mechanistic studies of synthesis of chlorine monofluoride and trifluoride from elements have shown the reactor wall material to produce a substantial effect on the reaction rate and the product composition [12,13]. For the synthesis of ClF with a minimal ClF$_3$ admixture, the most suitable reactor is hollow copper and the reagent's contact time in the heated zone (220 to 250 °C) is 40 to 60 s.

For the synthesis of ClF, the authors used laboratory equipment like that described in [1, p 24] and [14], but improved in accordance with the above data. The automatic dosing apparatus feeds into the reactor strictly equimolar amounts of chlorine and undiluted fluorine. ClF prepared by this method may be further used in organic reactions without preliminary purification. Work with chlorine monofluoride is much more convenient if the ClF current is produced by passing the chlorine and fluorine under excessive pressure but not by evacuation of the reactor at the outlet, as suggested in [14].

On a laboratory scale, chlorine monofluoride may also be obtained by passing the equimolar quantities of chlorine and chlorine trifluoride through a nickel tube at 300 to 400 °C [4,6].

$$Cl_2 + ClF_3 \xrightarrow[\text{300 to 400 °C}]{} 3ClF$$

The unchanged ClF_3 and Cl_2 are separated from ClF by condensation in a copper trap cooled to -78 °C. But most frequently synthesis according to the above scheme is carried out by heating a mixture of Cl_2 and ClF_3 for several hours in a stainless steel autoclave at 150 to 200 °C (40 atm) [15]. The autoclave is pre-passivated by chlorine trifluoride.

Preparation of chlorine monofluoride from ClF_3 is convenient, as ClF_3 is commercially available and may be stored and transported in steel cylinders [1].

There are several laboratory methods for the preparation of ClF without using F_2 or ClF_3. One of them is based on the endothermic reaction of chlorine with metal fluorides, resulting in a mixture of gases containing Cl_2, ClF and ClF_3. Chlorine is slowly passed through a melt of the mixture of alkali metal fluorides, and ClF is isolated by freezing [16].

$$Cl_2 + MF \xrightarrow[\text{800 to 1000 °C}]{} ClF + MCl$$

With silver fluoride or $AgBF_4$ the reaction proceeds under milder conditions. Another route to ClF is the reaction of chlorine, SbF_5 and anhydrous HF at -78 to -35 °C [17].

Chlorine monofluoride is easily prepared by the reaction of chlorine fluoro-sulfate with well-dried CsF [2,18].

$$ClOSO_2F + CsF \rightarrow ClF + CsOSO_2F$$

Table 1 presents some data on the properties of chlorine monofluoride and other halogen fluorides.

Chlorine trifluoride is a low-boiling liquid of light green colour. Its industrial production methods, apparatus design and material have been described in detail [1,6]. ClF_3 is synthesized by passing the necessary quantities of chlorine and fluorine through a copper or nickel tube heated to 250 to 280 °C. Chlorine trifluoride is formed as a result of a sequence of reactions [11].

$$Cl_2 + F_2 \xrightarrow{k_1} 2ClF$$

$$ClF + F_2 \xrightarrow{k_2} ClF_3 \quad k_1 \gg k_2$$

Table 1. Some properties of halogens and halogen fluorides [1, 4, 23–25]

Compound	B.p. (°C)	M.p. (°C)	Dipole moment (D)	Atomic distances (Å)	E_{diss} (kJ mol^{-1})	Heat of formation (kJ mol^{-1})
F_2	−188	−219	—	1.418	156	—
Cl_2	−35	−101	—	1.988	240	—
ClF	−100	−155	0.89	1.63	253	−50.5
ClF_3[a]	12	−76	0.65	1.598–1.698	105[b]; 160	−164
ClF_5[c]	−13	−93	—	1.58; 1.67		−244
Br_2	59	−7	—	2.284	295	—
BrF	(20)[d]	−33	1.29	1.756	254; 260	−75.5
BrF_3[a]	126	9	1.19	1.72; 1.81		−256
BrF_5[c]	41	−30	1.51	1.69; 1.78		−429
I_2	184	113	—	2.666	147	—
IF	1 (decomp.)		—	1.909	277	
IF_3[a]	−35 (decomp.)					
IF_5[c]	100	9	2.18	1.869; 1.844		
IF_7	5 (subl.)	6			122[b]	−824; −840

[a] Planar T-form structure

[b] The value of the heat of the reaction: $HlgF_n \rightarrow HlgF_{n-2} + F_2$

[c] Tetragonal pyramid

[d] The value was obtained by extrapolation

ClF_3 is formed much more slowly than ClF, therefore its complete synthesis from elements requires longer contact time in the reactor's heated zone than for ClF. The most suitable material for the reactor in this case is nickel [12,13]. Liquid chlorine trifluoride is purified by distillation and stored in steel cylinders. In Germany, during World War II, chlorine trifluoride was produced at 1000 tons per year [1].

Chlorine pentafluoride is a gas condensing upon cooling to a colourless liquid with b.p. $-13\,°C$. It is prepared by heating a mixture of chlorine trifluoride with fluorine under high pressure, or in a quartz reactor under UV irradiation for several hours [1,2,4].

$$ClF_3 + F_2 \xrightarrow[350\,°C,\ 250\ atm]{} ClF_5$$

$$ClF_3 + F_2 \xrightarrow[30\,°C]{hv} ClF_5$$

Bromine and iodine monofluorides unlike chlorine monofluoride, are unstable and have not been isolated in pure form. Nevertheless, passing of fluorine diluted with nitrogen through the cooled $CFCl_3$ or CF_2Cl_2 solution of the respective halogen leads to the BrF [19] or IF [20] solution.

$$Br_2 + F_2 \xrightarrow[-78\ to\ -45\,°C]{CFCl_3} 2BrF$$

$$I_2 + F_2 \xrightarrow[-45\,°C]{CF_2Cl_2} 2IF$$

At $-78\,°C$, BrF may be isolated from the solution as orange-red crystals. But even at low temperatures, crystalline BrF spontaneously decomposes to bromine and bromine trifluoride, being in equilibrium with them [4,19].

$$3BrF \rightleftarrows Br_2 + BrF_3$$

According to [21,22], BrF and IF are formed as a result of the reaction of bromine or idoine with AgF. The reaction of AgF with bromine was carried out in acetonitrile at $0\,°C$ and in the absence of pyridine which was thought by the authors to bind BrF to form the $BrF \cdot 2Py$ complex [21]. But the authors of a more recent work [19] were unable to record BrF or its pyridine complex in the above conditions or upon varying the reaction conditions. In the reactions of iodine with AgF, iodine monofluoride decomposes to $I_2\ IF_5$, or, in the presence of pyridine, gives complexes $IF \cdot 2Py$ and $IF \cdot Py$ [22].

$$AgF + I_2 \xrightarrow[-AgI]{} IF \begin{cases} \longrightarrow I_2 + IF_5 \\ \xrightarrow{Py} IF \cdot Py\ or\ IF \cdot 2Py \end{cases}$$

Formation of these complexes from IF and pyridine was confirmed in a recent work [20].

Bromine trifluoride is a stable compound formed in the reaction of fluorine with liquid bromine.

$$Br_2 + 3F_2 \rightarrow 2BrF_3$$

The reaction follows the radical chain mechanism forming simultaneously BrF, BrF_3, and BrF_5 [1]. Bromine trifluoride is separated from the pentafluoride by vacuum distillation. On a large scale BrF_3 was also produced by the reaction of fluorine with a stoichiometric amount of bromine vapours at 80 to 100 °C. The by-product of this reaction, BrF_5, reacts with bromine again, forming BrF_3. The apparatus design and the procedure have been described in detail [1, 6]. On a small scale BrF_3 may be obtained by the reaction of ClF_3 with liquid bromine [1].

$$Br_2 + 2ClF_3 \rightarrow 2BrF_3 + Cl_2$$

Bromine pentafluoride is also quite stable. It is produced in industry directly from bromine and fluorine [1].

$$Br_2 + 5F_2 \xrightarrow[200\ °C]{} 2BrF_5$$

Among iodine fluorides, IF_5 and IF_7 are reasonably stable. Their synthesis from elements has been described [1,4,6]. Iodine pentafluoride may be obtained from chlorine trifluoride and iodine which react with inflammation according to the scheme [1,4].

$$3I_2 + 10ClF_3 \rightarrow 6IF_5 + 5Cl_2$$

Iodine pentafluoride is a colourless liquid which may be stored in a quartz or steel vessel. Iodine heptafluoride, a colourless gas, was obtained from elements at high temperatures [1]. At first, iodine pentafluoride is formed, which afterwards reacts under more severe conditions with fluorine, giving IF_7.

$$I_2 + 5F_2 \rightarrow 2IF_5$$

$$IF_5 + F_2 \xrightarrow[250\ °C]{} IF_7$$

Halogen fluorides are polar compounds with high dipole moments. Chlorine, bromine, and iodine monofluorides have comparatively high dissociation energies (250–280 kJ mol^{-1}). This accounts for the liability of these compounds to heterolytic rather than homolytic reactions. Halogen polyfluorides have less stable Hlg–F bonds. Thus, the energy of dissociation of chlorine trifluoride to fluorine and the ClF_2^- species almost equals that of molecular fluorine, therefore the high activity of ClF_3 in radical reactions is not unexpected. The boiling points of many halogen fluorides are much higher than might be expected from their molecular masses, and in many cases decrease with the increased number of fluorines in a molecule (see Table 1, compare b.p. of chlorine and bromine trifluorides with those of the pentafluorides). The low volatility of these compounds is explained by the ability of these compounds to associate with formation of fluorine bridges [1,26].

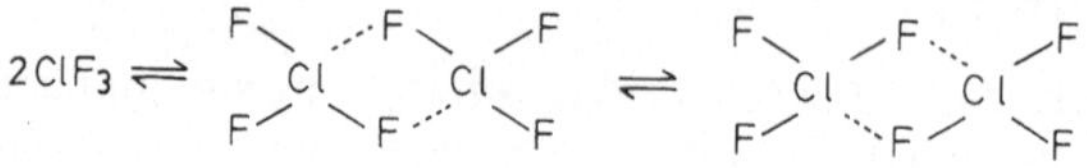

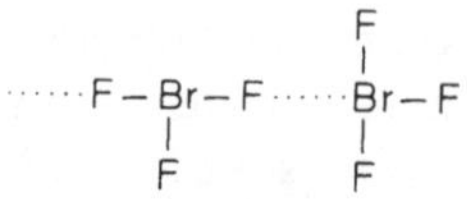

A specific feature of halogen fluorides is their amphoterism [4]. Thus bromine trifluoride is capable of self-ionisation.

$$2BrF_3 \rightleftarrows BrF_2^+ BrF_4^- \rightleftarrows BrF_2^+ + BrF_4^-$$

For the same reason, halogen fluorides react both with Lewis acids and bases, forming salt-like compounds with complex fluorine-containing anions and cations [2,27,28].

$$XF_n + YF_m \rightarrow XF_{n-1}^+ YF_{m+1}^-$$

$$XF_n + MF \rightarrow M^+ XF_{n+1}^-$$

Chlorine monofluoride reacts with Lewis acids giving complexes of type $Cl_2F^+YF_{m+1}^-$ [28], but not $Cl^+YF_{m+1}^-$, as considered earlier [4,17], since Hlg^+ cations cannot exist in solution or salt-like compounds because of their thermodynamic instability [29]. Anhydrous HF is used, which also forms highly polar and slightly stable 1:1 complexes with halogen fluorides. Halogen fluoride solutions in HF have a high electric conductivity at the expense of ionisation according to the scheme [1].

$$ClF_3 + HF \rightleftarrows ClF_3 \cdot HF \rightleftarrows ClF_2^+ + HF_2^-$$

In compounds with the complex fluorohalogenonium cation, the halogen atom bears a higher positive charge than in the starting molecule. Due to this, such compounds are even stronger oxidants than the starting halogen fluorides. The high electron affinity is demonstrated, e.g. by the oxidative fluorination of xenon by some complexes of this type [28,30].

$$Xe + 2ClF + AsF_5 \rightarrow XeF^+AsF_6^- + Cl_2$$

$$3Xe + 2BrF_2^+SbF_6^- + SbF_5 \rightarrow 3XeF^+SbF_6^- + Br_2$$

Being strong oxidants, halogen fluorides can undergo reactions with most elements, as well as organic and inorganic compounds [1]. Of few materials that are stable against these extremely corrosion-active substances at moderate temperatures, it is necessary to mention nickel, copper, Monel (70% Ni, 30% Cu), Kel-F and Teflon. The reactor intended for work with halogens may also be made of quartz. Glass is slowly destroyed under the action of halogen fluorides, especially in the presence of moisture which easily decomposes all compounds of this class, forming HF. Liquid halogen fluorides, especially chlorine and bromine polyfluorides, are rather dangerous to handle [5,6]. Chlorine trifluoride ignites wood, plastics, and in contact with ordinary solvents at $-100\,°C$ causes an explosion. Bromine trifluoride is less dangerous, though at room temperature it reacts explosively even with chlorocarbons. Iodine pentafluoride is not so aggressive. Only polyfluorocarbons may be quite safely mixed with halogen polyfluorides at low temperatures. While performing the reactions with low-reactive hydrogen-containing substrates, accumulation of halogen fluorides in a mixture should be avoided. Inadherence to this rule may cause an explosion [5]. Safe work with chlorine and bromine polyfluorides is best provided by diluting an organic substrate with a solvent inert relative to halogen fluoride before bringing it in contact with these reagents. It should also be remembered that in their toxicity, halogen fluorides are similar to elemental fluorine.

5.3 Reactions of Bromine and Iodine Monofluorides

Bromine and iodine monofluorides are unstable compounds, therefore they have been little studied, but the reactions of their stoichiometric equivalents of "BrF" and "IF" have been studied quite intensively.

5.3.1 Stoichiometric Equivalents of "BrF" and "IF"

The precursors of "BrF" and "IF" are the systems: *N*-bromo- or *N*-iodo-substituted amide—anhydrous HF [5,8,10], more rarely—Hlg_2–HF or Hlg_2–AgF [31–35]. As shown below, this does not mean that bromine and iodine monofluorides are really formed in these systems. *N*-Halogen-substituted amides or molecular halogens are the source of electrophilic halogen, whereas HF or AgF give a nucleophilic fluoride ion.

These systems are usually used for the addition of "BrF" and "IF" at a multiple bond. As a rule, halofluorination proceeds at -80 to $25\,°C$ in solution [5,36]. The electron-donating substituents at a multiple bond accelerate the reaction, the electron-accepting ones slow it down [37]. Halofluorination by such systems presents typical electrophilic addition reactions defined by the general equation.

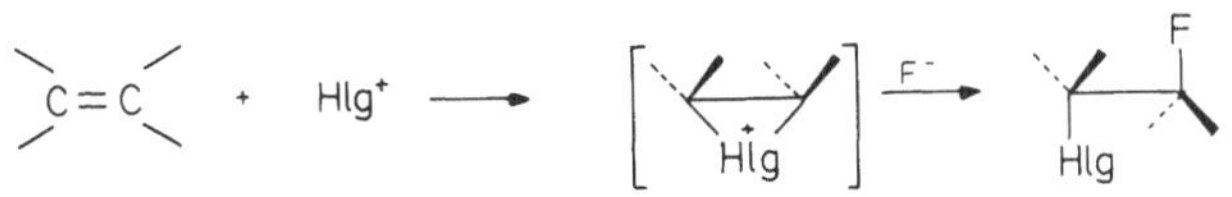

There are many preparative modifications of halofluorination, depending on the type of unsaturated compound and the source of electrophilic halogen. As a solvent, ether, tetrahydrofuran [5,36], HF(70%)–pyridine (30%) [31,38], or anhydrous HF [37] are used. Using oxygen-containing solvents is not always desirable, as they may participate in the addition reaction as nucleophiles [39].

Halofluorination in the HF-Py system gives good yields in the case of phenyl-substituted olefins [31,40,41]. Bromo- and iodofluorination of cyclohexene and its derivatives proceeds *anti*-stereospecifically [42,43]. The influence of substituents at a multiple bond on the regioselectivity and stereochemistry of bromofluorination of substituted alkenes has been well studied [8,10]. In general, the electrophilic and nucleophilic addition orientation is such as would be expected for the electrophilic mechanism of the reaction. As it is possible to subject unsaturated compounds with electron-deficient multiple bonds, which usually react with halogens via the free-radical mechanism in bromofluorination, it would be interesting to compare the competing effects of substituents on the addition orientation and stereochemistry of bromofluorination.

Bromofluorination of 1,3-dichloroprop-1-enes proceeds regiospecifically, and electrophilic bromine attaches to the central carbon atoms, whereas nucleophilic F^-—to the terminal carbon atom [44].

$$CHCl{=}CHCH_2Cl + {>}N{-}Br \xrightarrow{HF} CHClF{-}CHBrCH_2Cl$$

The chlorine atom at the multiple bond as a *p*-donor entirely controls the addition orientation. Introduction of substituents delocalising the positive charge at carbocation's 2-position reverses the addition orientation—electrophile is oriented to the terminal carbon atom, and nucleophile—to the central one [32].

$$CHCl{=}CRCH_2Cl + {>}N{-}\overset{+}{Br} \xrightarrow{HF} CHClBr{-}CRFCH_2Cl \qquad R{=}Me, Cl, F$$

The total orienting effect of the CH_2Cl with Me, F, and Cl substituents is stronger than that of chlorine.

Bromofluorination of acrylates also proceeds in accordance with the electronic interpretation of the Markownikoff rule. Therefore in the case of methyl acrylate, electrophilic bromine is predominantly oriented to the 2-, and for methyl methacrylate—to the 3-position [39,45]. Halogens as the *p*-donor substituents produce a much more stabilising effect on the intermediate carbocation than the CH_3-group [38,45]. Thus, bromofluorination of methyl *E*-2-methyl-3-chloropropenoate is regiospecific: the addition orientation is completely determined by the electronic influence of the chlorine atom.

The phenyl group is the π-donor and is a much more effective orientant than the halogen atom [40].

$$CHBr{=}CHPh + {>}N{-}Br \xrightarrow{HF} CHBr_2{-}CHFPh$$

Bromofluorination of multiple bonds proceeds chiefly stereospecifically, as *anti*-addition. In some cases this was strictly proved by the stereospecific dehydro-bromination of diastereomers to the respective fluorine-containing olefins [40,46]. The deviations observed in this case—incomplete *anti*-stereospecificity [32,40,46,47] or *syn*-addition [40]—may be explained in terms of the stability of bridged bromonium and open carbenium ions which are believed to be the intermediates in bromofluorination reactions. The incomplete *anti*-stereo-specificity of the addition of "BrF" to *E*-1,2-dichloroethylene and *E*-1,2,3-trichloropropene [32, 46] is attributable to the fact that the bromonium ions *1* with transoid chlorine atoms are less stable than ions *2* with cisoid chlorine atoms. This is in accordance with the greater stability of *cis*-dihaloethylenes than of *trans*-dihalaloethylenes [48,49].

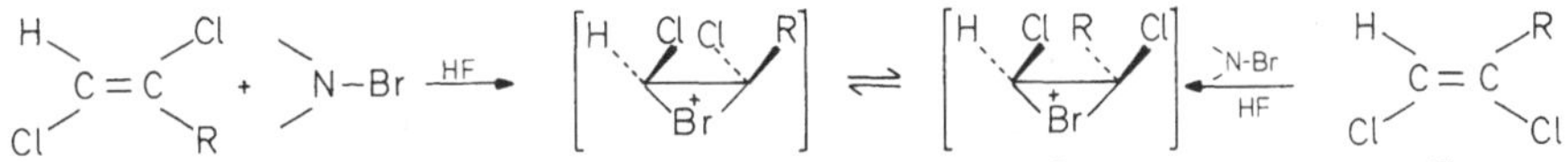

$R = H, CH_2Cl$

This results in the partial rearrangement of ion *1* to ion *2*. That is why bromofluorination of *Z*-olefins is completely *anti*-stereospecific, whereas bromo-fluorination of *E*-olefins is only stereoselective (85–60% of *anti*-adducts respectively) [32,46].

The non-stereospecific addition of "BrF" to ethyl *E*- and *Z*-3-chloro-crotonates is explained by other reasons [47].

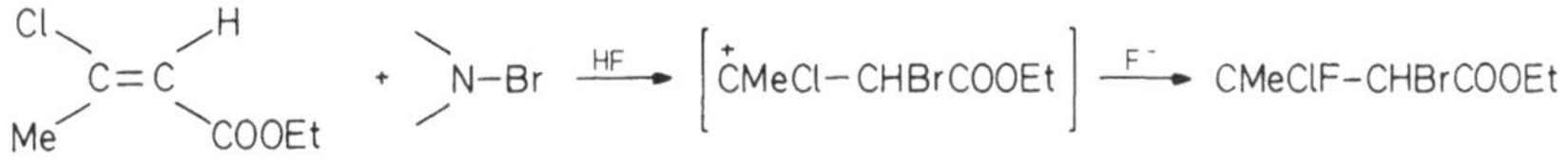

As a result of the addition of electrophilic bromine at a double bond of both *E*- and *Z*-3-chlorocrotonates, an intermediate is formed—the carbocation *3* rather than the bromonium ion. This occurs due to the presence of two stabilising substituents CH_3 and Cl at the carbocationic centre and the electron-accepting alkoxycarbonyl group. As a result, the reaction is regiospecific and non-stereospecific.

However it is enough to have at least a weak p-bridge bond in the non-symmetric bromonium ions *4*, *5*, *6* for the completely regiospecific reaction to proceed as an *anti*-stereospecific one [38,40,44,47].

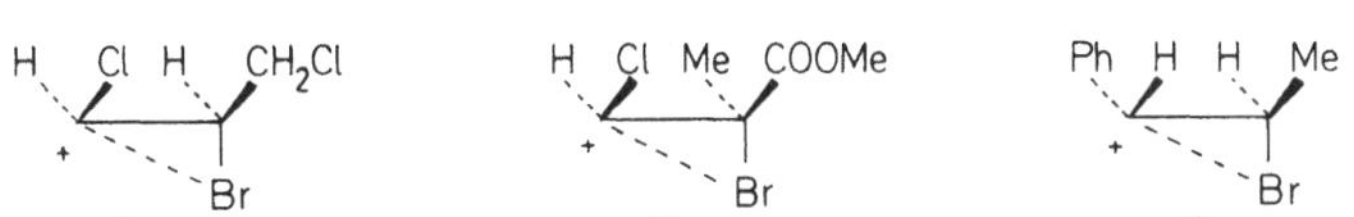

The deviations from *anti*-stereospecificity of bromofluorination of phenyl-substituted ethylenes results from the fact that in the process of the reaction, *cis*-olefins isomerise to more stable *trans*-olefins [40]. In the case of the sterically hindered *cis*-2-phenyl-1-*tert*-butylethylene, there occurs 100% *cis*-addition, and for the *trans*-isomer—100% *anti*-addition.

The unsaturated acetylated glucals have been reported [50,51] to undergo iodo- and bromofluorination by N-iodo and N-bromosuccinimide in HF with formation of *syn*-adducts. As shown in [33], the ^{1}H and ^{19}F NMR spectral data indicate that these reactions proceed predominantly as *anti*-addition, so the configuration assignment of halofluorinated sugars in [50,51] is erroneous.

For the effective halofluorination of unsaturated sugars and other compounds that are easily polymerisable or liable to undergo rearrangements in HF, it is recommended to use the system: bromine (or iodine)—finely divided silver fluoride in benzene or the benzene—acetonitrile mixture [33,34].

Bromofluorination of norbornene and the related compounds by NBS and HF in ether or pyridine is accompanied by the Wagner–Meerwein rearrangement [31,52,53]. Norbornene gives three main products: 2-*exo*-fluoro-7-*anti*-bromonorbornane *7*, 2-*exo*-fluoro-5-*exo*-bromonorbornane *8*, and 3-fluoronortricyclane *9*.

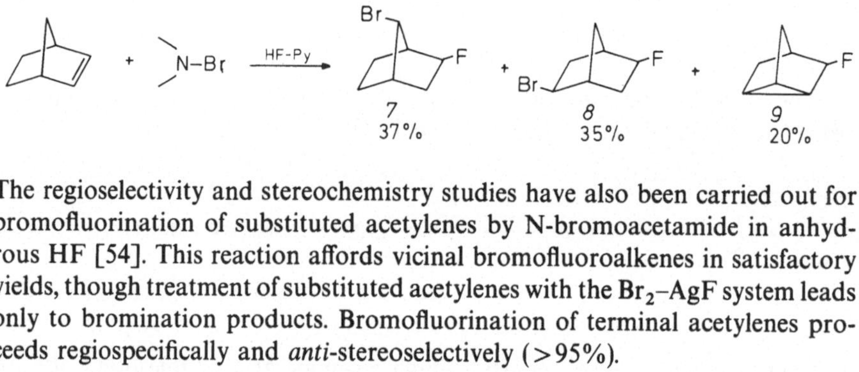

The regioselectivity and stereochemistry studies have also been carried out for bromofluorination of substituted acetylenes by N-bromoacetamide in anhydrous HF [54]. This reaction affords vicinal bromofluoroalkenes in satisfactory yields, though treatment of substituted acetylenes with the Br_2–AgF system leads only to bromination products. Bromofluorination of terminal acetylenes proceeds regiospecifically and *anti*-stereoselectively (>95%).

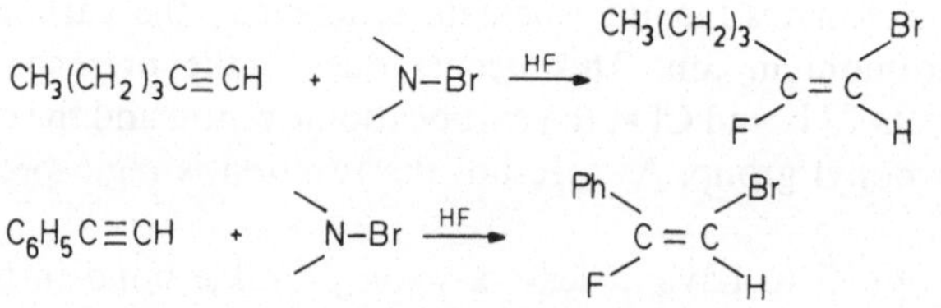

5.3.2 Bromine and Iodine Monofluorides

The mixtures of stoichiometric quantities of bromine trifluoride and molecular bromine, or iodine pentafluoride and molecular iodine were used as a source of bromine and iodine monofluorides in the addition reactions of perfluoro- and fluorochloroolefins, and alkyl perfluoroalkenyl ethers [55–58].

On the basis of the analysis of the products of addition at the multiple bond of olefins and the fact that BrF was really found in the Br_2-BrF_3 mixture, the following simple schemes were suggested for these reactions [59].

$$Br_2 + BrF_3 \rightleftarrows 3BrF$$

$$2I_2 + IF_5 \rightleftarrows 5IF$$

$$\text{>}C=C\text{<} + HlgF \rightarrow \text{>}CHlg-CF\text{/}$$

However recent studies have shown that the reactions proceed in a more complex way, and in some cases halogen polyfluorides may themselves be the main reagents in such mixtures. Therefore these reactions will be considered in detail in Sect. 5.5.

The addition of halogen monofluorides IF and BrF at multiple bonds has been reported in [60–63]. Iodine and bromine monofluorides were obtained by passing molecular fluorine diluted with nitrogen (8–10% v/v) into the strongly diluted (<1%) and cooled solutions of I_2 and Br_2 in the inert solvents ($CFCl_3$, $CHCl_3$). Addition of olefin to the resulting cool solution of HlgF gave the respective adducts with good yields. Some peculiarities of these reactions have been considered. Thus the reaction of olefins with iodine monofluoride in $CFCl_3$ easily proceeds according to the scheme.

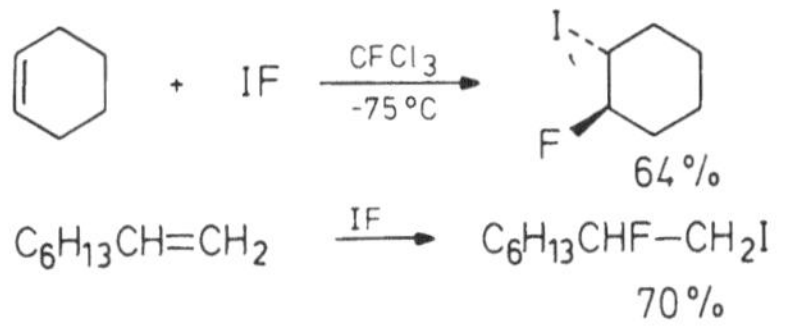

$$C_6H_{13}CH=CH_2 \xrightarrow{IF} C_6H_{13}CHF-CH_2I$$
$$70\%$$

Similar reactions with BrF proceed smoothly only in the presence of ethanol (traces) or any other alcohol. In the absence of alcohol, the reaction of BrF with olefins was very vigorous, and the bromofluoroadducts could not be isolated from the reaction mixture. The bromofluorination of olefins in the presence of an alcohol is suggested by the authors of [61] to proceed as follows.

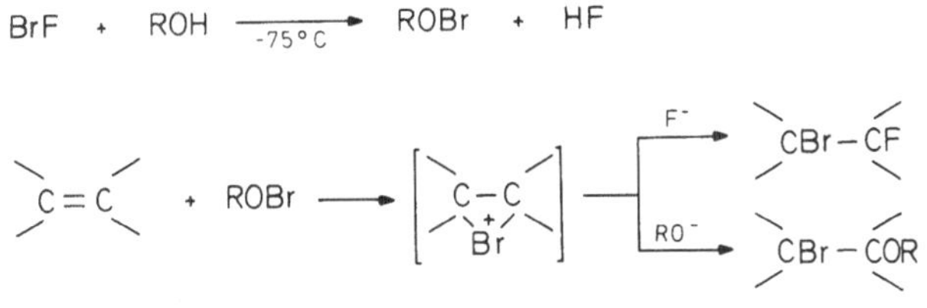

The bromoalkoxy adducts were really obtained as admixtures among the fluorinated products. Bromofluorination of *cis*- and *trans*-stilbenes by this method proceeds *anti*-stereospecifically, whereas iodofluorination [60] involves further substitution of iodine in the adducts by fluorine. In this case, *cis*-stilbene gave *d,l*-1,2-difluoro-1,2-diphenylethane, and *trans*-stilbene—a 1:1 mixture of *meso*- and *d,l*-diastereomers.

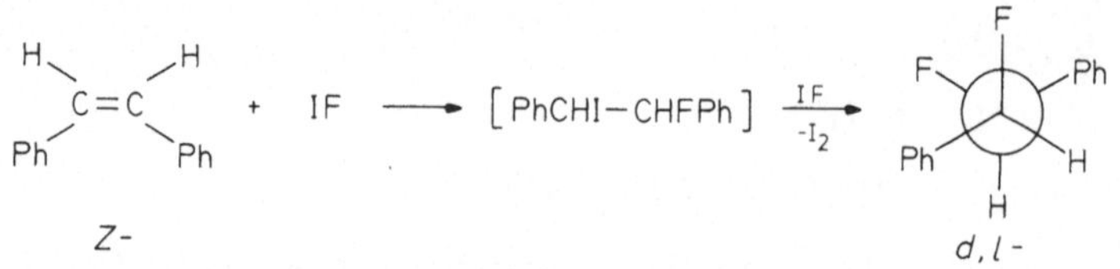

$$\underset{Z-}{\underset{Ph}{\overset{H}{>}}C=C\underset{Ph}{\overset{H}{<}}} \; + \; IF \; \longrightarrow \; [PhCHI-CHFPh] \xrightarrow[-I_2]{IF} \quad (d,l-)$$

Addition of IF and BrF to substituted acetylenes leads to formation of compounds containing the CF_2CX_2 group (X = Br, I) [62]. In the reactions of IF or BrF with phenylacetylene, one or two X atoms are substituted by fluorine.

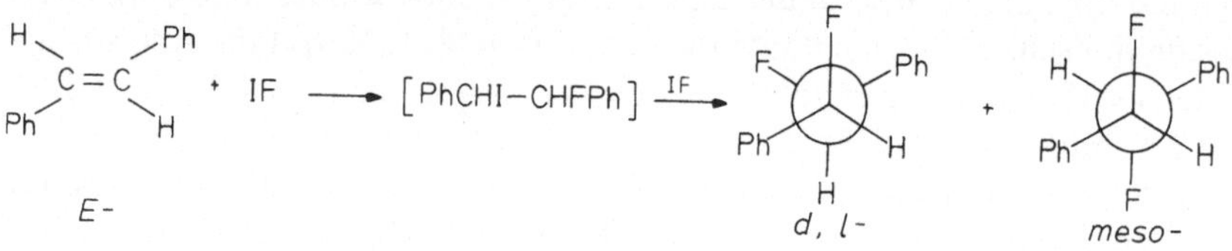

$$\underset{E-}{\underset{Ph}{\overset{H}{>}}C=C\underset{H}{\overset{Ph}{<}}} \; + \; IF \; \longrightarrow \; [PhCHI-CHFPh] \xrightarrow{IF} \quad (d,l-) \quad + \quad (meso-)$$

The alkyne esters react only with BrF.

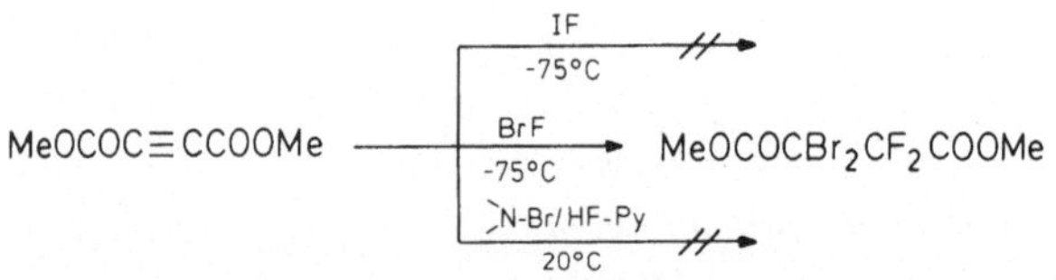

$$MeOCOC\equiv CCOOMe \quad \begin{array}{l} \xrightarrow[-75°C]{IF} \; /\!/ \\ \xrightarrow[-75°C]{BrF} \; MeOCOCBr_2CF_2COOMe \\ \xrightarrow[20°C]{>N-Br/HF-Py} \; /\!/ \end{array}$$

This clearly shows that BrF is much more reactive than IF and the N-bromoamide–HF–Py system.

Iodine monofluoride obtained by the above method in $CFCl_3$ reacts with hydrazones, forming geminal difluorides [63]. Possibly, at first IF undergoes electrophilic addition at the C=N bond of hydrazone, whereafter the heterolytic fission of the C–N bond occurs under the action of excess IF.

$$R^1R^2C=NNR^3R^4 + IF \rightarrow [R^1R^2CFNINR^3R^4] \xrightarrow{IF} R^1R^2CF_2$$

$$R^1, R^2 = Me, Ar; \; R^3, R^4 = H, Me$$

Unsubstituted hydrazones ($R^3 = R^4 = H$) react with IF much faster than their methyl-substituted analogues.

Fluorination of unsaturated compounds by IF and BrF described in [60–63] expand to some extend the possibilities of fluoroorganic synthesis, but the authors of these works obviously overestimate the preparative possibilities and convenience of the method. In this respect, the reactions of fluorination by chlorine monofluoride (see Sect. 5.4) are certainly more advantageous, the more so that ClF is a stable and available product.

5.4   Reactions of Chlorine Monofluoride

Gaseous chlorine monofluoride and its stoichiometric equivalents proved to be extremely interesting and highly selective fluorinating agents which allow one to synthesize mono-, di-, and polyfluoroorganic compounds, and this was the first suitable method of their synthesis. The regio- and stereochemistry studies of fluorination of organic compounds by these reagents and comparison of the results with reactions of the related compounds facilitated critical consideration of some theoretic aspects of electrophilic reactions which were open to discussion.

5.4.1   Stoichiometric Equivalents of "ClF"

The sources of stoichiometric equivalents of "ClF" are the N-chloroamide–HF [31], hexachloromelamine–HF [64] or molecular chlorine–HF [31] systems. Methyl hypochlorite in the presence of boron trifluoride can also chlorofluorinate unsaturated compounds [65].

The conditions of chlorofluorination of unsaturated compounds only slightly differ from those of bromofluorination; the yields (considering the starting alkene) are 30–80%, depending on the type of a substrate, reagent ratio and the solvent [8,31,37,46,47,64–69]. The regioselectivity and stereochemistry have been most intensively studied in the case of chlorofluorination of substituted olefins by hexachloromelamine–HF [8,10].

As in the case of bromofluorinations by the stoichiometric equivalents of "BrF", the expected regio- and stereoselectivity of the electrophilic addition of "ClF" at multiple bonds is observed. Thus allyl alcohol and its derivatives are chlorofluorinated non-regiospecifically. The regioisomer ratio depends on the electron nature of substituent in the allyl position [8]. In the case of allyl bromide, substantial migration of bromine from the allyl position to the central carbon atom has been shown.

$$CH_2=CHCH_2Br \; + \; R-NCl_2 \; \xrightarrow{HF} \; \left[\begin{array}{c} Br \\ CH_2 \overset{+}{-} \overset{.}{CH} - CH_2 \\ Cl \end{array} \right] \xrightarrow{F^-} \begin{array}{l} CH_2ClCHFCH_2Br \\ \quad 60\% \\ CH_2FCHClCH_2Br \\ \quad 15\% \\ CH_2FCHBrCH_2Cl \\ \quad 25\% \end{array}$$

Introduction of carbocation-stabilising substituents into the 2-position of the propene system leads to the regiospecific chlorofluorination [8].

$$CH_2=CHlgCH_2X + R-NCl_2 \xrightarrow{HF} CH_2ClCHlgFCH_2X$$

$$Hlg = Cl, F; \; X = OH, Cl$$

A similar substituent effect is observed in chlorofluorination of methylacrylate and its 2-substituted derivatives [64,66–68]. Chlorofluorination of E- and Z-1,3-dichloroprop-1-enes is regiospecific and highly *anti*-stereoselective (93–100%)

[44]. As in bromofluorination of these alkenes, the chlorine atom at a double bond controls the addition orientation: electrophilic chlorine is attached to the central carbon atom, and fluoride ion—to the terminal carbon. The addition of "ClF" to methyl E-2-methyl-3-chloro- and Z-2-fluoro-3-chloropropenoates also proceeds regioselectively and *anti*-stereospecifically [37,47]. The electrophilic chlorofluorination of ethyl E- and Z-3-chlorocrotonates proceeds non-stereospecifically due to the reason considered previously (see Sect. 5.3.1).

Using the systems involving chlorine donors and anhydrous HF, it is possible to carry out at -10 to $10\,°C$ the substitutive fluorination of some organic bromoderivatives [20]. As a source of electrophilic chlorine, N-chloroamines (hexachloromelamine, trichloroisocyanuric acid) may be used, as well as some hypochlorites (trifluoroacetyl hypochlorite). The reaction follows the route.

$$-\underset{|}{\overset{|}{C}}Br \quad + \quad \overset{\diagdown}{\underset{\diagup}{N}}-Cl\,(or-OCl)\,+\,HF \longrightarrow -\underset{|}{\overset{|}{C}}F \quad + \quad \overset{\diagdown}{\underset{\diagup}{N}}-H\,(or-OH)\,+\,BrCl$$

Hypochlorites as a rule easily decompose in HF, leading to a decreased yield of fluoro-derivatives. NBS is unreactive in these conditions. For effective fluorination, the solubility of either the reagent or substrate in anhydrous HF is necessary. The best results (the yield of fluoro-derivatives 50–70%) have been obtained for polyhaloalkanes containing bromine at a secondary or tertiary carbon atom. The substitutive fluorination of 1,2-dibromo-3-chloroisobutane with hexachloromelamine–HF (one equivalent of N-chloroamine per one mole of haloalkane) gives exclusively 1-bromo-2-fluoro-3-chloroisobutane (10).

$$CH_2BrCBrCH_2Cl + R-NCl_2 \xrightarrow{\;HF\;} [BrCH_2\overset{+}{C}CH_2Cl] \xrightarrow{\;F^-\;} CH_2BrCFCH_2Cl$$
$$\quad\;\;\underset{Me}{|} \qquad\qquad\qquad\qquad\;\; \underset{Me}{|} \qquad\qquad\qquad \underset{\underset{10}{Me}}{|}$$

With an excess of electrophilic reagent, the primary bromine atom in alkane 10 is not substituted by fluorine. The rate-determining stage of the reaction is obviously the electrophilic elimination of bromine from the substrate, forming the intermediate carbocation which is stabilised by capturing the fluoride ion. The substitutive fluorination by these systems is not always regiospecific. The vicinal dibromides may give both fluoro-containing isomers, possibly due to formation of bridged bromonium ions.

$$CH_2BrCHBrCH_2Cl \;+\; R-NCl_2 \xrightarrow{\;HF\;} \left[\underset{Br}{\overset{+}{CH_2-CHCH_2Cl}}\right] + BrCl$$

$$11$$

$$CH_2BrCHFCH_2Cl \longleftarrow \qquad\qquad\Big| F^- \qquad\qquad \longrightarrow CH_2FCHBrCH_2Cl$$
$$\underset{12}{} \qquad\qquad\qquad\qquad\qquad\qquad\qquad\qquad\qquad \underset{13}{}$$
$$45\% \qquad \Big|\underset{HF}{\overset{R-NCl_2}{\longrightarrow}} \nrightarrow CH_2FCHFCH_2Cl \longleftarrow \underset{HF}{\overset{R-NCl_2}{\Big|}} \quad 15\%$$
$$\underset{14}{}$$

When 1,2-dibromo-3-chloropropane *11* was treated with one equivalent of hexachloromelamine in HF, a 3:1 mixture of isomeric monofluorides *12, 13* was obtained with a total yield of 60%. With an excess of electrophilic reagent, only secondary bromide *13* is converted to difluorochloropropane *14*. As seen from the above examples of substitutive fluorinations, the chlorine atoms in the molecule remain intact.

This indicates that in the $>$N–Cl—HF systems, chlorine monofluoride is not formed. As will be shown (see Sect. 5.4.4), ClF easily substitutes bromine by fluorine in the primary bromides such as *10* and *12*, and in anhydrous HF it substitutes chlorine in the primary chloro-derivatives.

5.4.2 Chlorination with Chlorine Monofluoride

Chlorine monofluoride readily reacts both with organic and inorganic hydroxyl-containing compounds and some of their derivatives forming the respective hypochlorites according to the general scheme [71–76].

$$ROH + ClF \rightarrow ROCl + HF$$

Such reactions usually proceed vigorously, with heat evolution, and the end products—hypochlorites—are in most cases explosive. Therefore the reactions should be carried out with caution: gradual introduction of the reagent, the use of inert solvents, and strict control over the temperature are required. This procedure afforded high yields of chlorine nitrate [71,72], perfluoroalkyl hypochlorites [73], trifluoroacetyl hypochlorite [74], acyl hypochlorites [9,75], and chlorine triflate [76]. These compounds were used in subsequent syntheses as effective electrophilic reagents [9]. Acyl hypochlorites are also easily formed in the reaction of ClF with acyl anhydrides [75].

$$(CH_3CO)_2O + ClF \xrightarrow[0\ °C]{CCl_4} CH_3COOCl + CH_3COF$$

Chlorine monofluoride reacts with water very vigorously, forming a mixture of $ClFO_2$, Cl_2, O_2, and HF [77]. At low temperatures the main reaction products are Cl_2O and HF [71]. The organic and inorganic compounds with the N–H bond react with ClF in a similar way as the hydroxyl-containing compounds, forming the respective N-chloroamides [2,78].

$$RR'NH + ClF \rightarrow RR'NCl + HF$$

Chlorine monofluoride chlorinates organic compounds—benzene, chlorobenzene, and even nitrobenzene [79]. From chlorobenzene and toluene, *ortho*- and *para*-derivatives have been obtained in the ratio of 2:1, from nitrobenzene—*meta*-chloronitrobenzene. These are the examples of typical electrophilic aromatic substitution.

5.4.3 Addition of Chlorine Monofluoride at Multiple Bonds

Chlorine monofluoride adds at heteroatomic multiple bonds $C\equiv N$ [80–82], $N\equiv S$ [83], $S=O$ [18,84], $C=S$ [85], $S=N$ [86], $C=N$ [9, 78,87], $C=O$ [88–91]. In some cases such addition proceeds without catalysts. Thus sulfur trioxide easily reacts with ClF, forming chlorine fluorosulfate.

$$SO_3 + ClF \rightarrow ClOSO_2F$$

Difluorophosgene, perfluoroacyl fluorides and perfluoroketones react with ClF only in the presence of catalysts—alkali metal fluorides or strong Lewis acids. The reactions catalysed by alkali metal fluorides (KF, RbF, CsF) are said to produce perfluoroalkoxides as intermediates, which are further transformed to the respective hypochlorites with regeneration of the alkali metal fluoride [88,89].

$$(R_F)_2C=O + CsF \rightarrow (R_F)_2CFO^-Cs^+ \xrightarrow[-CsF]{ClF} (R_F)_2CFOCl$$

In the presence of Lewis acids, there occurs the electrophilic addition of chlorine monofluoride to carbonyl [90]. In this case, the catalyst enhances the electrophilicity of chlorine monofluoride by forming the respective complexes (see Sect. 5.2).

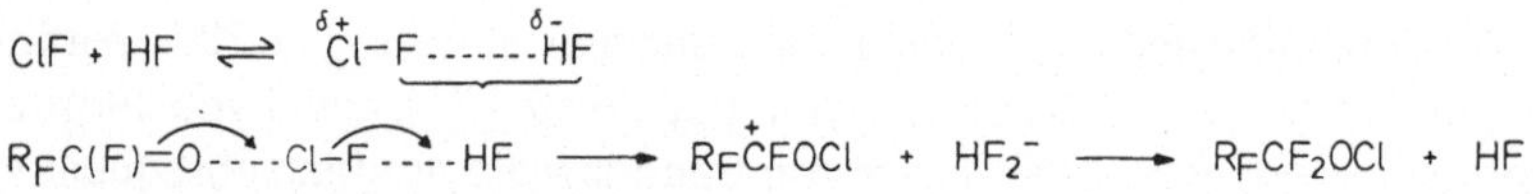

Fluorine does not react with perfluorocarbonyl compounds in the same conditions and in the presence of Lewis acids.

The perfluoroalkanenitriles and fluorocyan add ClF at the nitrile group, forming adducts in high yields (70–90%) [80–82].

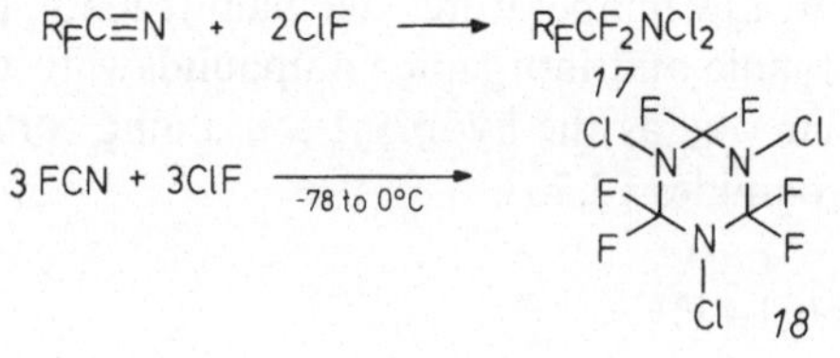

N-Chloroamines 17, 18 are themselves mild fluorinating agents [81]. Chlorocyan also adds 2 moles of ClF, but the chloramine produced is then dechlorinated to give N-chlorimine [92].

$$ClCN + 2ClF \rightarrow ClCF_2NCl_2 \rightarrow CF_2=NCl + Cl_2$$

Fluorinated isocyanates readily add ClF at room temperature [87].

$$R_FN{=}C{=}O + ClF \rightarrow R_FNCl{-}COF$$
$$75\text{–}90\%$$

$$R_F = CF_3,\ CF_3CO,\ FCO,\ Cl$$

Similar reactions of ClF proceed with perfluoroacylimines, haloimines, and perfluoroalkylimines [9].

A distinction of the above examples is that the substrates in the reactions with chlorine monofluoride are, as a rule, inorganic or perfluoroorganic compounds. The experimental procedures used in the above-mentioned works do not allow to perform syntheses with non-halogenated compounds. The reactions were carried out in small autoclaves or tubes which were charged with a substrate, then the calculated amount of ClF was condensed at $-196\,°C$ and the temperature was gradually raised. No wonder that in such experimental conditions the reactions of ClF with such substrates as acetonitrile [82] and *tert*-butanol [73] should proceed with an explosion. But if chlorine monofluoride diluted with nitrogen $(1:3)$ is passed through the solution of an unsaturated compound in an inert solvent $(CCl_4, C_2H_4Cl_2)$, then the selective reaction of ClF addition to the C=C bond proceeds smoothly even for butadiene and styrene. Good results have also been obtained in the reactions with haloolefins: allyl and vinyl chloride, 1,2-dichloroethylene, and trichloroethylene. The respective ClF addition products have been obtained with good yields -70 to 80% [14, 78, 93].

$$CHCl{=}CHCl + ClF \rightarrow CHCl_2CHFCl$$

$$CH_2{=}CHCH_2Cl + ClF \rightarrow CH_2ClCHFCH_2Cl + CH_2FCHClCH_2Cl$$

$$PhCH{=}CH_2 + ClF \rightarrow PhCHFCH_2Cl$$

In more recent works, it was shown that chlorine monofluoride may be effectively used for chlorofluorination of α,β-unsaturated esters, acyl chlorides, as well as allyl esters. The chlorofluoroadducts are formed in satisfactory, and frequently in very high, yields [94, 95]. The chlorofluorinations are carried out in inert solvents $(CCl_4, CHCl_3, 1,1,2\text{-}C_2F_3Cl_3)$ at room temperature. In most cases the reactions proceed smoothly with undiluted ClF.

The regioselectivity and stereochemistry studies have been carried out for chlorofluorination of substituted olefins by chlorine monofluoride in non-polar solvents and in anhydrous HF (Table 2). The orientation of ClF addition to the multiple bond of substituted alkenes depends on the substituents and the medium. Vinyl chloride and trichloroethylene are chlorofluorinated regiospecifically in $1,1,2\text{-}C_2F_3Cl_3$, according to the Markownikoff's rule: the electrophilic chlorine is oriented to the most hydrogenised carbon atom (Table 2, Nos. 1, 4). Chlorofluorination of vinylidene chloride proceeds non-regiospecifically in these conditions -12% of the adduct is formed against the Markownikoff rule (Table

Table 2. Chlorofluorination of alkenes with ClF (1:1, mol)

N	Alkene	Products (%)				Ref.
		In CHCl$_3$ or 1,1,2-C$_2$Cl$_3$F$_3$[a] (in HF[b])				
		Yield	Regioselectivity		Stereoselectivity (% *anti*-adduct)	
			C^{1}Cl–C^{2}F	C^{1}F–C^{2}Cl		
1	C^{1}H$_2$=C^{2}HCl[c]	68	100	0		79
2	CHCl=CHCl[c]	75				14, 95
3	C^{1}H$_2$=C^{2}Cl$_2$[c]	70	88	12		95
4	C^{1}HCl=C^{2}Cl$_2$[c]	83	100	0		79, 95
5	CCl$_2$=CCl$_2$[c,d]	92				95
6	C^{1}H$_2$=C^{2}HCH$_2$Cl[c]	73	57	43		79
7	C^{1}H$_2$=C^{2}HCH$_2$Br[c,e]	84	55	45		95
8	C^{1}H$_2$=C^{2}HCH$_2$OCOH	48(38)	64(75)	36(25)		94
9	C^{1}H$_2$=C^{2}HCH$_2$OCOOMe	70(44)	60(68)	40(32)		94
10	C^{1}H$_2$=C^{2}HCH$_2$OCOCCl$_3$	82[f]	60	40		94
11	Z-C^{1}HCl=C^{2}HCH$_2$Cl[c]	95	2	98	90	94, 95
12	E-C^{1}HCl=C^{2}HCH$_2$Cl[c]	90	20	80	90	94, 95
13	C^{1}H$_2$=C^{2}HCCl$_3$[c]	96	0	100		94
14	C^{1}F$_2$=C^{2}FCF$_3$[c]	99	0	100		95, 96
15	E-CF$_3$CF=CFCF(CF$_3$)$_2$[c]	no reaction				95
16	C^{1}H$_2$=C^{2}HCOOMe	89(60)	30(9)	70(91)		94

No.	Alkene					Ref.
17	C¹H₂=C²MeCOOMe	78(82)	53(95)	47(5)		94
18	C¹H₂=C²MeCOCl	72	35	65		94
19	Z-C¹HCl=C²FCOOMe[d]	71(70)	22(0)	78(100)	50(100)	94, 95
20	Z-C¹HF=C²FCOOMe	[f](59)	(0)	(100)		95
21	C¹F₂=C²(CF₃)COOMe	60(55)	0(0)	100(100)		95
22	Z-C¹ClMe=C²HCOOEt	[f](72)	(0)	(100)	(50)	94
23	E-C¹ClMe=C²HCOOEt	[f](72)	(0)	(100)	(50)	94
24	Z-C¹H(COOMe)=C²HCOOMe	66(45)			100(100)	94
25	E-C¹H(COOMe)=C²HCOOMe	68(48)			100(100)	94
26	Z-C¹H(COOMe)=C²ClCOOMe	[d]50(86)	100(100)	0(0)	2:1(1:4)[g]	95
27	E-C¹H(COOMe)=C²ClCOOMe	[d]50(80)	100(100)	0(0)	1:4(1:1)[g]	95
28	C¹Cl(COOMe)=C²FCOOMe	60(70)	40(10)	60(90)	15:1(1:8)	95

[a]15%(v/v) alkene
[b]50%(v/v) alkene
[c]Because of insolubility of haloalkenes in HF, their chlorofluorination was not investigated
[d]ClF was passed through the solution of alkene in $C_2Cl_3F_3$ (2–10 ml of ClF per mol of alkene)
[e]Product of the bromine migration (25%) was also formed
[f]Complex mixture of unknown products
[g]Reaction was non-stereoselective. Configuration of diastereomers was not determined. One of them dominated in non-polar solvent, and another—in HF.

2, No. 3), and the reaction is accompanied by considerable proton elimination leading to trichloroethylene.

$$CH_2=CCl_2 + ClF \longrightarrow \left[\begin{array}{c} CH_2-\overset{+}{C}Cl_2 \\ | \\ Cl \end{array} \right]_{F^-} \begin{array}{c} \longrightarrow CH_2ClCCl_2F + CH_2FCCl_3 \\ \\ \xrightarrow{-HF} CHCl=CCl_2 \end{array}$$

Chlorofluorination of the allyl derivatives proceeds non-regiospecifically both in non-polar media and HF (see Table 2, Nos. 6–10); the adduct formed according to the Markownikoff rule predominate in the mixture of isomers, irrespective of the type of allyl substituent.

In the chlorofluorination of allyl bromide, bromine migrates from the allyl position to the central carbon atom, the regioisomer ratio being the same as in the chlorofluorination by hexachloromelamine–HF (see Sect. 5.4.1). Addition of ClF to Z- and E-1,3-dichloropropenes (Table 2, Nos. 11 and 12) proceeds regioselectively, as opposed to the complete regiospecificity in the conjugated chlorofluorination of the olefins, and anti-stereoselectively.

The electrophilic mechanism of chlorofluorination of polyhaloalkenes, including tetrachloroethylene and perfluoropropene, by chlorine monofluoride in non-polar media has been confirmed by participation of ethyl acetate as an external nucleophile [95]. Chlorofluorination of dichloroethylene and perfluoropropene in 1,1,2-$C_2F_3Cl_3$ and in the presence of ethyl acetate gave, apart from alkanes 19 and 21, up to 30% of acetates 20 and 22.

$$CHCl=CHCl + ClF \xrightarrow{AcOEt} \left[\begin{array}{c} CHCl-CHCl \\ \diagdown \overset{+}{Cl} \ _{F^-} \end{array} \right] \begin{array}{c} \longrightarrow \begin{array}{c} CHCl_2CHClF \\ 19 \\ 50\% \end{array} \\ \\ \xrightarrow[-EtF]{AcOEt} \begin{array}{c} CHCl_2CHClOAc \\ 20 \\ 30\% \end{array} \end{array}$$

$$CF_2=CFCF_3 + ClF \longrightarrow [\overset{+}{C}F_2CFClCF_3] \begin{array}{c} \longrightarrow (CF_3)_2CFCl \\ 21 \quad 65\% \\ \\ \xrightarrow[-EtF]{AcOEt} CF_3CFClCF_2OAc \\ 22 \quad 26\% \\ \downarrow EtOH \end{array}$$

$$CF_3CFClCOOEt$$

The α,β-unsaturated esters react with ClF in non-polar solution and in HF with varying degrees of regio- and stereoselectivity (Table 2, Nos. 16–28). Methyl acrylate (Table 2, No. 16) in $CHCl_3$ forms regioisomers in the ratio of 1:2.5, in HF—1:1; for methyl metacrylate (Table 2, No. 17) the figures are respectively 1:1 and 19:1. Nevertheless these unsaturated compounds react with ClF in both media via the electrophilic mechanism: the reactions are very fast and give the products of external nucleophile participation in addition to chlorofluoro-adducts in the presence of AcOEt [94].

Dimethyl maleate and fumarate readily add ClF both in CHCl$_3$ and HF. The addition proceeds *anti*-stereospecifically in both media, also indicating the electrophilic mechanism.

The α,β-unsaturated esters containing more than two electron-accepting substituents at the multiple bond, are smoothly chlorofluorinated only by the system ClF–HF (Table 2, Nos. 19–23, 26–28). The reactions studied are non-stereospecific. To complete the reaction of unsaturated esters with electron-deficient multiple bonds in a non-polar medium, a large excess of the reagent is required, and the stereochemical result here is absolutely different than in HF (Table 2, Nos. 26–28).

Chlorine monofluoride activated by HF is a highly effective electrophilic agent. Methyl perfluorometacrylate, for which no electrophilic addition reactions were known [97], smoothly reacts with the stoichiometric quantity of ClF in HF.

5.4.4 Substitutive Fluorodehalogenation in Organic Halogenoderivatives

Chlorine monofluoride proved to be an effective agent for the selective substitutive fluorodebromination of alkyl bromides and esters. The reactions are carried out by passing the stoichiometric quantity of the ClF gas through liquid alkyl bromide or a solution of alkyl bromide in CCl$_4$ or CFCl$_2$CF$_2$Cl at -50 to $50\,°$C [98,99]. In these conditions the chlorine atoms and alkoxycarbonyl groups of the substrate remain intact. Yields in most cases are very high. This type of interaction may be represented by the following elementary scheme[1].

$$RBr + ClF \xrightarrow[-BrCl]{} [R^+] \xrightarrow{F^-} RF$$

$$BrCl \rightleftarrows Br_2 + Cl_2$$

Chlorine monofluoride eliminates the bromine atom from organic substrate to form bromine monochloride and the respective carbocation, which is stabilised by capturing the fluoride ion. The reactivity of bromides changes in the same series as usually observed for the carbocationic reactions: tertiary > secondary > primary bromide.

The most favourable temperatures for substitutive fluorodebromination are those above $0\,°$C. In this case chlorine monofluoride reacts practically with any bromo-derivatives except for compounds of type R$_F$CF$_2$Br and RCHBrCOOR'. For the substitutive fluorodebromination of alkyl bromides, especially for those carried out at low temperatures (-20 to $-50\,°$), the ClF to substrate ratio should be $1:1$. Excessive ClF reacts with the BrCl or Br$_2$ product giving BrF$_3$,

---

[1]The mechanistic details are discussed in Sect. 5.6

which may crystallise from the reaction mixture. Increased temperature of the reaction mixture may bring about an explosion [99].

The high selectivity of the substitutive fluorination is seen in the following example.

CH$_2$BrCBrCH$_2$Cl + ClF $\xrightarrow{-BrCl}$ $\left[\text{CH}_2\text{Br}\overset{+}{\text{C}}\text{CH}_2\text{Cl}\right]$ $\longrightarrow$ CH$_2$BrCFCH$_2$Cl
    | Me F$^-$ |
   Me   Me

CH$_2$BrCFCH$_2$Cl + ClF $\longrightarrow$ $\left[\overset{+}{\text{C}}\text{H}_2\text{CFCH}_2\text{Cl}\right]$ $\longrightarrow$ $\left[\text{MeCH}_2\overset{+}{\text{C}}\text{FCH}_2\text{Cl}\right]$ $\longrightarrow$ MeCH$_2$CF$_2$CH$_2$Cl
  | Me F$^-$ F$^-$
 Me   25   26

Treatment of 1,2-dibromo-3-chloroisobutane with one mole of ClF leads to regioselective formulation of tertiary fluoride *10*; upon treatment with a second mole of ClF, the primary bromide elimination also proceeds easily and is accompanied by methyl migration. This results in the formation of a more stable α-fluorocarbocation *25* which adds the fluoride ion, giving the geminal difluoride *26*.

This example vividly shows gaseous ClF in non-polar solution to be a more powerful electrophilic agent than hexachloromelamine in HF. The latter is unable to eliminate primary bromine from haloalkane *10* (see Sect. 5.4.1). Substitution of bromine atoms in primary bromides proceeds in most cases with migration of vicinal substituents—methyl groups and chlorine atoms. Fluorination of 1-bromo-2-chloropropane (*27*) proceeds readily with chlorine migration. Hence the primary alkyl bromide *27* and its isomer *28* give the same fluorination product—1-chloro-2-fluoropropane *29*.

$$\text{MeCHClCH}_2\text{Br} \xrightarrow{\text{ClF}} \text{MeCHFCH}_2\text{Cl} \xleftarrow{\text{ClF}} \text{MeCHBrCH}_2\text{Cl}$$
      27   29   28

The driving force of all migrations in substitutive fluorinations is formation of a stable carbocation. The 1,2-hydrogen shift occurs even in non-polar media when a stable tertiary carbocation is formed.

CH$_2$BrCHFMe + ClF $\longrightarrow$ $\left[\overset{+}{\text{C}}\text{H}_2\text{CFMe}\right]$ $\longrightarrow$ $\left[\text{Me}\overset{+}{\text{C}}\text{FMe}\right]$ $\longrightarrow$ MeCF$_2$Me
              H   F$^-$ 100%

MeCHCH$_2$Br + ClF $\longrightarrow$ $\left[\text{MeC}\overset{+}{\text{H}}\text{CH}_2\right]$ — $\Big\{$ $\begin{array}{l}\left[\text{t–Bu}^+\right] \longrightarrow \text{t–BuF} \\ \text{F}^-\quad\quad\quad 90\% \\ \\ \left[\text{Me}\overset{+}{\text{C}}\text{HCH}_2\text{Me}\right] \longrightarrow \text{MeCHFCH}_2\text{Me} \\ \text{F}^-\end{array}$
 |
Me   Me

Fluorination of propyl bromide in 1,1,2-C$_2$F$_3$Cl$_3$ at 0 to 10 °C also proceeds with hydrogen migration, but apart from 2-fluoropropane, it gives 1-fluoropropane.

$$CH_3CH_2CH_2Br \;+\; ClF \longrightarrow \left[\begin{array}{c} CH_3CH_2CH_2^+ \\ F^- \end{array}\right] \longrightarrow \left[\begin{array}{c} CH_3\overset{+}{C}HCH_3 \\ F^- \end{array}\right]$$

$$\downarrow \qquad\qquad\qquad\qquad \downarrow$$

$$\begin{array}{cc} CH_3CH_2CH_2F & CH_3CHFCH_3 \\ 30\% & 70\% \end{array}$$

The reactions in non-polar solutions may lead to primary fluorides if the alternative hydrogen migration would lead to a carbocation that is only slightly more stable, or if the adjacent carbon is bonded with fluorine [98,99].

$$CH_2BrCH_2F \;+\; ClF \longrightarrow \left[\begin{array}{c} \overset{+}{C}H_2CH_2F \\ F^- \end{array}\right] \overset{\longrightarrow\!\!/\!\!/\longrightarrow [CH_3CHF^+]}{\underset{\longrightarrow CH_2FCH_2F}{}}$$

Comparison of the relative rates of fluorination of propyl and *iso*-propyl bromides showed the difference in their reactivities to be small (1:2) [99]. Nevertheless 1,2-dibromopropane reacts with one mole of ClF, selectively forming 1-bromo-2-fluoropropane.

$$CH_2BrCHBrCH_3 + ClF \longrightarrow \begin{array}{l} \longrightarrow [CH_2Br\overset{+}{C}HCH_3] \xrightarrow{\;F^-\;} CH_2BrCHFCH_3 \\ \qquad\qquad\quad 30 \\ \qquad\qquad\quad \uparrow \\ \longrightarrow [\overset{+}{C}H_2CHBrCH_3] \\ \qquad\qquad\quad 31 \end{array}$$

Due to the liability of halogens (Cl, Br) to migration, the primary cation *31* formed by bromine elimination rearranges to a more stable secondary cation *30*, which accounts for the high selectivity of substitution.

Studies of the relative rates of substitutive fluorination of alkyl bormides in non-polar solution by the competing reactions method [99] have shown vicinal halogens to strongly slow down the fluorination. For secondary bromides, the following reactivity sequence is observed.

$$MeCHBrMe \gg MeCHBrCH_2Cl \gg CH_2ClCHBrCH_2Cl$$

The same is observed for primary bromides.

$$MeCH_2CH_2Br \gg MeCHClCH_2Br$$

The geminal halogens slow down fluorination only slightly. Thus ethyl bromide and 1-bromo-1-chloroethane are fluorinated at about equal rates, and in the series of dibromoethanes the ratio of fluorination rates is the following.

$$CH_2BrCH_2Br : CH_2BrCHClBr : CH_2BrCCl_2Br = 10:3:1$$

In this case, with the stoichiometric amount of ClF, only α-halogenfluorides are formed. Such a high selectivity obviously results from the fact that, e.g., in compound *32* two vicinal halogens deactivate the bromine atom of the CH_2Br group to a greater extent than when one geminal and one vicinal halogens deactivate the bromine atom of the CHBrCl group. For the same reason MeCHClBr is fluorinated many times faster than CH_2ClCH_2Br. As halogen atoms show electronic effects of a different nature—the negative inductive effect and the positive mesomeric effect, the resonance delocalisation of the positive charge of α-halocarbocation should already occur at the transition stage of bromine elimination, whereas vicinal halogens do not produce the anchimeric effect on electrophilic dehalogenation.

The fluorine atoms show a strong resonance effect and do not have migratory aptitude, therefore the substitutive fluorination of vicinal dibromochloroalkanes and polybromoalkanes leads to geminal difluorides.

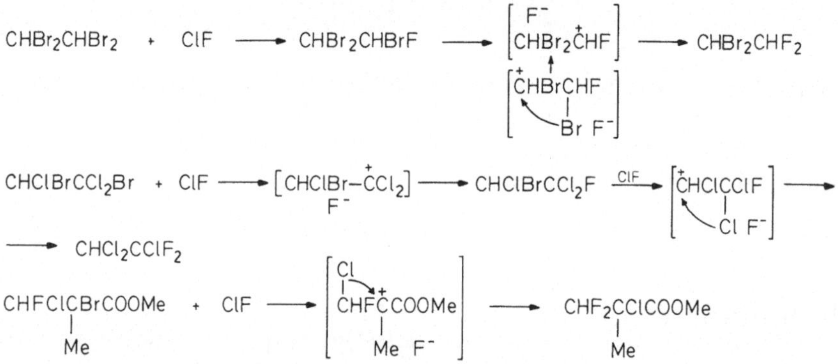

The high selectivity is not to be expected for the fluorination of bromoalkanes where the CH_2Br and $CHBrHlg$ ($CBrHlg_2$) groups occupy not vicinal but more remote positions.

The ester group deactivates the α-bromine atom. Methyl 2,3-dibromopropionate reacts with ClF in a non-polar solution, giving the product of fluorodebromination predominantly at the 3-position [98].

$$CH_2BrCHBrCOOMe + ClF \xrightarrow{CHCl_3} CH_2FCHBrCOOMe$$

$$\xrightarrow{ClF} \!\!\!\!\!\!/\!\!\!\!\!\!\rightarrow CH_2FCHFCOOMe$$

Introduction of carbocation-stabilising α-substituent activates α-halogen, therefore methyl 2,3-dibromoisobutyrate reacts with one mole of ClF, forming a 3:1 mixture of 2- and 3-fluorine-containing regioisomers, and treatment of this mixture with a second mole of ClF leads to methyl 2,3-difluoroisobutyrate.

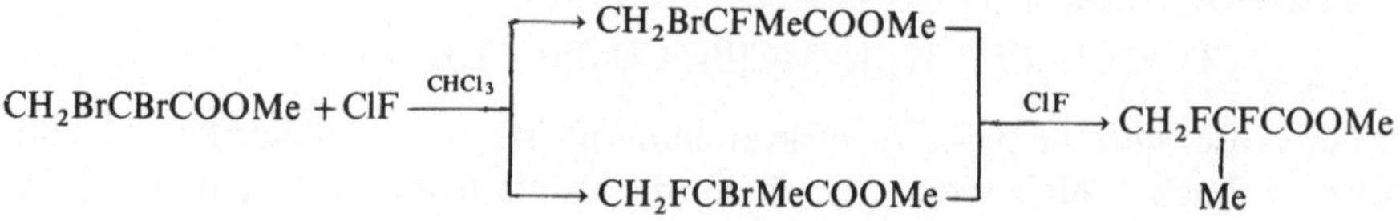

The reactions of substitutive fluorination by chlorine monofluoride possess low stereoselectivity. Fluorination of *parf*- and *pref*[2]-1-bromo-2-fluoro-1,2-dichloro-ethanes *33* by chlorine monofluoride in the absence of solvent proceeds non-stereospecifically [98]. From *parf-33*, *pref*-1,2-chloro-1,2-difluoroethane *34* is predominantly formed (85% in the diastereomer mixture); from *pref-33*, a mixture of equal amounts of *pref*- and *parf-34*.

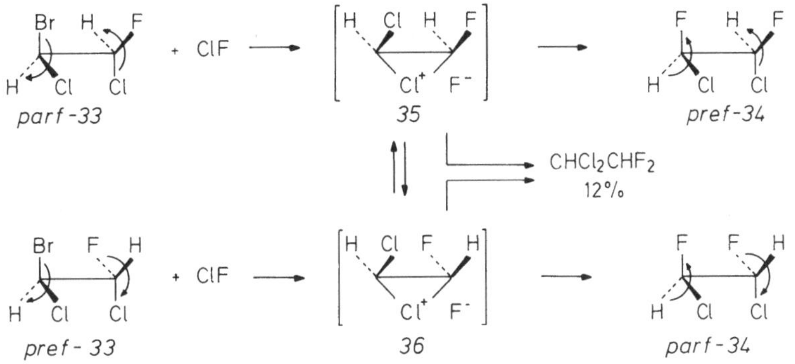

parf-33 + ClF → [35] → *pref-34*

CHCl$_2$CHF$_2$ 12%

pref-33 + ClF → [36] → *parf-34*

The predominant presentation of configuration in *parf-33* may be attributed to stabilisation of the intermediate carbocation *35* by interaction between the cisoid chlorine and fluorine atoms; in the case of *pref-33*, there is no such stabilisation in carbocation *36*, and the substitution reaction proceeds non-stereoselectively.

The substitutive fluorination of *pref*- and *parf*-1,2-dibromo-1,3-dichloro-propanes *37* proceeds regiospecifically at the terminal bromine atom and predominantly (85–90%) with preservation of configuration, which may be explained by formation of the bromonium ion *38* as a result of participation of the adjacent bromine atom maintaining the carbocation configuration.

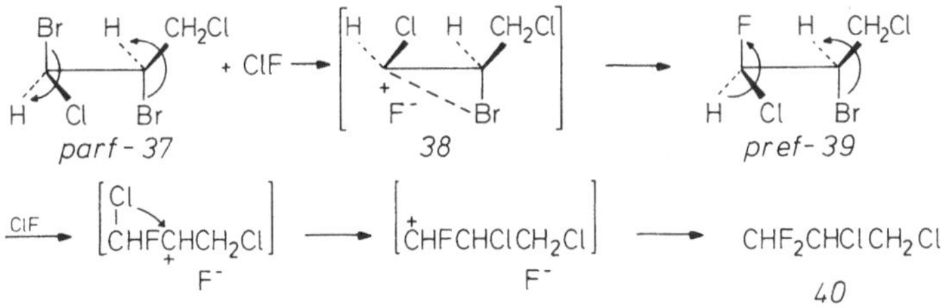

parf-37 + ClF → [38] → *pref-39*

$\xrightarrow{ClF}$ [CHFCHCH$_2$Cl with Cl, F$^-$] → [CHFCHClCH$_2$Cl, F$^-$] → CHF$_2$CHClCH$_2$Cl *40*

When 2-bromo-1-fluoro-1,3-dichloropropane *39* reacts with a second mole of ClF, bromine elimination proceeds with chlorine migration, therefore the single fluorination product is the geminal difluoride *40*.

Chlorine monofluoride can substitute chlorine by fluorine without special catalysts in such compounds which lead via chlorine elimination to rather stable carbocations [102]. Such reactions are carried out by passing ClF into the

---

[2]The Carey-Kuehne notation system is used [100,101]

appropriate substrate or its solution in inert solvent at -50 to $-10\,°C$. In this way, *tert*-butyl chloride, 1,2-dichloropropane, and 1,2,3-trichloropropane are easily fluorinated, giving nearly quantitative yields. Even chloroform is fluorinated in these conditions to a high degree.

$$CH_3CHClCH_2Cl + ClF \xrightarrow{C_2F_3Cl_3} CH_3CHFCH_2Cl$$

$$CH_2ClCHClCH_2Cl + ClF \rightarrow CH_2ClCHFCH_2Cl$$

$$CHCl_3 + ClF \rightarrow CHFCl_2$$

The above examples show the high selectivity of fluorination: the primary chlorine atoms remain intact. For deeper fluorination, activation of ClF by HF or Lewis acids is necessary [102].

Chlorine monofluoride in anhydrous HF substitutes chlorine in organic chloro-derivatives by fluorine at -60 to $10\,°C$. The reaction proceeds readily when gaseous ClF is passed into the HF solution or stirred emulsion of a chloro-derivative, according to the scheme.

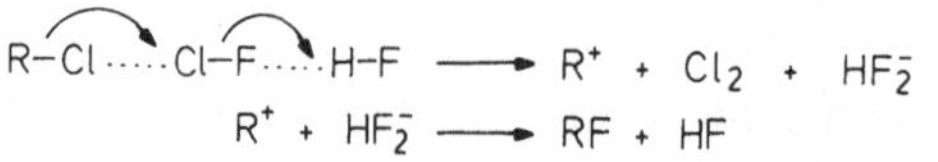

$$R\overset{\frown}{-}Cl\cdots Cl\overset{\frown}{-}F\cdots H-F \longrightarrow R^+ + Cl_2 + HF_2^-$$
$$R^+ + HF_2^- \longrightarrow RF + HF$$

This reaction has the same limitation as the substitution of bromine by fluorine: the chlorine atom of the CHCl group in α-position to the alkoxycarbonyl group is not substituted. When one mole of ClF reacts with one mole of methyl 2,3-dichloropropanoate, only the β-chlorine atom is substituted.

$$CH_2ClCHClCOOMe + ClF \xrightarrow{HF} CH_2FCHClCOOMe$$

$$\xrightarrow{ClF/HF} \!\!\!\not\!\!\!\rightarrow CH_2FCHFCOOMe$$

Upon passing of a second mole of ClF, the α-chlorine atom remains intact, and the electrophilic attack of ClF is directed at the carbonyl oxygen (see Sect. 5.4.5). Introduction of carbocation-stabilising substituents into the α-position leads to substitution of α-chlorine by fluorine.

$$CH_2ClCCl_2COOMe + ClF \xrightarrow{HF} CH_2ClCFClCOOMe$$

$$\xrightarrow[HF]{ClF} CH_2ClCF_2COOMe$$

$$\underset{\underset{Me}{|}}{CHClFCClCOOMe} + ClF \xrightarrow{HF} \underset{\underset{Me}{|}}{CHF_2CClCOOMe}$$

$$\xrightarrow{ClF} \underset{\underset{Me}{|}}{CHF_2CFCOOMe}$$

The HF solution promotes hydrogen migrations with formation of more stable carbocations in substitutive fluorination. In the reaction of 1,2,3-trichloropropane with one mole of ClF, the central chlorine atom is selectively substituted, and further attack on the primary chlorine atom of 1,3-dichloro-2-fluoropropane is accompanied by hydrogen migration. This leads to 1-chloro-2,2-difluoropropane, in which the chlorine atom remains intact in these conditions.

$$CH_2ClCHClCH_2Cl \; + \; ClF \; \xrightarrow{HF} \; CH_2ClCHFCH_2Cl \; \xrightarrow[HF]{ClF} \; \left[\overset{+}{C}H_2CFCH_2Cl \atop \underset{H \quad F^-}{\diagdown \diagdown |} \right] \longrightarrow CH_3CF_2CH_2Cl$$

$$\xrightarrow[HF]{ClF} \!\!\!\!\!\!/\!\!/\!\!\!\!\!\!\to CH_3CF_2CH_2F$$

The catalytic amount of antimony halide (3–5%) enhances the fluorinating ability of chlorine monofluoride in a non-polar solution [102]. This may be explained by the formation of fluorodichloronium hexafluoroantimonate [28], in which the electrophilic activity of chlorine is so high that such complex can fluorinate 1,1,2-trichlorotrifluoroethane to 1,2-dichlorotetrafluoroethane at -40 to $-60\,°C$.

$$CClF_2CCl_2F + Cl_2F^+SbF_6^- \xrightarrow[-Cl_2, \, -ClF]{} [CClF_2\overset{+}{C}ClF]$$
$$SbF_6^-$$

$$\xrightarrow[-SbF_5]{} CF_2ClCF_2Cl$$

The antimony salt-catalysed substitutive fluorination by chlorine monofluoride proceeds less selectively than the non-catalysed or ClF–HF-catalysed ones. Thus the reaction of 1,2,3-trichloropropane with an equivalent amount of ClF in the presence of 5% of SbCl$_5$ gave, apart from the main product, 1,3-dichloro-2-fluoropropane *41*, substantial amounts of 3-chloro-1,2-difluoropropane *14*, 3-chloro-1,1-difluoropropane *42*, and 1,1,3-trifluoropropane *43*.

$$CH_2ClCHClCH_2Cl \; + \; ClF \; \xrightarrow{(SbCl_5)} \; \left[\underset{Cl^{'}}{\overset{Cl}{CH_2 \!-\! \overset{+}{C}H \!-\! CH_2}} \atop SbF_6^- \right] \longrightarrow \left[\overset{+}{C}HClCH_2CH_2Cl \atop SbF_6^- \right]$$

$$[CH_2FCHClCH_2Cl] \quad + \quad CH_2ClCHFCH_2Cl \qquad\qquad [CHFClCH_2CH_2Cl]$$
$$\downarrow \qquad\qquad\qquad 41 \qquad\qquad\qquad\qquad \downarrow$$
$$CHF_2CHFCH_2Cl \qquad\qquad\qquad\qquad\qquad\qquad CHF_2CH_2CH_2Cl$$
$$14 \qquad\qquad\qquad\qquad\qquad\qquad\qquad 42$$
$$\qquad\qquad\qquad\qquad\qquad\qquad\qquad\qquad\qquad CHF_2CH_2CH_2F$$
$$\qquad\qquad\qquad\qquad\qquad\qquad\qquad\qquad\qquad 43$$

Thus fluorination with chlorine monofluoride possesses wide possibilities for the selective synthesis of various mono- and polyfluorides, with variation of the reaction temperature and catalysing additions.

Chlorine monofluoride can also substitute chlorine of the S–Cl bond in perfluoroalkyl chlorosulfites [103].

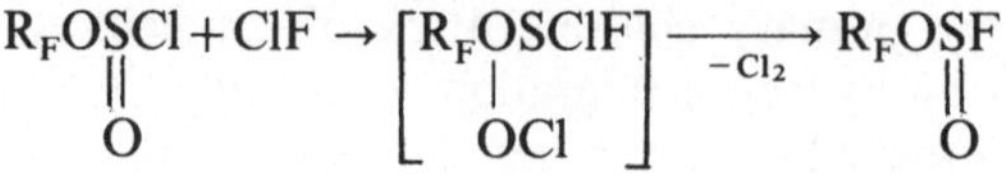

$$R_FOSCl + ClF \rightarrow \left[\begin{matrix} R_FOSClF \\ | \\ OCl \end{matrix} \right] \xrightarrow[-Cl_2]{} R_FOSF$$

with the O double bonds on the sulfur groups (shown as ‖ O under R_FOSCl and under R_FOSF).

$$44$$

As no such substitution occurs in the reactions with alkali metal fluorides at high temperatures, it is reasonable to suppose that the reaction proceeds by the electrophilic addition at the S=O bond and subsequent dechlorination of the resulting hypochlorite 44.

A similar chlorofluorination with subsequent dechlorination is suggested to take place in the reaction of chlorine monofluoride with cyanur chloride, hexachloromelamine [104] and some chlorine-containing polyfluoroalkylimines [105].

$$CF_3CCl_2N=C(CF_3)_2 + ClF \xrightarrow{CsF} CF_3CF_2NClCF(CF_3)_2 + Cl_2$$

5.4.5 Transformation of the Carbonyl Group to the Difluoromethylene Group

When a solution of an ester in anhydrous HF is treated with chlorine monofluoride at -70 to $-20\,^\circ$C, there smoothly proceeds transformation of the carbonyl group to the difluoromethylene one [106]. Presumably, at first there occurs the known reaction of addition of ClF at the carbonyl group to form hypochlorite 47. Further reaction of hypochlorite 47 with ClF seems to proceed via the electrophilic elimination of the ClO group with formation of a stable carbocation 48, which adds the fluoride ion forming difluoroether 49.

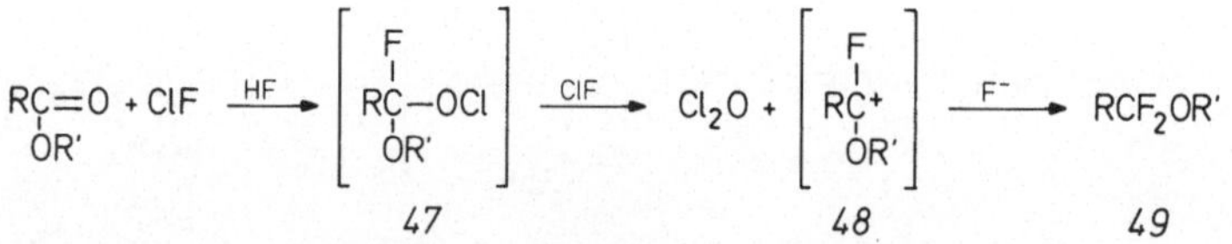

$$RC{=}O + ClF \xrightarrow{HF} \left[\begin{matrix} F \\ | \\ RC{-}OCl \\ | \\ OR' \end{matrix} \right] \xrightarrow{ClF} Cl_2O + \left[\begin{matrix} F \\ | \\ RC^+ \\ | \\ OR' \end{matrix} \right] \xrightarrow{F^-} RCF_2OR'$$

$$\qquad\quad 47 \qquad\qquad\qquad\qquad 48 \qquad\quad 49$$

$$R = Alk,\ RCHCl;\ R' = CH_3,\ R_FCH_2$$

The requirements to the stability of cation 48 are so high that introduction of two electron-accepting substituents into the α-position to carbonyl suppresses the reaction. Thus in these conditions, the RCF_2COO group may not be transformed to the RCF_2CF_2O group. As many compounds of type 49, such as $CH_3CF_2OCH_2CF_2CF_2H$, are hydrolytically unstable, their contact with water upon isolation from the reaction mixture should be avoided.

In this way, the hydrolytically stable fluoroethers may be synthesized, e.g. $CH_2ClCF_2OCH_2CF_2CF_2H$ *50*.

When ethyl alkanoates are used as substrates, there occurs deeper fluorination. Ethyl chloroacetate reacts with three moles of ClF giving 1-chloro-2,2,4-trifluoro-3-oxapentane *51* with a good yield.

$$CH_2ClCOOCH_2CH_3 + 3ClF \xrightarrow{\text{HF}} CH_2ClCF_2OCHFCH_3$$
$$51$$

5.4.6 Oxidative Fluorination of Sulfur- and Selenium-Containing Compounds

Chlorine monofluoride reacts with sulfur-containing compounds with sequential oxidation of S(II) to S(IV) and S(VI). These reactions have been well studied in the case of perfluoroalkyl sulfur-containing compounds and discussed in reviews [9,107].

The oxidation degree of sulfur in the products of the reaction of perfluoroalkyl sulfides with ClF chiefly depends on the mole ratio of the starting sulfide to ClF. At the 1:2 mole ratio, the respective difluorosulfuranes are readily formed with good yields [108].

$$R_F-S-R_F' + 2ClF \rightarrow R_F-SF_2-R_F'$$
$$R_F = CF_3; R_F' = CF_3, C_2F_5, C_3F_7$$

The reactions are carried out in a Monel autoclave. The ClF—perfluoroalkyl sulfide mixture is allowed to stand at $-75\,°C$, thereafter the temperature is slowly raised to $25\,°C$.

The reaction of perfluoroalkyl sulfide with a 4-fold excess of ClF at $-75\,°C$ and subsequently at room temperature gives a mixture of the respective *cis*- and *trans*-derivatives of S(VI). The reactions may be accompanied by the C–S bond cleavage [109,110].

$$CF_3SCH_3 + 4ClF \xrightarrow[25\,°C]{} CF_3SF_4CH_3 + 2Cl_2$$
$$70\%$$

$$CF_3SCH_2SCF_3 + 8ClF \rightarrow (CF_3SF_4)_2CH_2$$
$$30\%$$

$$CF_3SCF_3 + ClF \rightarrow CF_3SF_4CF_3 + CF_3SF_4Cl$$
$$48\%$$

In a similar way proceed the reactions of oxidative fluorination of tetrafluoro-1,3-dithiane *52* and octafluoro-1,4-dithiane [9] by chlorine monofluoride at room temperature.

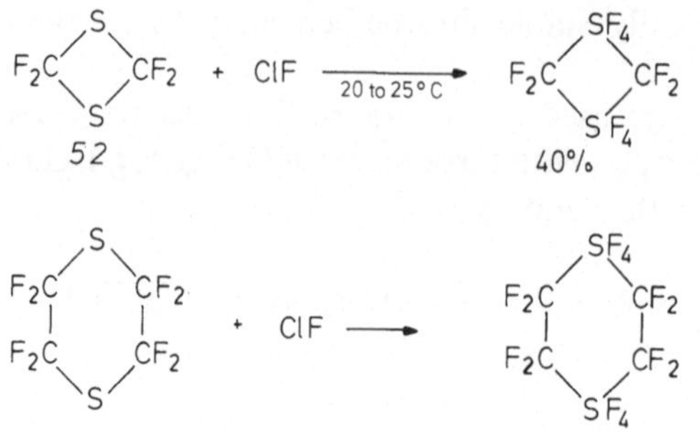

It is interesting to note that in the reaction of dithiane 52 with chlorine monofluoride, upon gradual increase of the temperature from -196 to $0\,°C$, the chlorofluorinated derivative 53 is formed initially, which is further converted to the perfluorinated product. Both steps give quantitative yields [111].

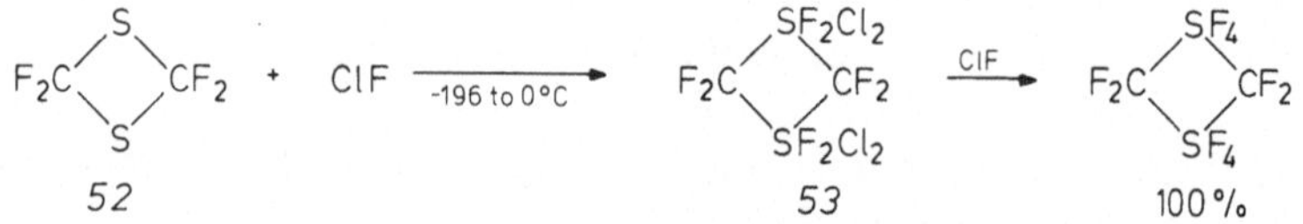

Perfluoroalkylsulfenyl chlorides and disulfides also readily undergo the oxidative fluorination by chlorine monofluoride, forming $trans$-R_FSF_4Cl [112].

$$R_FSCl \atop R_FSSR_F' \quad \xrightarrow{\quad ClF \quad} R_FSF_4Cl$$

$R_F, R_F' = $ perfluoroalkyl

The perfluoroalkyl derivatives $R_FSF_4R_F'$ and R_FSF_4Cl are chemically inert, whereas the fluorosulfuranes may be used in organic synthesis. Thus the reaction of $R_FSF_2R_F'$ with anhydrous HCl in a tube, in the presence of glass powder, gives perfluoroalkyl sulfoxides, which in their turn may undergo oxidative fluorination by chlorine monofluoride.

$$R_FSF_2R_F' \xrightarrow[\text{2) } H_2O]{\text{1) HCl}} R_FSOR_F'$$

$$R_FSOR_F' + ClF \xrightarrow{-78\,°C,\ 3\,h} R_FSOF_2R_F'$$

$R_F, R_F' = CF_3, C_2F_5$

In this way perfluoroalkylsulfuryl difluorides were first obtained [113].

The reactions of chlorine monofluoride with some bis(perfluoroalkyl) selenides and bis(perfluoroalkyl) diselenides have been studied in detail in one of recent works [114]. The oxidative fluorination of perfluoroalkyl selenides to $(R_F)_2SeF_2$ smoothly proceeds in the conditions described above for the reactions with analogous sulfides. The reactions give very high yields of products.

$$(R_F)_2Se + 2ClF \xrightarrow[-78\,°C]{} (R_F)_2SeF_2 + Cl_2$$
$$98-100\%$$

$$R_F = CF_3, C_2F_5$$

But the Se(IV) derivatives produced in these reactions do not react with ClF, as opposed to the sulfur-containing analogues, even after keeping the reaction mixture at room temperature for a long time. Compounds $(R_F)_2SeF_4$ are not formed in more rigid conditions either.

The reactions of chlorine monofluoride with bis(perfluoroalkyl) diselenides also have some peculiarities. The reaction of bis(perfluoroethyl) diselenide with ClF in the mole ratio of 1:6 at $-78\,°C$ gives perfluoroethylselenium trifluoride *54* with a quantitative yield.

$$(C_2F_5)_2Se + 6ClF \xrightarrow[-78\,°C]{} 2C_2F_5SeF_3 + 3Cl_2$$

Compound *54* at room temperature consumes one more mole of ClF during several hours, forming *trans*-perfluoroethylselenium chlorotetrafluoride *55*.

$$C_2F_5SeF_3 + ClF \xrightarrow[20\,°C]{} C_2F_5SeF_4Cl$$
$$\textit{54} \qquad\qquad\qquad \textit{55} \quad 40\%$$

By contrast with the chemically inert derivatives of S(VI), compound *55* is itself a vigorous oxidant, as indicated by the reaction of this compound with metallic mercury, in which HgF_2 and $HgCl_2$ are formed [114].

The oxidative fluorination of bis(trifluoromethyl) telluride by chlorine monofluoride leads to bis(trifluoromethyl)tellurium difluoride, which is a white, easy-hydrolysable substance [115].

5.5 Fluorination with Halogen Polyfluorides

Section 2 considered in detail methods of synthesis and properties of halogen polyfluorides. Among these, the most intensively studied are ClF_3, BrF_3, and IF_5. Being liquids, these compounds are preferable to gaseous chlorine mono-fluoride, as they do not require any special equipment. They may be stored in copper, steel, nickel or Teflon vessels and used when required. The most aggressive of all halogen fluorides—chlorine trifluoride—has not found use in controlled fluoroorganic synthesis. Bromine trifluoride and iodine pentafluoride as far less aggressive and dangerous reagents attracted greater attention, so this Section discusses the results of their preparative uses and recent mechanistic studies of their reactions with organic substrates.

5.5.1 Organic Derivatives of Iodine (III), Iodine (V) and Bromine (III) Fluorides

Chlorine and bromine trifluorides, bromine pentafluoride, and elemental fluorine oxidise the iodine atom in perfluoroalkyl iodides to I(III) and I(V). The reaction is carried out in perfluorohexane or without solvents, at -60 to $-80\,°C$ [116]. With ClF_3 the reaction proceeds according to the following schemes.

$$3R_FI + 2ClF_3 \rightarrow 3R_FIF_2 + Cl_2$$

$$3R_FI + 4ClF_3 \rightarrow 3R_FIF_4 + 2Cl_2$$

$$R_F = C_2F_5, C_4F_9, C_6F_{13}, C_{10}F_{21}$$

Iodine pentafluoride is ineffective in this reaction. The use of elemental fluorine leads to formation of three products at a time.

$$C_4F_9I + F_2 \rightarrow C_4F_9IF_2 + C_4F_9IF_4 + (C_4F_9)_2I^+IF_6^-$$

The freshly prepared (difluoroiodo)perfluoroalkanes are the low-melting solids which are transformed upon standing to unmelting and insoluble products shown by the X-ray structure analysis to have the structure of $(R_F)_2I^+Y^-$, where Y^- may be IF_4^-, IF_6^-, or IF_8^-. Treatment of tetramethoxysilane in SO_2 with iodine pentafluoride at low temperatures yielded methoxyiodo(V) fluorides 56 and 57 as unstable, low-melting crystalline substances [117].

$$IF_5 + Si(OMe)_4 \xrightarrow[-70\,°C]{SO_2} IF_4OMe$$
$$56$$

$$IF_5 + Si(OMe)_4 \xrightarrow[-15\,°C]{SO_2} IF_2O(OMe)$$
$$57$$

The reaction of tetrakis(perfluorophenyl)silane with iodine pentafluoride gave the stable crystalline (tetrafluoroiodo)pentafluorobenzene [118].

$$(C_6F_5)_4Si + IF_5 \xrightarrow[80\,°C]{Py,\ MeOH} C_6F_5IF_4 + SiF_4 2Py$$
$$49\%$$

Hexafluorobenzene and pentafluorobenzene do not react with IF_5 even at $120\,°C$. The reactions of 1,3,5-trifluorobenzene and pentafluorobenzene with $IF_4^+SbF_nCl_{6-n}^-$ in HF, IF_5, and SO_2FCl at 15 to $50\,°C$ with subsequent hydrolysis of the reaction mixture gave the aromatic compounds of I(V) [119].

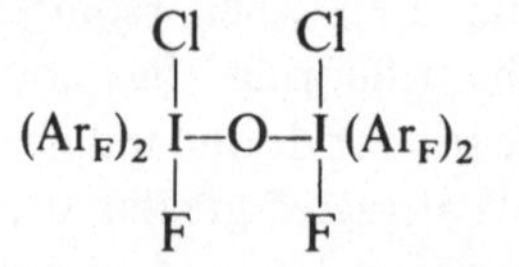

$$(Ar_F)_2 \overset{\displaystyle Cl}{\underset{\displaystyle F}{I}} - O - \overset{\displaystyle Cl}{\underset{\displaystyle F}{I}} (Ar_F)_2 \qquad Ar_F = C_6F_5,\ 2,4,6\text{-}C_6F_3H_2$$

Stable perfluoroaromatic Br(III) derivatives have been obtained by the reaction of bromine trifluoride with pentafluorophenyltrifluorosilane [120].

$$C_6F_5SiF_3 + BrF_3 \xrightarrow[-5 \text{ to } 0\,°C]{CFCl_3} C_6F_5BrF_2 + SiF_4$$
$$58$$

Compound *58* is a thermally stable solid, melting at 36 °C. Its treatment with $(CF_3CO)_2O$ affords bis-trifluoroacetoxybromopentafluorobenzene.

$$C_6F_5BrF_2 + 2(CF_3CO)_2O \xrightarrow[0\,°C]{CFCl_3} C_6F_5Br(OCOCF_3)_2 + 2CF_3COF$$
$$91\%$$

An interesting organic Br(III) derivative was obtained by the reaction of aryl bromide *59* with bromine trifluoride [121].

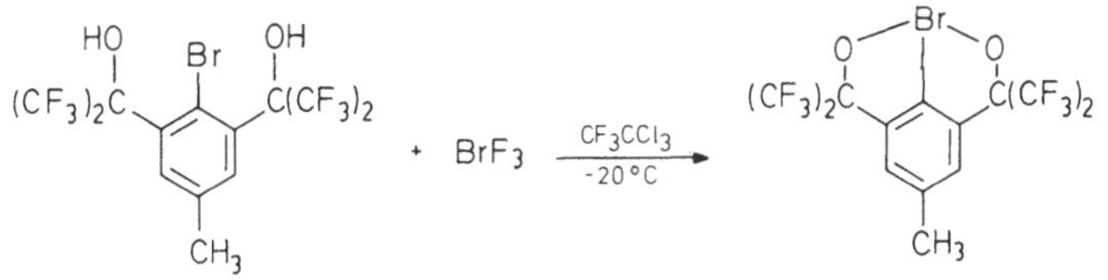

The authors of [122] have suggested, in this and earlier works, a method for the synthesis of diarylbromonium salts $(XC_6H_4)_2Br^+BF_4^-$ by the reactions of BrF_3 and boron trifluoride etherate with arenes, organomercury compounds or tetraarylstannanes in CH_2Cl_2 and acetonitrile at $-70\,°C$.

The reaction of chloropentafluorobenzene with elemental fluorine was reported to give stable perfluoroaromatic derivatives of Cl(III), $C_6F_5ClF_2$, and $C_6F_5ClF–ClFC_6F_5$ [123]. However, Frohn [120] considers these data to be doubtful.

5.5.2 Reactions of Halogen Polyfluorides and Their Inorganic Complexes with Unsaturated Compounds

The mixture of stoichiometric quantities of BrF_3 and Br_2, IF_5 and I_2, react with $C=C$ bonds as 3 equivalents of BrF and 5 equivalents of IF respectively.

$$BrF_3 + Br_2 \rightleftarrows 3BrF$$

$$IF_5 + 2I_2 \rightleftarrows 5IF$$

The reactions of these systems with polyfluoroolefins were carried out in a steel equipment in the range of 20 to between 150 and 170 °C. The adducts were obtained in some cases in very high yields [55–57]. The orientation of addition to some olefins indicates the electrophilic mechanism of the reaction.

$$CH_2=CF_2 + 1/5(IF_5 + 2I_2) \xrightarrow[100\,°C]{} CF_3CH_2I$$
$$86\%$$

$$CF_2=CFCF_3 + 1/3(BrF_3 + Br_2) \xrightarrow[20\ °C]{} (CF_3)_2CFBr$$
$$46\%$$

The reaction of perfluoroalkenyl ethyl ethers with the BrF_3–Br_2 system proceeds both with the addition of BrF at the multiple bond and substitution of α-hydrogen in the ethyl group by fluorine.

$$(CF_3)_2C=CFOCH_2CH_3 + BrF_3 + Br_2 \xrightarrow[60\ °C]{} (CF_3)_2CBrCF_2OCHFCH_3$$
$$28\%$$

It had long been thought impossible to carry out the selective fluorination of polyhydro-substrates by halogen polyfluorides: as a rule, these reactions gave a complex mixture of the products of fluorination and chlorination (bromination) of different degrees [5,6,124,125].

$$CHCl=CHCl + ClF_3 \xrightarrow[55\ °C]{} C_2H_2Cl_3F + C_2H_2Cl_2F_2$$
$$24\% \qquad\qquad 40\%$$

$$\begin{array}{c} CHCl=CCl \\ | \\ CHCl=CCl \end{array} + ClF_3 \xrightarrow[20\ to\ 150\ °C]{} C_4F_6Cl_4 + C_4F_5Cl_5 + C_4F_4Cl_6 + C_4F_3Cl_7$$
$$19\% \qquad\quad 45\%$$

The mixture of octafluorotetrabromocyclohexane isomers $C_6F_8Br_4$ was obtained by treatment of hexabromobenzene with ClF_3, IF_5 or IF_7 in bromine at -10 to $60\ °C$ [126]. One of the first examples of selective fluorination of unsaturated polyhydro-substrates was the reaction of bromine trifluoride with lower nitriles and acetone in anhydrous HF [127].

$$RCN + BrF_3 \xrightarrow[-20\ to\ 15\ °C]{HF} RCF_3$$

$$R = Me,\ Et,\ CH_2Cl$$

$$CH_3COCH_3 + BrF_3 \xrightarrow{HF} CH_3CF_2CH_3$$

These reactions possibly proceed via the intermediate addition at the $C\equiv N$ and $C=O$ bonds. However in the case of ethyl methyl ketone, there occurs the predominant fragmentation of the molecule at the C–C bonds. The conditions have been found, in which BrF_3 without molecular bromine addition smoothly reacts with mono- and dihaloalkenes with a short carbon chain, giving the equimolar mixture of bromofluorides and difluorides in a high total yield [128]. The reaction is carried out by gradually adding liquid BrF_3 (0.5 mole per one multiple bond) to the diluted (3 to 5% v/v) alkene solution in $C_2F_3Cl_3$.

$$CHBr=CHBr + 1/2\ BrF_3 \xrightarrow[10\ to\ 20\ °C]{C_2F_3Cl_3} CHBr_2CHBrF + CHBr_2CHF_2$$
$$38\% \qquad\qquad 45\%$$

$$CH_2=CClCH_2Cl + 1/2\,BrF_3 \rightarrow CH_2BrCClFCH_2Cl + CH_2ClCF_2CH_2Cl$$
$$\qquad\qquad\qquad\qquad\qquad\qquad 35\% \qquad\qquad\qquad 33\%$$

The equimolar BrF_3–Br_2 mixture may be used for bromofluorination of multiple bonds of $\alpha,\beta,$-unsaturated esters [128].

$$CH_2=CMeCOOMe + 1/3(BrF_3+Br_2) \xrightarrow[\text{10 to 20 °C}]{C_2F_3Cl_3} CH_2BrCFMeCOOMe$$
$$\qquad\qquad\qquad\qquad\qquad\qquad\qquad\qquad 46\%$$

$$+ CH_2FCBrMeCOOMe$$
$$36\%$$

$$CH_2=CClCOOMe + 1/3(BrF_3+Br_2) \rightarrow CH_2BrCClFCOOMe$$
$$\qquad\qquad\qquad\qquad\qquad\qquad\qquad 42\%$$

The reaction of unsaturated esters with pure BrF_3 fails to give satisfactory yields of the bromofluoro- and difluoro-adducts.

Detailed studies have been carried out on the products of the reaction of polyfluoroaromatic compounds with the BrF_3–Br_2 system [129,130], pure BrF_3 [131], and polyfluorohalogenonium salts [131–133].

In the reactions of hexafluorobenzene, pentafluorobenzene, bromopentafluorobenzene, octafluorotoluene, decafluoro-p-xylene, 2,3,4,5,6-pentafluorotoluene, and 2,3,4,5,6-pentafluoroanisole with an equimolar BrF_3–Br_2 mixture (in the mole ratio of substrate: BrF_3 : $Br_2 = 1:1:1$, i.e. 3 equivalents of "BrF" per 1 mole of polyfluoroarene) in inert solvents at $0\,°C$, at first there occurs fluorination of the aromatic ring with formation of substituted polyfluoro-1,4-cyclohexadienes, which further add BrF at one or two multiple bonds, giving bromopolyfluorocyclohexenes and -cyclohexanes (the products were isolated after decomposition of the BrF_3 excess with water).

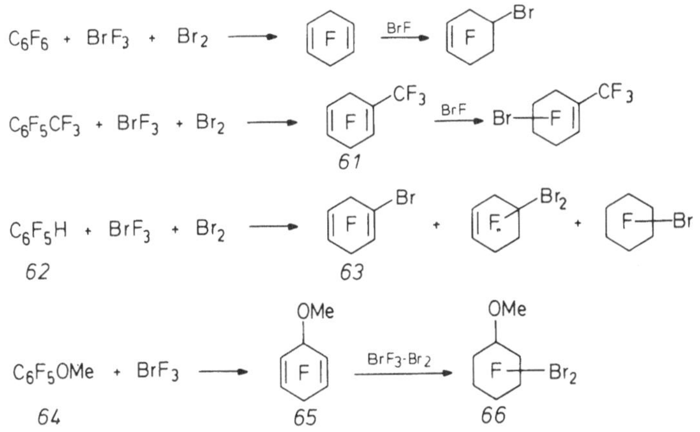

If the starting perfluoroarene has an electron-accepting substituent (except fluorine atoms), for example, the CF_3 group, then in the diene product it occupies the 1-position of *61*, and an electron-donating substituent occupies the 3-position of *65*. Fluoroaromatic compounds with electron-donating substituents, such as pentafluoroanisole *64*, react with BrF_3 at $0\,°C$ in the absence of Br_2, giving substituted dienes in good yields. Diene *65* is bromofluorinated to a mixture of isomeric methoxynonafluorodibromocyclohexanes *66* only by the BrF_3-Br_2 system. Pentafluorobenzene *62* gives mainly the diene *63*.

The products of the reaction of octafluoronaphthalene *67* with BrF_3-Br_2 depend on the molar ratio of the substrate to reagents [130].

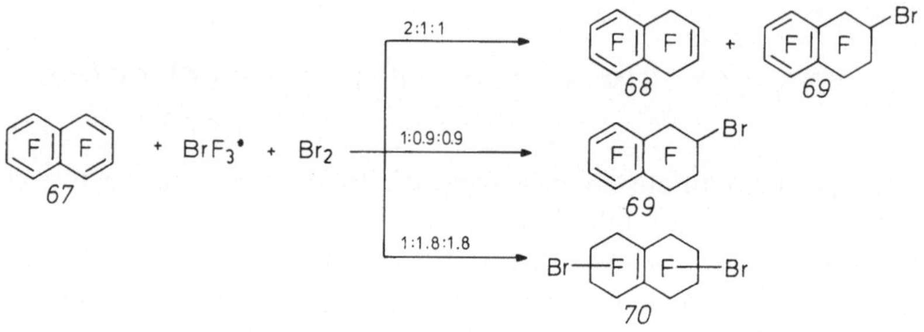

At the molar ratio of substrate: BrF_3: $Br_2 = 2:1:1$, the products were perfluoro-1,4-dihydronaphthalene *68* and 2-bromoundecafluorotetralin *69* wth a good yield, at the ratio $1:0.9:0.9$—bromopolyfluorotetralin *69*, with a large excess of reagents—a mixture of isomeric dibromotetradecafluorobicyclo[4.4.0]decenes *70*. In the case of perfluoronaphthalene, as for perfluorobenzene and its derivatives, first there occurs fluorination by BrF_3-Br_2, and subsequently bromofluorination of the diene product *68*.

Thus molecular bromine undoubtfully enhances the reactivity of bromine trifluoride towards perfluoroaromatic compounds.

The reactivity of halogen fluorides (BrF_3, BrF_5, and ClF_3) is increased much more by transferring them to complexes $BrF_2^+BF_4^-$, $BrF_2^+SbF_6^-$, $BrF_2^+Sb_2F_{11}^-$, $ClF_2^+BF_4^-$, and $ClF_2^+SbF_6^-$ [131–133]. Some fluoroaromatic compounds react with these complexes in SO_2FCl at -50 to $-90\,°C$.

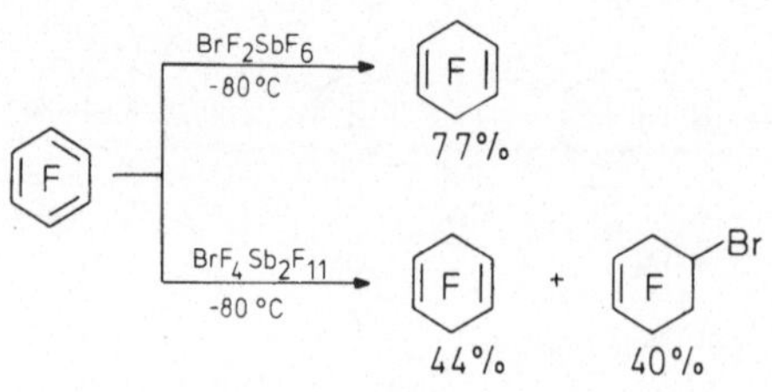

Pentafluoropyridine, 3-chlorotetrafluoropyridine, and nitropentafluorobenzene are more stable against the difluorobromonium cation and react with it only at -50 to $20\,°C$.

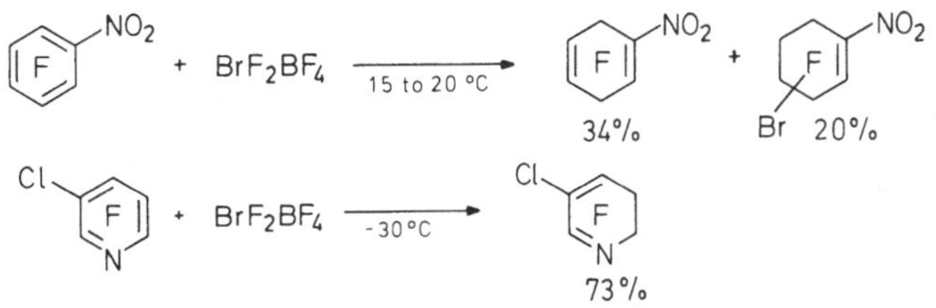

The reaction of hexafluorobenzene and octafluorotoluene with ClF_2BF_4 and ClF_2SbF_6, unlike the reactions with the difluorobromonium cation, yielded the poducts of direct chlorination of the aromatic ring, apart from the fluorination products.

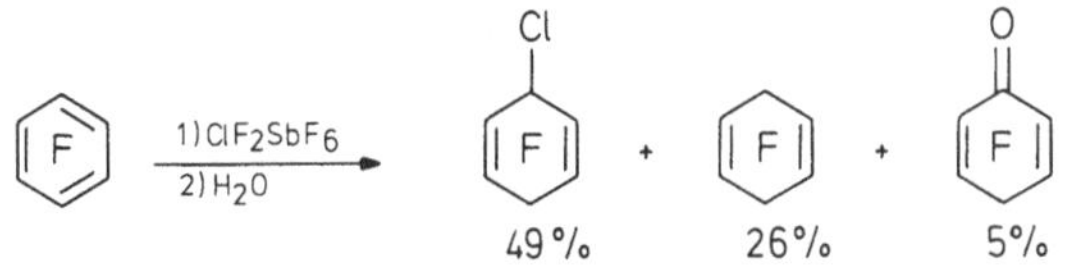

Anhydrous HF also catalyses the reaction of halogen polyfluorides with fluoroaromatic compounds, owing to formation of highly polar complexes forming the halogenonium cations upon dissociation.

$$BrF_3 + HF \rightleftharpoons BrF_3 \cdot HF \rightleftharpoons BrF_2^+ + HF_2^-$$

The suggested mechanisms of the reactions of BrF_3, BrF_3–Br_2 system and fluorohalogenonium cations $HlgF_n^+$ with polyfluoroaromatic compounds, and the reactions of BrF_3 with alkenes will be discussed in Sect. 5.6.

5.5.3 Substitutive Fluorination Reactions of Saturated Haloderivatives with Halogen Polyfluorides

The very first attempts to use chlorine trifluoride for fluorination of saturated chlorohydrocarbons showed ClF_3 to be hardly suitable for this purpose [6]. Simultaneously with partial substitution of chlorine by fluorine there possibly occurred the radial fluorination and chlorination at the C–H bonds.

Studies on the substitution of iodine and bromine (less frequently—chlorine) atoms by fluorine in completely halogenated lower alkanes under the action of IF_5 and BrF_3 were more successful [5,7].

$$CI_4 + IF_5 \xrightarrow[90\ to\ 100\ °C]{} CF_3I + I_2$$
$$90\%$$

$$CF_2ICF_2I + IF_5 \rightarrow CF_3CF_2I$$
$$85\%$$

$$CCl_4 + BrF_3 \xrightarrow[0\ to\ 10\ °C]{} CFCl_3 + CF_2Cl_2$$

The exchange of bromine and chlorine for fluorine in haloalkanes in the reaction with IF_5 proceeds at high temperatures with difficulty, and has no preparative value [7]. The interaction of chloro- and bromo-deivatives with bromine trifluoride usually led to the formation of a mixture of products of various fluorination degrees [1,6]. The reaction of BrF_3 with α,α,α'-trichlorohexafluoropiperidine at $-50\,°C$, however, gave one product—N-bromoperfluoropiperidine [134].

An efficient reagent for the substitutive fluorination of bromopolyfluoroalkanes is the BrF_3–Br_2 system [135]. The reaction is carried out by gradually adding bromofluoroalkane to a mixture of BrF_3 (1/3 mole per one g-atom of bromine) and molecular bromine at 0 to 50 °C, with simultaneously distilling off the low-boiling fluorination product.

$$CF_3CHBr_2 + BrF_3 + Br_2 \rightarrow CF_3CHFBr + CF_3CHF_2$$
$$92\% \qquad\quad 5\%$$

$$CF_2BrCHBr_2 + BrF_3 + Br_2 \rightarrow CF_3CHBrF$$
$$77\%$$

$$CF_2BrCBr_3 + BrF_3 + Br_2 \rightarrow CF_2BrCFBr_2 + CF_2BrCF_2Br$$
$$79\% \qquad\qquad 4\%$$

This procedure may be used to fluorinate lower bromofluoroalkanes where the C:H ratio is not higher than 1:1, at a small time of contact of substrate with reagent, as the reaction is carried out with a large excess of BrF_3 relative to substrate in the beginning and in the process of the reaction. Molecular bromine as a solvent surely plays an essential role in these reactions, promoting ionisation of the BrF_3 molecule, as indicated by the relatively high electric conductivity of the BrF_3–Br_2 system [1]. In addition, liquid bromine as medium in the reactions involving the carbocation intermediates is similar in its action to such a well-investigated electrophilic solvent as H_2SO_4 [136]. In more recent works it has been shown that pure BrF_3 may be used as an efficient agent of selective fluorodebromination of bromo-derivatives (with perfluorocarbons as solvent, 10 to 25 °C [137,138]. The reaction follows the scheme.

$$RBr + 1/3BrF_3 \rightarrow RF + 2/3Br_2$$

The reaction is carried out by gradually adding bromine trifluoride to a solution of the bromo-derivative (5 to 30% v/v), with vigorous stirring. The higher is the content of halogens in a substrate, the more concentrated solution may be used in the synthesis. In the process of fluorination, all the three fluorines of the BrF_3 molecule at the given stoichiometry are effectively used. The chlorine atoms of the molecule usually remain intact. The fluorodebromination of bromoalkanes by bromine trifluoride is non-stereospecific and proceeds with carbocationic rearrangements considered in Sect. 5.4.4. The typical examples of substitutive fluorination by BrF_3:

$$CH_2ClCH_2Br + BrF_3 \rightarrow CH_2FCH_2Cl$$
$$80\%$$

$$CH_2BrCHBrCH_2Cl + BrF_3 \rightarrow CH_2FCHFCH_2Cl$$
$$85\%$$

$$CH_2BrCFMeCH_2Cl + BrF_3 \rightarrow CH_2FCFMeCH_2Cl + MeCH_2CF_2CH_2Cl$$
$$6\% \qquad\qquad 67\%$$

$$CH_3(CH_2)_3CHBrCH_2Br + BrF_3 \rightarrow CH_3(CH_2)_3CHFCH_2Br$$
$$72\%$$

$$\begin{array}{c} pref\text{-CHClBrCHFCl} \\ parf\text{-CHClBrCHFCl} \end{array} \bigg] \xrightarrow{BrF_3} pref\text{-CHClFCHFCl}$$
$$70\text{–}80\%$$

In the reactions under discussion, bromine trifluoride is much less reactive than chlorine monofluoride. Thus ClF readily and completely reacts with 1-bromo-2,3-dichloropropane at -30 to $-40\,°C$, whereas with BrF_3 at the same temperature there is practically no reaction [137].

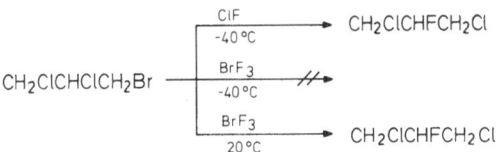

The difference in the reactivities of BrF_3 and ClF is still more vividly seen in the reactions with bromine-containing esters. Thus the reaction of methyl dibromo-isobutyrate 72 with ClF proceeds to an end when the stoichiometric amount of ClF is passed through this solution for 1 h, whereas the reaction with BrF_3 takes about 30 h.

$$CH_2BrCBrMeCOOMe + 1/3\,BrF_3 \xrightarrow[20\,°C,\ 30h]{} CH_2BrCFMeCOOMe$$
$$72 \qquad\qquad\qquad\qquad\qquad\qquad ·55\text{–}60\%$$
$$+ CH_2FCHBrMeCOOMe$$
$$15\text{–}20\%$$

In this case, such a low reactivity of BrF_3 is also explained by the fact that the COOMe group deactivates BrF_3 in the substitutive fluorination reaction. Fluorination of methyl 2,3-dibromo-2-chloropropanoate 73 in similar conditions proceeds much faster, which may be explained by the decreased basicity of the COOMe group of ester 73 [138].

$$CH_2BrCBrClCOOMe + BrF_3 \xrightarrow[20\,°C,\ 10h]{} CH_2BrCFClCOOMe$$
$$73 \qquad\qquad\qquad\qquad\qquad\qquad 74$$

Of course, this is not due to the fact that chlorine at the reaction centre promotes substitution more than the CH_3 group in ester 72, as in the reaction of BrF_3 will a mixture of esters 72 and 73, compounds 72 is fluorinated first.

The fluorinating ability of BrF_3 sharply increases in the presence of some Lewis acids, such as $SbCl_5(SbF_5)$ and $SnCl_4(SnF_4)$ [139]. Especially good results have been obtained with the use of $SnCl_4$, which does not decrease the selectivity of fluorination. In the presence of 1 to 3% of $SnCl_4$, fluorination of ester *73* proceeds rapidly and gives ester *74* in a good yield. Ester *74* is a convenient substrate for the synthesis of methyl 2-fluorometacrylate *75* a valuable but almost unavailable monomer [138]. Ester *74* is dehalogenated by zinc powder in acidified water.

$$CH_2BrCFClCOOMe \xrightarrow[H_3O^+]{Zn} CH_2=CFCOOMe$$
$$\qquad\quad 74 \qquad\qquad\qquad\qquad\qquad 75$$

Apart from BrF_3, for the preparation of intermediate products in the synthesis of 2-fluoroacrylates, chlorine monofluoride may be used. Syntheses using ClF and BrF_3 are attractive not only due to high yields of the fluoro-derivatives but also due to the possibility to utilise the evolving chlorine and bromine for regeneration of ClF and BrF_3.

Recently, conditions for the selective fluorination of some saturated hydrocarbons by chlorine trifluoride have been found [140]. The reactions are carried out at $-75\,°C$ in $1,2\text{-}C_2Cl_2F_4$ or liquid CO_2 (under low pressure). Substitution of hydrogen atoms by fluorine follows the general scheme.

$$RH + ClF_3 \rightarrow RF + HF + ClF$$

Selectivity of the reactions substantially depends on the structure of hydrocarbons. Most easily substituted are the hydrogen atoms at the tertiary carbon atom. Thus 2,3-dimethylbutane undergoes 80% conversion, giving a mixture of mono- and difluoroalkanes with a high total yield.

$$Me_2CHCHMe_2 + ClF_3 \xrightarrow[-75\,°C]{} Me_2CFCHMe_2 + Me_2CFCFMe_2$$
$$\qquad\qquad\qquad\qquad\qquad\qquad\quad 55\% \qquad\qquad 20\%$$

On the other hand, 2,2,3,3-tetramethylbutane (Me_3CCMe_3), which contains hydrogen atoms only in CH_3 groups, and methane do not react in the given conditions.

The reaction of ClF_3 with some hydrocarbons (e.g. 2-methylbutane) give alkenes. They are possibly formed in the process of product isolation due to easy dehydrofluorination of monofluorohydrocarbons.

Fluorination of 2,2-dimethylbutane with ClF_3 is accompanied by the methyl group migration.

$$Me_3CCH_2Me + ClF_3 \rightarrow Me_2CFCHMe_2 + Me_2CFCFMe_2$$
$$\qquad\qquad\qquad 76 \quad 69\% \qquad\qquad 26\%$$

The formation of alkyl fluoride *76* indicates the carbocationic mechanism of the reaction. Brower suggested the following route involving the three-centre transition state or intermediate 77.

$$RH + ClF_3 \xrightarrow[-F^-]{-ClF} \left[R-\!\!\!\begin{array}{c} H^+ \\ \diagdown F \end{array} \right] \xrightarrow[-HF]{F^-} [R^+, F^-] \rightarrow RF$$

77

$$RF + HF \rightarrow R'^+ + HF_2^- \rightarrow R'F + HF$$

$$R^+ = Me_3C\overset{+}{C}HMe; \; R'^+ = Me_2\overset{+}{C}CHMe_2$$

However in our view this substitution of hydrogen atoms by chlorine trifluoride proceeds in a similar way as the substitution of halogens (Cl, Br) by chlorine monofluoride or bromine trifluoride according to the mechanism considered in Sect. 5.6, the only difference being in orientation of the electrophilic attack of oxidising agent on the electrons of the C–H bond. The few examples of α-hydrogen substitution in the Et and Me groups of haloesters by ClF [106] or the Br_2–BrF_3 system [58] seem to refer to the same type of reactions.

5.6 Halogen Fluorides as Electrophilic Agents

Investigation of the reactions of halogen monofluorides HlgF (Hlg = Cl, Br, I) with organic unsaturated compounds has shown that these interhalogens may be referred to as typical electrophilic agents. All the observed facts—rate and orientation of addition at multiple bonds depending on the electronic nature of substituents and on the medium, stereochemistry of addition, skeletal rearrangements and migrations accompanying these reactions in the polar and nonpolar media, may be easily interpreted in terms of the carbocation chemistry.

The reactions of HlgF with most unsaturated compounds in polar and nonpolar media proceed as the electrophilic anti-addition.

The action of halogen polyfluorides on unsaturated compounds has been studied in less detail. Most data on the mechanism of addition to alkenes and arenes have been obtained for BrF_3. As considered above (see Sect. 5.5.2), the reactions of BrF_3 with alkenes and perfluoroarenes form the additive fluorination and bromofluorination products. In the case of the reactions of BrF_3 with E- and Z-1,2-dichloroethylenes, and with E- and Z-1,3-dichloropropenes, it was shown that fluorination and bromofluorination by bromine trifluoride proceed with widely varying stereo- and regioselectivities [128]. Thus the bromofluorination anti-stereospecifically gives vicinal bromofluorides, whereas the fluorination proceeds non-stereospecifically and is accompanied by halogen migration.

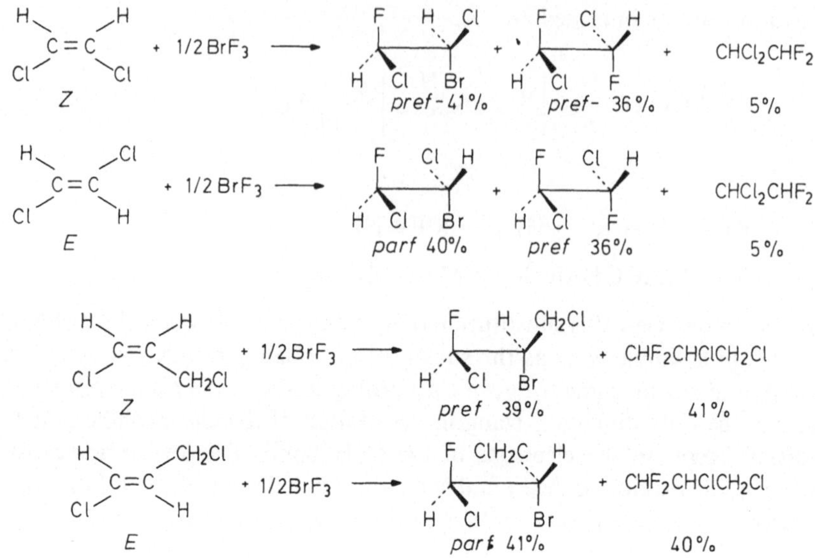

Z: $+ 1/2\,BrF_3 \longrightarrow$ pref-41%   pref-36% $CHCl_2CHF_2$ 5%

E: $+ 1/2\,BrF_3 \longrightarrow$ parf 40%   pref 36% $CHCl_2CHF_2$ 5%

Z: $+ 1/2\,BrF_3 \longrightarrow$ pref 39% $+ CHF_2CHClCH_2Cl$ 41%

E: $+ 1/2BrF_3 \longrightarrow$ parf 41% $+ CHF_2CHClCH_2Cl$ 40%

On the basis of these data, the reactions of fluorination and bromofluorination of alkenes by bromine trifluoride were suggested [128] to proceed via different independent transition states or intermediates (Scheme 1).

Scheme 1

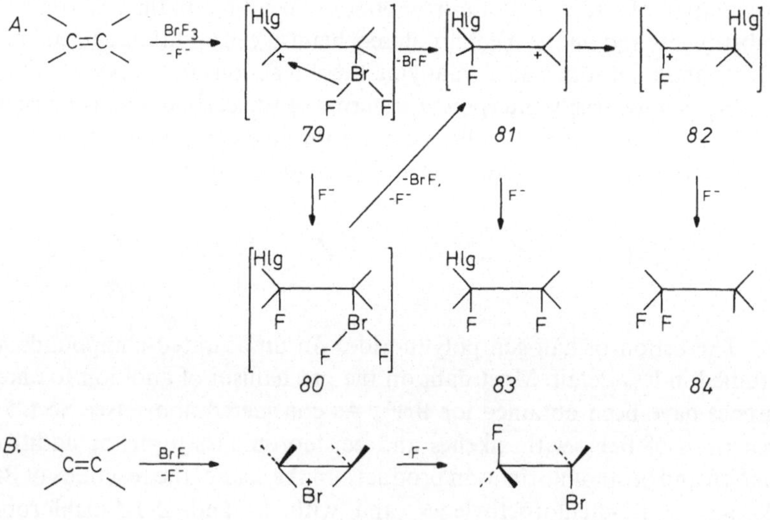

A. $\ce{C=C}$ $\xrightarrow[-F^-]{BrF_3}$ 79 $\xrightarrow{-BrF}$ 81 $\longrightarrow$ 82

80   83   84

B. $\ce{C=C}$ $\xrightarrow[-F^-]{BrF}$ $\xrightarrow{-F^-}$

The electrophilic attack at the olefin double bond in non-polar medium is effected by the BrF_3 molecule. The bromine (III)-containing carbocation 79 formed in this process eliminates the BrF molecule to form the β-fluorocarbocation 81. The latter is stabilised by capturing the fluoride ion, forming vicinal difluoride *83*, or is rearranged to a more stable α-fluorocarbocation *82*, which is further transformed to the more geminal difluoride *84*. Carbocation *79* possibly manages to capture the fluoride ion, giving an unstable intermediate *80*, which further also decomposes to bromine monofluoride and carbocation *81*. A relatively stable adduct of type *80* with trivalent iodine was obtained in the addition reaction of iodine tris(fluorosulfate) with perfluoroolefins [141].

Thus the initial stage (stage *A*) of the reaction involves the addition of two fluorine atoms at the C=C bond, and the C=C bond is bromofluorinated by bromine monofluoride liberated at the first stage, according to the ordinary scheme of electrophilic *anti*-addition (stage *B*) (cf. Sect. 5.3.1).

The suggested reaction mechanism adequately accounts for the unfavourable (from the first sight) migration of vinyl halogen (Cl, Br) leading to geminal difluorides *84*, and the sharp difference in the stereochemistry of formation of vicinal difluoro- and bromofluoro-adducts. As mentioned in Sect. 5.5.2, the reaction of perfluoroaromatic compounds with BrF_3 initially gives the products of fluorination of the aromatic ring—perfluorocyclohexadienes, and then—their adducts with BrF [129]. Bastock and his co-workers [129] carried out the semi-empirical thermodynamic calculations of the reactions of BrF_3–Br_2 with per-fluoroarenes considering ionisation potentials of fluoroaromatic compounds, and suggested a general scheme of one-electron oxidation of these substrates to radical cations. This opinion is shared by the scientists who studied the reactions of fluorohalogenonium cations (ClF_2^+, BrF_2^+, BrF_4^+, and IF_4^+) with perfluoroaro-matics [131–133].

All unmarked bonds to fluorine
Radical cation *85* was recorded by ESR spectroscopy in the reaction of perfluoronaphthalene with ClF_3, BrF_3, BrF_5, and XeF_2 at $-50\,°C$ [133].

$$C_{10}F_8 + BrF_3 \xrightarrow[-50\,°C]{SO_2FCl} C_{10}F_8^{+\cdot}$$

Fluorohalogenonium cations XF_n^+ (X = Cl, Br, I) have a higher electron affinity than the starting halogen fluorides, being thus more reactive. For example, $EA(BrF_2^+)$ and $EA(ClF_2^+)$ are equal to 11–12 eV. The ionisation potential of most perfluoroarenes ranges from 9 eV (octafluoronaphthalene) to 10.6 eV (nitro-pentafluorobenzene).

Thus halogen fluorides as electrophilic agents towards unsaturated compounds differ from "normal" electrophiles such as Hlg_2, $HHlg$, HgX_2, $RSHlg$, etc., only in their extremely high reactivity, as they are simultaneously very strong oxidants. Nevertheless many of their reactions may be controlled by usual methods—dilution with inert solvents (polychlorofluorocarbons, SO_2FCl, etc.), mixing the reagents in small portions, etc.

Halogen fluorides are most close in reactivity to the agents with an electropositive halogen atom, such as chlorine, bromine and iodine fluorosulfates $HlgOSO_2F$ [9,142], chlorine and bromine triflates [76,143,144], chlorine perchlorate $ClOClO_3$ [145], chlorine nitrate $ClONO_2$ [146], chlorine chlorosulfate $ClOSO_2Cl$, and chlorine ethoxysulfate [147,148], as well as iodine and bromine tris(fluorosulfates) [141,142]. These reagents easily add at the multiple bond of alkenes according to the electrophilic mechanism (as substrates, perfluoroolefins were most frequently used). Currently there is not enough evidence to make up a reactivity sequence for heteroatomic electrophiles including halogen fluorides. Nevertheless it would be reasonable to expect the reactivity of compounds with electropositive halogen to directly depend on the electronegativeness of group Y in the molecule $Hlg^{\delta+} - Y^{\delta-}$.

On the basis of stereochemistry studies of the addition of $ClOSO_2CF_3$ to E- and Z-1,2-difluoroethylenes, the authors of [76,144] suggested a concerted mechanism of cyclic syn-addition of this reagent. But, as shown in review [10], the configuration assignment of the diastereomers produced may not be considered reliable. The reactions of alkenes with $ClOSO_2CF_3$ and related compounds including ClF, BrF, and IF, apparently, proceed as the ordinary electrophilic $anti$-addition.

Among the reactions of halogen fluorides, the most interesting seem to be the substitutive fluorodebromination and fluorodechlorination of organic chloro- and bromo-derivatives by chlorine monofluoride and bromine trifluoride. This must be a new and prospective method of selective fluorination.

$$RBr + ClF \rightarrow RF + 1/2Cl_2$$

$$RBr + 1/3BrF_3 \rightarrow RF + 2/3Br_2$$

$$RCl + ClF \xrightarrow{\text{Cat.}} RF + Cl_2$$

Organic iodides RI as substrates are of no interest, as the released iodine is oxidised, as a rule, by chlorine monofluoride or bromine trifluoride to IF_5, leading to uneconomical waste of the fluorinating agent.

Interest in these reactions stems essentially from the fact that they represent an unusual type of nucleophilic substitution at a saturated carbon atom. Many examples of these reactions are now scattered in the literature. Halogens in alkyl halides may be substituted by various oxidising agents: molecular bromine, chlorine, ICl [149,150], nitric acid, $KMnO_4$, $K_2Cr_2O_7$, $KClO_3$, $NaNO_2$, H_2O_2, $Ca(OCl)_2$ [151,152], organic peroxy acids [153,154], chlorine and iodine fluorosulfates, bromine and iodine tris(fluorosulfates) [142,155–157], chlorine

and bromine triflates [76,158], chlorine perchlorate [159], nitronium fluorobor-
ate [160], the N-chloroamine–HF system [70], and halogen fluorides. The
peculiarity of these reactions is in the oxidation of the halogen to Hlg_2 (or to the
higher oxidation state).

Let us consider this type of nucleophilic substitution with reference to
fluorination of alkyl bromides by chlorine monofluoride and bromine trifluoride. As shown in Sects. 5.4.4 and 5.5.3, the rearrangements and migrations
which frequently occur in these syntheses, as well as the dependence of the
reactivity of organic halides on the electronic nature of substituents at the
reaction centre, indicate the carbocationic mechanism. But as the substitutive
fluorination by halogen fluorides easily occurs in non-polar media, it is unlikely
that the heterolysis of C–Hlg bond in the electrophilic elimination of Hlg atom
would lead to formation of carbocations as the kinetically independent species,
as it was presented for the sake of simplicity in the previous Sections.

Interaction of ClF with organic halides possibly proceeds in the following
way (Scheme 2).

$$Scheme\ 2$$

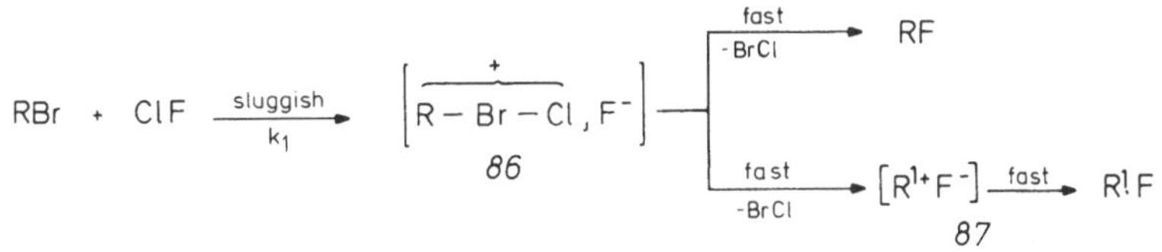

$$2BrCl \rightleftarrows Br_2 + Cl_2$$

$$BrCl + 3ClF \xrightarrow{k_2} BrF_3 + 3/2\,Cl_2$$

$$Br_2 + 6ClF \xrightarrow{k_3} 2BrF_3 + 3Cl_2$$

$$k_1 \gg k_2;\ k_1 \gg k_3$$

The first, rate-determining stage of the reaction involves the electrophilic
attack of ClF at the electron lone pairs of bromine to form ion pair 86 containing
the linear alkylbromonium cation. Under the action of internal nucleophile F^-,
ion pair 86 releases BrCl to form quickly an organic fluoride, or quickly
rearranges to ion pair 87 containing a more stable carbocation R'^+ to give
organic fluoride R'F. Thus the role of ClF as an oxidating agent is to form an
easy-leaving group—the neutral BrCl molecule. In this respect the reactions
considered are similar to the deamination reactions where the leaving group is
the N_2 molecule. Formation of ion pairs such as 86 occurs in the substitutive
chlorination of iodoalkanes by iodine monochloride [150].

Bromine monochloride is an unstable compound decomposing at room
temperature to Cl_2 and Br_2 [4]. The latter can react with ClF forming BrF_3
[1,99], which is also a fluorinating agent. Special experiments have shown that

oxidation of Br_2 or $BrCl$ by chlorine monofluoride to BrF_3 is generally much more sluggish than the oxidation of bromine in organic bromides ($k_1 \gg k_2$ and $k_1 \gg k_3$) (Scheme 2). Consequently, BrF_3 does not transfer fluorine in the fluorination of organic bromo-derivatives by chlorine monofluoride [137]. Nevertheless it is not unlikely that in the reactions of ClF with low-reactive bromo-derivatives, such as perfluoroalkyl bromides, the k_1 and k_2 (k_3) values may become comparable. As a result, bromine trifluoride will accumulate in the reaction mixture, because of its weaker fluorinating ability than that of ClF [137]. This occurs in the reactions of chlorine perchlorate with perfluoroalkyl bromides [159].

$$R_FBr + 2ClOClO_3 \rightarrow R_FOClO_3 + Cl_2 + BrOClO_3$$

In the reactions of bromine trifluoride with organic bromides, the same products are mostly formed as in the fluorination by chlorine monofluoride. The reaction mechanism may be represented by a similar scheme.

Scheme 3

$$Br_2F_2 \longrightarrow 2BrF \rightleftarrows 2/3BrF_3 + 2/3Br_2$$
$$RBr + BrF \longrightarrow RF + Br_2$$

The electrophilic attack of BrF_3 at the bromine atom of organic bromide leads to formation of ion pair *88*, which further releases the Br_2F_2 molecule to form organic fluoride RF or to rearrange to ion pair *89* containing a more stable carbocation R'^+, and is further transformed to isomeric organic fluoride R'F. It should be noted that for the reactions of substitutive fluorination of bromo-ethanes by the BrF_3–Br_2 system, the authors of [135] suggested the carbocationic mechanism involving formation of the hypothetic unstable Br_2F_2 molecule as a result of Br^- elimination under the action of the BrF_2^+ cation. Fragmentation of the Br_2F_2 molecule yields two molecules of bromine monofluoride, which, by analogy with ClF, may possibly also fluorinate the organic bromides. But, in contrast to ClF, bromine monofluoride easily and almost completely decomposes to Br_2 and BrF_3, which makes it impossible to judge about participation of BrF in the fluorodebromination by bromine trifluoride.

This mechanism accounts for the main regularities of substitutive fluorination by halogen fluorides:

1. The electron-accepting substituents at the reaction centre slow down the fluorination, whereas the electron-donating ones accelerate it. Vicinal halogens strongly slow down the reaction, while the geminal ones only slightly affect the fluorination rate.

2. Cationoid intermediates easily rearrange by the 1,2-shift of chlorine, bromine, hydrogen, and methyl. Fluorine atoms do not migrate. The migrating atoms (groups) produce no anchimeric effect upon substitution.
3. The selectivity of fluorination is determined not only by the selectivity of halogen (Cl, Br) elimination but also by the high selectivity of 1,2-migrations of groups in the intermediate forming more stable carbocations. Hence the reactions predominantly form the secondary, tertiary or α-halo-substituted organic fluorides.

It is interesting to compare the reactions of halogen fluorides and some Lewis acids (SbF_5, SbF_3, HgF_2, AgF, etc.) with alkyl halides. At the first, rate-determining stage, the action of the reagents of both types involves the transformation of the halogen to be substituted to an easier-leaving group.

$$RHlg + Hlg'F \rightarrow [\overset{\delta+}{R}-\overset{\delta+}{Hlg}-\overset{\delta+}{Hlg'}, F^-] \rightarrow RF + HlgHlg'$$

$$RHlg + MF_n \rightarrow [\overset{\delta+}{R} \ldots \overset{\delta-}{Hlg} \ldots \overset{\delta-}{MF_n}] \rightarrow RF + MF_{n-1}Hlg$$

$$Hlg = Cl, Br; M = Sb, Hg, Ag$$

There is a typical difference between these reactions: under the action of halogen fluorides the halogen is eliminated in the oxidised form (Cl_2, Br_2, $BrCl$), whereas Lewis acids eliminate halogen as a negatively charged species being part of metal halide (SbF_nCl, $HgFCl$, $AgCl$, $AgBr$, etc.). For that reason, in the reaction with alkyl halides, a Lewis acid usually attacks the halogen whose elimination would give the most stable carbocation, whereas the halogen fluoride attack is oriented to the most easily oxidisable halogen atom. Thus in the fluorination of 1,2-dibromo-1,1,2-trichloroethane and 1,2-dibromo-1,1-dichloroethane by chlorine monofluoride or HgF_2 (HgO in HF) or by SbF_5(SbF_3/SbF_5/HF), haloethanes 90 and 91 respectively are formed as intermediates, irrespectively of the fluorinating agent. But in further fluorinations of compounds 90 and 91 with antimony or mercury fluorides, fluorine is substituted for the chlorine atoms of the $CFCl_2$ group [3,124,161], whereas treatment of these compounds with chlorine monofluoride proceeds with bromine substitution accompanied by the chlorine 1,2-migration [99].

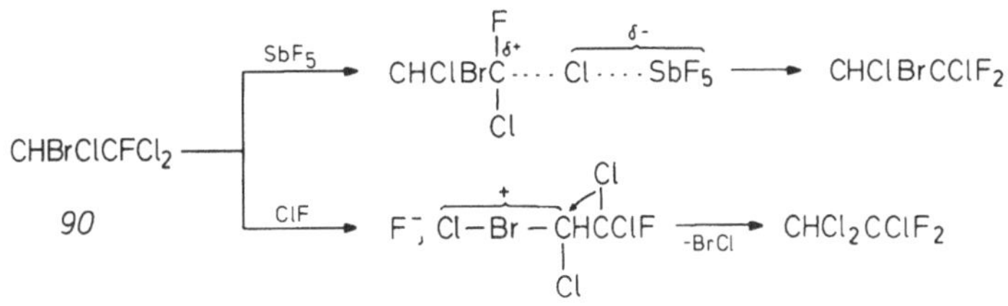

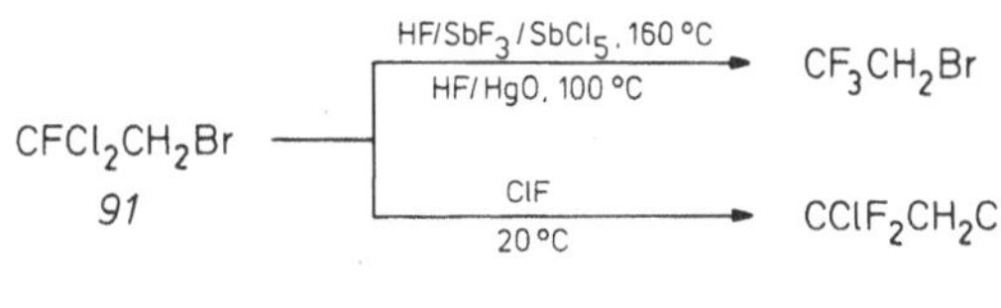

Another peculiarity of the oxidative nucleophilic substitution reactions is in the fact that halogen is substituted only by a more electronegative atom or group, e.g. F, $OClO_3$, or OTf. In view of this, it may be concluded that eventually not only the substituted halogen is oxidised but also the carbon atom of the reaction centre.

In [162], a quantitative estimation is given of the oxidative ability of a series of halogen fluorosulfates: $IOSO_2F$, $I(OSO_2F)_3$, $BrOSO_2F$, $Br(OSO_2F)_3$, $ClOSO_2F$, and $(FSO_2O)_2$. The reactivity of these compounds in the reactions of halogen (Cl, Br) substitution at the tetrahedric carbon atom is shown to vary, as a rule, parallel with their oxidative ability. It is also shown that the reactivity of these compounds increases with the increased acidity of medium (from HSO_3F to $HSO_3F–SbF_5$) [162, 163]. The results obtained in this work and the reaction routes suggested certainly present interest for understanding the mechanism of halogen substitution in organic halides by oxidising agents, including halogen fluorides—ClF and BrF_3 as compounds related to fluorosulfates $ClOSO_2F$ and $Br(OSO_2F)_3$. Nevertheless the term "oxidative electrophilic substitution at the carbon atom" [162] seems unacceptable, as in the electrophilic substitution at the tetrahedric carbon atom, the electrophilic attack is directed at the carbon with which the electrophile further forms the σ-bond. Considering this, it would seem better to use the term "electrophilic substitution at the halogen" suggested in [158] for the reactions of Cl and Br substitution in perfluoroalkyl halides by $ClOSO_2CF_3$ and $BrOSO_2CF_3$. But this definition is also hardly acceptable for organic reactions where the object of consideration is first of all the carbon-containing fragment of a substrate. Moreover, with such an approach, the numerous nucleophilic substitution reactions proceeding under the electrophilic catalysis by Brönsted or Lewis acids would have to be referred to as the electrophilic substitution at the heteroatom. For the same reason, these reactions may not be considered as "halophilic" [164]. This term suggested in [165] is altogether unacceptable, as classification of the reactions according to the initial attack of reagent at the substrate in many cases convey nothing to an organic chemist about the result of the reaction. Interpretation of the reactions as an instance of the nucleophilic substitution at the tetrahedric carbon allows one to plan syntheses and forecast their results.

The heterolytic substitution of hydrogen atom by halogen fluorides [58, 106, 140] or the related compounds such as $ClOSO_2F$ and $Br(OSO_2F)_3$ [142, 157] under mild conditions also seems to refer to the oxidative nucleophilic substitution at a saturated carbon atom. The low-temperature fluorination of hydrocarbons by chlorine trifluoride and halogenated ethers by chlorine monofluoride may be represented by the following scheme.

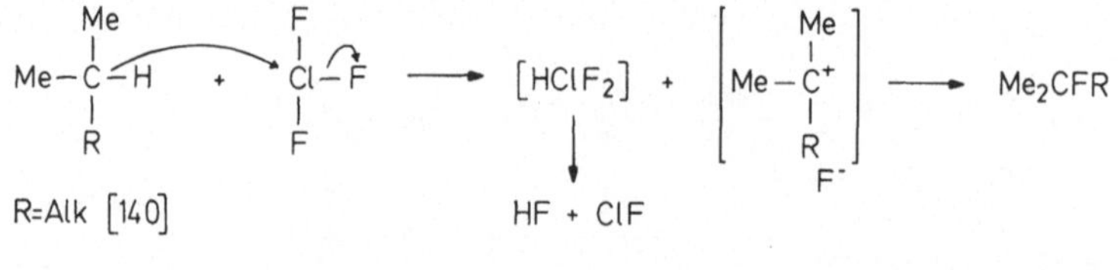

$$RCF_2OC \overset{\underset{\displaystyle Me}{|}}{\overset{\displaystyle H}{|}} H \;+\; Cl{-}F \longrightarrow HCl \;+\; \left[RCF_2\overset{+}{\underset{\underset{\displaystyle F^-}{|}}{O}}{-}\overset{\underset{\displaystyle Me}{|}}{\overset{\displaystyle H}{|}}C^+ \right] \longrightarrow RCF_2OCHFMe$$

$$HCl + ClF \longrightarrow HF + Cl_2$$

R = haloalkyl [106]

A similar mechanism of oxidative substitution of α-hydrogen atoms has been suggested for the bromination reactions of ethers [166]. The reactions of selective hydrogen substitution by halogen fluorides have been little studied, and their mechanism is not quite clear as yet, and it is not clear whether they will have the same preparative possibilities as the halogen substitution reactions. Nevertheless the authors of this chapter believe that in the very near future studies on the reactions of this type may yield both theoretically and synthetically interesting results.

5.7 Preparations

1. 1-Bromo-2-fluorohexane [36]
In a polyethylene bottle, 130 g (6.5 mol) of anhydrous HF was mixed, under deep cooling, with 240 ml of diethyl ether. Then 69 g (0.5 mol) of N-bromoacetamide and 42 g (0.5 mol) of hexene-1 were alternately added in portions, for 27 min at $-80\,°C$. The mixture was stirred at this temperature for 2 h, then carefully poured into a mixture of Na_2CO_3 (690 g), water (1.2 l), crushed ice (500 g), and ether (300 ml). The ether layer was dried with $MgSO_4$ and distilled to give 71.8 g (79%) of 1-bromo-2-fluorohexane; b.p. 62 °C (18 mm), d_4^{20} 1.287, n_D^{20} 1.4370. The product contained 5% of 2-bromo-1-fluorohexane.
2. Methyl 2-chloro-3-fluoropropanoate *15* and methyl 3-chloro-2-fluoro-propanoate *16* [66]
To a solution of 73 g (0.22 mol) of hexachloromelamine in 100 ml of anhydrous HF was added 86 g (1 mol) of methyl acrylate with stirring and cooling to between -30 and $-20\,°C$. The mixture was stirred at this temperature for 1 h, then warmed to room temperature and allowed to stand overnight. Then 250 ml of chloroform was added and the solution was cooled to between 0 and $-5\,°C$. An excess of hexachloromelamine was decomposed by dry sodium sulfite at this temperature and with stirring, then excessive HF was neutralised by dry sodium bicarbonate. The resulting mixture was filtered and the filtrate distilled to give 74 g (52.5%) of 81 : 19 mixture of esters *15* and *16*; b.p. 60 to 67 °C (35 mm), n_D^{20} 1.4160. The mixture was rectified on a column (20 th.pl.). Ester *15*: b.p. 148 to 150 °C, d_4^{20} 1.2990, n_D^{20} 1.4150; ester *16*: b.p. 165 to 166 °C, d_4^{20} 1.3120, n_D^{20} 1.4190.
3. 1-Bromo-2-fluoro-3-chloroisobutane *26* [70]
To a solution of 14 g (0.04 mol) of hexachloromelamine in 80 ml of anhydrous HF was added dropwise 50 g (0.20 mol) of 1,2-dibromo-3-chloroisobutane with stirring and cooling (-10 to $0\,°C$). The mixture was stirred for 4 h at this

temperature, then poured onto ice. The organic layer was washed with water, then with the sodium sulfite solution, dried with $MgSO_4$ and distilled to give 22 g (59%) of haloalkane *10*: b.p. 55 °C (13 mm), d_4^{20} 1.5994, n_D^{20} 1.4630.

4. 1*R**,2*R**-1-fluoro-1,2,3-trichloropropane *23* [94]

Into a solution of 134 g (1.2 mol) of *Z*-1,3-dichloropropene in 500 ml of chloroform, 30 l of gaseous ClF was passed for 60 to 70 min at 20 to 30 °C. After the reaction was completed, chlorine monofluoride was not absorbed (KI probe at the outlet of the reactor). The product was distilled to give 184 g (92%) of fluorotrichloropropane *23*: b.p. 143 °C, d_4^{20} 1.4818, n_D^{20} 1.4554.

5. Dimethyl *threo*-2-chloro-3-fluorosuccinate *24* [94]

Into a solution of 85 g (0.71 mol) of dimethyl maleate in 350 ml of chloroform, 0.8 mol of ClF was passed by the above-described method. The product was distilled to give 77.3 g (66%) of *threo*-diastereomer *24*: b.p. 88 to 90 °C (2 mm), d_4^{20} 1.3645, n_D^{20} 1.4357.

6. 1,2,3-Trifluoropropane *45* [98]

A glass cylinder reactor with a reflux condenser, thermometer, and polyethylene bubbler was charged with 281 g (1 mol) of 1,2,3-tribromopropane, and ClF was passed at 0 to -5 °C and at 25 l h^{-1}, for 3 h. The reaction mixture was treated with NaOH solution under cooling to disappearance of bromine colouring, then dried over $MgSO_4$ and distilled to give 69 g (71%) of trifluoropropane *45*: b.p. 70 °C, d_4^{20} 1.2680, n_D^{20} 1.3250.

7. Methyl 2,3-difluoroisobutyrate *46* [98,102]

A. Gaseous ClF (50 l) was passed through 257 g (1 mol) of methyl 2,3-dibromo-isobutyrate during 2 h, as described above. Then chloroform (200 ml) was added to the reaction mixture, and the mixture was washed with Na_2SO_3 solution to give 125 g (93%) of ester *46*: b.p. 49 °C (18 mm), d_4^{20} 1.1877, n_D^{20} 1.3850.

B. A polyethylene cylinder reactor was charged with a solution of 86 g (0.5 mol) of methyl 2,3-dichloroisobutyrate in 60 ml of anhydrous HF, and 26 l of ClF was passed at -20 to -40 °C for 2 h. The reaction mixture was poured onto ice, the organic layer was separated, washed with $NaHSO_3$ solution, dried over $MgSO_4$ and distilled to give 57 g (83%) of ester *46*: b.p. 50 °C (18 mm), n_D^{20} 1.3850.

8. 1,3-Dichloro-2-fluoropropane *41* [102]

In a glass cylinder reactor were placed 147.5 g (1 mol) of 1,2,3-trichloropropane and 50 ml of 1,1,2-trichlorotrifluoroethane, and ClF was passed at -25 to -30 °C for 3 h (10 l h^{-1}). The reaction mixture was washed with Na_2SO_3 solution, dried and distilled to give 111 g (84%) of haloalkane *41*: b.p. 127 °C, d_4^{20} 1.4350, n_D^{20} 1.4353.

9. 1-Chloro-2,2-difluoro-3-oxabutane *50* [106]

A polyethylene cylinder reactor was charged with 27 g (0.25 mol) of methyl chloroacetate and 20 ml of anhydrous HF and cooled to -60 to -40 °C. Chlorine monofluoride (15 l, 0.6 mol) was passed into the solution for 25–30 min. The reaction mixture was poured onto ice. The organic layer was washed with ice water, dried and distilled to give 26 g (80%) of ether *50*: b.p. 76 °C, d_4^{20} 1.2850, n_D^{20} 1.3530.

10. 1-Chloro-2,2,4-trifluoro-3-oxapentane *51* [106]

Into a solution of 24.5 g (0.2 mol) of ethyl chloroacetate in 20 ml of HF, 18 l (0.7 mol) of ClF was passed under the above conditions, for 40–50 min. The procedure gave 19.6 g (61%) of ether *51*: b.p. 47 °C (115 mm), d_4^{20} 1.3045, n_D^{20} 1.3530.

11. *Pref*-2-bromo-1-fluoro-1,3-dichloropropane   *39*   and   2,3-dichloro-1,1-difluoropropane *40* [128]

In a Teflon reactor with a nickel stirrer, nickel réflux condenser and quartz dropping funnel, were placed 24.4 g (0.22 mol) of Z-1,3-dichloropropene and 380 ml of 1,1,2-$C_2F_3Cl_3$. With stirring and cooling the reactor with ice water 15.2 g (0.11 mol) of BrF_3 was gradually added for 30–40 min at 5 to 10 °C. The reaction mixture was washed with aqueous sodium sulfite, dried with $CaCl_2$, and distilled on a column (5 th.pl.). This procedure gave 13.3 g (40%) of halopropane *40*: b.p. 55 °C (100 mm), d_4^{20} 1.4460, n_D^{20} 1.4040. Then 19.1 g (41%) of *pref*-diastereomer *39* was isolated: b.p. 40 °C (5 mm), d_4^{20} 1.8233, n_D^{20} 1.4908.

12. 4-Bromononafluorocyclohexene *71* [129]

A nickel reactor with a stirrer, dropping funnel, and copper reflux condenser was charged with 10 g (0.054 mol) of hexafluorobenzene, 40 ml of perfluoromethylcyclohexane (solvent), and 8.6 g (0.054 mol) bromine. Then 7.3 (0.054 mol) of BrF_3 was slowly added with stirring at 0 °C. The mixture was stirred for 1 h at 0 °C, whereupon water was added to decompose unchanged BrF_3. The organic layer was washed with $NaHSO_3$ solution and dried over $MgSO_4$. Distillation of the reaction mixture gave 7.6 g (43%) of cyclohexene *71*: b.p. 99 °C, and 1.1 g (9%) of octafluorocyclohexa-1,4-diene.

13. 2-Bromo-1,1,1,2-tetrafluoroethane *78* [135]

A Monel flask provided with a nickel condenser, nickel dropping funnel and nickel stirrer was charged with 54 g (0.4 mol) of BrF_3 and 236 g of bromine. To this mixture, 247 g (1 mol) of 2,2-dibromo-1,1,1-trifluoroethane was added for 15 min, with cooling of the flask to 25 to 35 °C and stirring. The gaseous products were simultaneously collected in a glass trap cooled to −78 °C. Then the reaction mixture was heated to 50 °C to complete the reaction. The reaction products were washed with the cooled NaOH solution to remove bromine. This procedure gave 156 g (85%) of haloethane *78*: b.p. 8 to 9 °C (with a 0.5% admixture of CF_3CHF_2).

14. 1-Bromo-3-chloro-2-fluoropropane *12* and 3-chloro-1,2-difluoropropane *14* [138]

A Teflon reactor with a nickel stirrer, nickel reflux condenser, thermometer in a nickel pocket and quartz dropping funnel was charged with a solution of 170 g (0.67 mol) of 1,2-dibromo-3-chloropropane in 400 ml of 1,1,2-$C_2F_3Cl_3$. Then 50.4 g (0.37 mol) of BrF_3 was added dropwise, with vigorous stirring for 3 h at 5 to 10 °C. The reaction mixture was stirred for 1 h more, then treated with the $NaHSO_3$ solution, dried and distilled to give 24.7 g (25%) of halopropane *14*, b.p. 96 °C (750 mm) and 69.5 g (60%) of halopropane *12*, b.p. 125 to 130 °C (100 mm), d_4^{20} 1.7408, n_D^{20} 1.4700.

A similar reaction of 170 g of 1,2-dibromo-3-chloropropane with 61.5 g (0.45 mol) of BrF_3 gave 70 g (65%) of chlorodifluoropropane *14*, b.p. 96 °C, d_4^{20} 1.3000, n_D^{20} 1.3813.

15. Methyl 3-bromo-2-chloro-2-fluoropropanoate *74* and methyl 2-fluoroacrylate *75* [138]

In a Teflon reactor was placed 95 g (0.34 mol) of methyl 2,3-dibromo-2-chloropropanoate in 100 ml of 1,1,2-$C_2F_3Cl_3$, then 1 g of $SnCl_4$ was added, and 19.5 g (0.145 mol) of BrF_3 was added with stirring for 30 min at 25 to 35 °C. The reaction mixture was stirred for a further 30 min, whereupon it was treated with $NaHSO_3$ solution, and dried over $MgSO_4$. After removing the solvent, the mixture was distilled in vacuum to give 61 g (81%) of ester *74*: b.p. 62 to 63 °C (10 mm), d_4^{20} 1.6940, n_D^{20} 1.455.

A mixture of 49 g of zinc dust, 170 ml of water, 1 ml of H_2SO_4, and 0.05 g of N-nitroso-N-phenylhydroxylamine sodium salt as an inhibitor was heated to 100 °C, and 126 g of ester *74* was added with vigorous stirring. Simultaneously the azeotrope (b.p. 82 to 83 °C) was distilled off. The latter was washed with aq. $CaCl_2$ (30 ml), the organic layer was dried and distilled to give 53 g (82%) of ester *75*: b.p. 91 °C (750 mm), d_4^{20} 1.1140, n_D^{20} 1.3880.

5.8 References

1. Nikolaev NS, Sukhoverkhov VF, Shishkov YD, Alenchikova NF (1968) Khimiya galoidnykh soedinenii ftora. Nauka, Moscow
2. Christe KO (1974) In: 24th Intern. Congress Pure Appl. Chemistry, 4:115
3. Ishikawa N, Kobayashi E (1982) Ftor. Khimiya i primenenie. Russ. per. Mir, Moscow
4. Meinert H (1967) Z. Chem. 7:41
5. Sheppard WA, Sharts CM (1969) Organic fluorine chemistry. Benjamin, New York
6. Musgrave W (1960) Adv. Fluorine Chem. 1:1
7. Methoden der organische Chemie (Houben-Weyl) (1962) 4th edn, vol 5, part 3, Thieme, Stuttgart, p 72
8. Boguslavskaya LS (1972) Usp. Khimii 41:1591
9. Schack CJ, Christe KO (1978) Isr. J. Chem. 17:20
10. Boguslavskaya LS (1984) Usp. Khimii, 53:2024
11. Fletcher EA, Dahnek BE (1969) J. Amer. Chem. Soc. 91:1603
12. Petrov YP, Mukhametshin FM, Stepukhovich AD (1976) Kinetika i kataliz, 17:1347
13. Mukhametshin FM, Petrov YuP, Shumikhin AG (1975) In: 4th Vses. Symposium po khimii neorg. ftoridov, 29 Sept.–30 Oct. 1975. Dushanbe, USSR, p 134
14. Gambaretto GP (1973) Chim. Ind. 55:18
15. Schack CJ, Wilson RD (1971) Synth. Inorg. Metalorg. Chem. 3:393
16. US Pat 3446592 (1969); (1969) Chem. Abs. 71:23371
17. US Pat 3451775 (1969); (1969) Chem. Abs. 71:62669
18. Schack CJ, Wilson RD (1970) Inorg. Chem. 9:311
19. Naumann D, Lehman E (1975) J. Fluor. Chem. 5:307
20. Schmeisser M, Sartory P, Naumann D (1970) Chem. Ber. 72:109

21. Schmidt H, Meinert H (1959) Angew. Chem. 71:126
22. Schmidt H, Meinert H (1960) Angew. Chem. 72:109
23. Krasnov KS (ed) (1979) Molekulyarnye postoyannye neorganicheskikh soedinenii. Khimiya, Leningrad
24. King RC, Armstrong GT (1970) J. Res. Nat. Bur. Stand. 74A:769
25. Wool AA (1980) J. Fluor. Chem. 15:533
26. Frey RA, Redington RL, Aljibury ALK (1971) J. Chem. Phys. 54:344
27. Opalovskii AA (1973) Usp. Khimii 36:1673
28. Christe KO, Sawodny W (1969) Inorg. Chem. 8:212
29. Gillespie RJ (1974) In: 24th Intern. Congress Pure Appl. Chem. 4:143
30. Stein L (1973) In: 7th Intern. Symposium Fluorine Chem. 15–20 July 1973, Santa Cruz, California, p I-30
31. Olah GA, Nojima M, Kerekes I: Synthesis 1973:779
32. Mel'nikova NB, Yasman YB, Boguslavskaya LS, Kartashov VR (1980) Zh. Org. Khim. 16:2026
33. Hall LD, Manvill JF (1969) Can. J. Chem. 47:361
34. Hall LD, Manvill JF (1973) Can. J. Chem. 51:2902
35. Owen GR, Verheyden JPH, Moffatt JC (1976)·J. Org. Chem. 41:3010
36. Pattison FLM, Peters DAV, Dean FN (1965) Can. J. Chem. 43:1689
37. Boguslavskaya LS, Morozova TV, Voronin AP (1978) Zh. Org. Khim. 14:1442
38. Hamman S, Beguin CG (1979) J. Fluor. Chem. 13:163
39. Bose AK, Das G, Funke FT (1964) J. Org. Chem. 29:1202
40. Zupan M, Pollak A: J. Chem. Soc. Perkin Trans. I 1976:971
41. Weber FG, Giese H, Westphal G (1975) Z. Chem. 15:475
42. Bowers A, Denot F, Becerra R (1960) J. Amer. Chem. Soc. 82:4007
43. Gregorcic A, Zupan M (1984) J. Org. Chem. 49:333
44. Boguslavskaya LS, Mel'nikova NB, Voronin AP, Kartashov VR (1978) Zh. Org. Khim. 14:1401
45. Dean FH, Pattison FLM (1965) Can. J. Chem. 43:2415
46. Boguslavskaya LS, Voronin AP, Yarovykh KV, Sineokov AP (1975) Zh. Org. Khim. 11:257
47. Boguslavskaya LS, Chernov AN, Krom EN (1980) Zh. Org. Khim. 16:1640
48. Carig NC, Ewans DA, Piper LG, Wheeler VL (1970) J. Phys. Chem. 74:4520
49. Henderson GS, Gajar A (1971) J. Org. Chem. 36:3834
50. Wood KR, Kent PW, Fisher D: J. Chem. Soc. 1966:912
51. Kent PW, Barnett JEG: J. Chem. Soc. 1964:6196
52. Dean FH, Marshall DR, Warnhoff EW, Pattison FLM (1967) Can. J. Chem. 45:2279
53. Limasova TI, Rumyanzeva AT, Kollegova MI, Chernyakhovskaya SG, Berus EI, Barkhash VA (1971) Zh. Org. Khim. 7:751
54. Dear BEA (1970) J. Org. Chem. 35:1703
55. Hauptschein M, Braid M (1961) J. Amer. Chem. Soc. 83:2383
56. Chambers RD, Musgrave WR, Savari J: J. Chem. Soc. 1961:3779
57. Lo ES, Readio JD, Iserson H (1970) J. Org. Chem. 35:2051
58. Yuminov VS, Pushkina LN, Sokolov SV (1967) Zh. Org. Khim. 37:375
59. Stein L (1959) J. Amer. Chem. Soc. 81:1273
60. Rozen S, Brand MA (1980) Tetrahedron Lett. 21:4543
61. Brand MA, Rozen S (1982) J. Fluor. Chem. 20:419
62. Rozen S, Brand MA (1986) J. Org. Chem. 51:222
63. Zamir D, Brand MA, Rozen S (1987) J. Fluor. Chem. 35:62

64. Knunyants IL, German LS: Izv. Akad. Nauk SSSR. Ser. Khim. 1966:1065
65. Heasley VL, Gipe RK, Martin JL (1983) J. Org. Chem. 48:3195
66. Boguslavskaya LS, Brovkina GV, Yarovykh KV, Sineokov AP (1974) Zh. Org. Khim. 10:2067
67. Aronovich DA, Boguslavskaya LS, Trofimov NN, Voronin AP (1975) Zh. Org. Khim. 11:695
68. Boguslavskaya LS, Yarovykh KV, Sineokov AP, Etlis VS, Bulovyatova AB (1972) Zh. Org. Khim. 8:1139
69. Boguslavskaya LS, Yarovykh KV, Sineokov AP (1973) Zh. Org. Khim. 9:231
70. Chuvatkin NN, Morozova V, Boguslavskaya LS (1983) Zh. Org. Khim. 19:1107
71. Christe KO (1972) Inorg. Chem. 11:1120
72. Schack CJ (1967) Inorg. Chem. 6:1938
73. Young DE, Anderson LR, Gould DE, Fox WB (1970) J. Amer. Chem. Soc. 92:2313
74. Ratcliffe CT, Hardin CV, Anderson LR, Fox WB (1971) J. Amer. Chem. Soc. 93:3886
75. Avt. Svid. SSSR 789503 (1980); (1981) Chem. Abs. 94:120854
76. Katsuhara Y, Hammaker RM, DesMarteau DD (1980) Inorg. Chem. 19:607
77. Bougon R, Carles M, Aubert J (1967) C.R. Acad. Sci., Ser. C 265C:179
78. Moldavskii DD, Temchenko VG (1969) Zh. Org. Khim. 39:1393
79. Gambaretto GP, Napoli M (1976) J. Fluor. Chem. 7:569
80. Hynes JB, Austin TE (1966) Inorg. Chem. 5:488
81. Kirchmeer RL, Sprengler GH, Shreeve JM (1975) Inorg. Nucl. Chem. Lett. 11:699
82. Sehiya A, DesMarteau DD (1979) J. Amer. Chem. Soc. 101:7640
83. Clifford AF, Zailenga GB (1969) Inorg. Chem. 8:979
84. Schack CJ, Wilson RD, Muirhead JS, Cohz SN (1969) J. Amer. Chem. Soc. 91:2907
85. Dahms G, Didernich G, Haas A, Yazdanbakhch M (1974) Chem. Ztg. 98:109
86. Yu SL, Shreeve JM (1976) Inorg. Chem. 15:14
87. Sprenger GH, Wright KJ, Shreeve JM (1973) Inorg. Chem. 12:2890
88. Gould DE, Anderson LR, Young DE, Fox WB (1969) J. Amer. Chem. Soc. 91:1310
89. Schack CJ, Maya V (1969) J. Amer. Chem. Soc. 91:2902
90. Young DE, Anderson LR, Fox WB (1970) Inorg. Chem. 9:2602
91. Mukhametshin FM (1980) Usp. Khimii 49:1260
92. Young DE, Anderson LR, Fox WB: J. Chem. Soc. Chem. Commun. 1970:395
93. Gambaretto GP, Napoli M (1974) Gazz. Chim. Ital. 105:1291
94. Boguslavskaya LS, Chuvatkin NN, Panteleeva IY, Ternovskoi LA, Krom EN (1980) Zh. Org. Khim. 16:2525
95. Boguslavskaya LS, Chuvatkin NN, Panteleeva IY (1982) Zh. Org. Khim. 18:2082
96. Moldavskii DD, Temchenko VG, Slesareva VI, Antipenko GL (1973) Zh. Org. Khim. 9:673
97. Rokhlin EM, Abduganiev EG, Utebaev U (1976) Usp. Khimii 45:1177
98. Boguslavskaya LS, Chuvatkin NN, Panteleeva IY, Ternovskoi LA (1982) Zh. Org. Khim. 18:938
99. Morozova TV, Chuvatkin NN, Panteleeva IY, Boguslavskaya LS (1984) Zh. Org. Khim. 20:1379
100. Carey FA, Kuehne ME (1982) J. Org. Chem. 47:3811
101. Boguslavskaya LS (1986) Zh. Org. Khim. 22:1568
102. Chuvatkin NN, Panteleeva IY, Boguslavskaya LS (1982) Zh. Org. Khim. 18:946
103. De Marco RA, Kovacina TA, Fox WB (1975) J. Fluor. Chem. 6:93
104. Shaw GC, Seaton DL, Bissel ER (1961) J. Org. Chem. 26:4765
105. Peterman KE, Shreeve JM (1975) Inorg. Chem. 14:1223

106. Boguslavskaya LS, Panteleeva IY, Chuvatkin NN (1982) Zh. Org. Khim. 18:222
107. Shreeve JM (1973) Accounts Chem. Res. 6:387
108. Sauer DT, Shreeve JM(1971) An. Asoc. Quim. Argent. 59:157
109. Abe T, Shreeve JM (1973) Inorg. Nucl. Chem. Lett. 9:465
110. Yu SL, Shreeve JM (1975) J. Fluor. Chem. 6:259
111. Kitazume T, Shreeve JM: J. Chem. Soc. Chem. Commun. 1978:154
112. Abe T, Shreeve JM (1973) J. Fluor. Chem. 3:187
113. Sauer DT, Shreeve JM (1971) Z. anorg. allg. Chem. 385:113
114. Lau C, Passmore J (1976) J. Fluor. Chem. 7:261
115. Naumann D, Herberg S (1982) J. Fluor. Chem. 19:205
116. Rondesvelt CS (1969) J. Amer. Chem. Soc. 91:3054
117. Frohn HJ, Pahlmann W (1983) J. Fluor. Chem. 24:219
118. Frohn HJ (1984) Chem. Ztg. 108:146
119. Bardin VV, Furin GG, Yakobson GG (1980) Zh. Org. Khim. 16:1256
120. Frohn HJ, Giesen N (1984) J. Fluor. Chem. 24:9
121. Shuda M (1981) Kagaku to Kogyo 34:251
122. Nesmeyanov AN, Vanchikov AN, Lisichkina IN, Grushin VV, Tolstaya TP (1980) Doklady Akad. Nauk SSSR 255:1386
123. Obaleye JA, Rahbarnooli R, Sams LS (1982) J. Fluor. Chem. 21:52
124. Barbour AK, Belf LJ, Buxton MW (1963) Adv. Fluorine Chem. 3:181
125. Brit. Pat 878385 (1961); (1962) Chem. Abs. 57:11018
126. US Pat 3651155 (1972); (1972) Chem. Abs. 76:140046
127. Stevens TE (1961) J. Org. Chem. 26:1627
128. Boguslavskaya LS, Chuvatkin NN, Kartashov AV, Ternovskoi LA (1987) Zh. Org. Khim. 23:262
129. Bastock TW, Harley ME, Pedler AE, Tatlow JC (1975) J. Fluor. Chem. 6:331
130. Bastock TW, Pedler AS, Tatlow JC (1976) J. Fluor. Chem. 8:11
131. Bardin VV, Furin GG, Yakobson GG (1981) Zh. Org. Khim. 17:999
132. Bardin VV, Furin GG, Yakobson GG (1982) J. Fluor. Chem. 23:67
133. Yakobson GG, Bardin VV, Furin GG (1983) In: 3rd Regular Meeting of Soviet-Japanese Fluorine Chemists, 7–8 May 1983, Tokyo, p 141
134. Avt. svid. SSSR 172816 (1965); (1966) Chem. Abs. 64:716
135. Davis RA, Larsen ER (1967) J. Org. Chem. 32:3478
136. Baklan VF, Antipenkova LS, Zakharchenko LI, Fesenko TE, Kukhar' VA (1984) Zh. Org. Khim. 20:1212
137. Chuvatkin NN, Kartashova AV, Morozova TV, Boguslavskaya LS (1987) Zh. Org. Khim. 23:269
138. Boguslavskaya LS, Chuvatkin NN, Morozova TV, Panteleeva IY, Kartashov AV, Sineokov AP (1987) Zh. Org. Khim. 23:1173
139. Chuvatkin NN, Kartashov AV, Morozova TV, Boguslavskaya LS (1986) In: 5th Vses. Conf. po Khimii ftororgan. soedinenii, 20–22 May 1986, Moscow, USSR, p 92
140. Brower KR (1987) J. Org. Chem. 52:798
141. Fokin AV, Studnev YN, Rapkin AI, Tatarinov AS: Izv. Akad. Nauk SSSR, Ser. Khim. 1984:1916
142. Fokin AV, Studnev YN, Kuznetsova LD, Krotovich IN (1982) Usp. Khimii 51:1258
143. Katsuhara Y, DesMarteau DD (1979) J. Amer. Chem. Soc. 101:1039
144. Katsuhara Y, DesMarteau DD (1980) J. Org. Chem. 45:2441
145. Schack CJ, Pilipovich D, Hon JF (1973) Inorg. Chem. 12:897
146. Fink W (1961) Angew. Chem. 73:466

147. Zefirov NS, Kozmin AS, Sorokin VD (1984) J. Org. Chem. 49:4086
148. Zefirov NS, Kozmin AS, Sorokin VD, Zhdankin VV (1986) Zh. Org. Khim. 22:898
149. Beringer FM, Schultz HS (1955) J. Amer. Chem. Soc. 77:5533
150. Schmid GH, Gordon JW (1983) J. Org. Chem. 48:4010
151. Svetlakov NV, Moisak IE, Averko-Antonovich IG (1969) Zh. Org. Khim. 5:2105
152. Svetlakov NV, Moisak IE, Shafigulin NK (1971) Zh. Org. Khim. 7:1097
153. MacDonald TL, Narasimhan N, Burka LT (1980) J. Amer. Chem. Soc. 102:7760
154. Davidson RI, Kropp PJ (1982) J. Org. Chem. 47:1904
155. Schack CJ, Christe KO (1980) J. Fluor. Chem. 16:63
156. Fokin AV, Studnev YN, Rapkin AI, Tatarinov AS, Seryanov YV: Izv. Akad. Nauk SSSR. Ser. Khim. 1985:1635
157. Fokin AV, Studnev YN, Rapkin AI, Chelikin VG, Verenikin OV: Izv. Akad. Nauk SSSR. Ser. Khim. 1985:659
158. Katsuhara Y, DesMarteau DD (1980) J. Amer. Chem. Soc. 102:2681
159. Schack CJ, Pilipovich D, Christe KO (1975) Inorg. Chem. 14:145
160. Bach RD, Holubka JW, Taaffee TH (1979) J. Org. Chem. 44:35
161. Brit. Pat 805764 (1953); (1953) Chem. Abs. 53:10035
162. Fokin AV, Rapkin AI, Seryanov YV, Studnev YN: Izv. Akad. Nauk SSSR. Ser. Khim. 1986:2734
163. Arapov OV, Rudenko AP, Zarubin MY (1985) Zh. Org. Khim. 21:168
164. Fokin AV, Semin TK, Raevskii AM, Gyshchin SI, Rapkin AI, Tatarinov AS, Studnev YN: Izv. Akad. Nauk SSSR. Ser. Khim. 1986:244
165. Zefirov NS, Makhonkov DI (1982) Chem. Rev. 82:615
166. Barter RM, Littler JS: J. Chem. Soc. (B) 1967:205

6 New Uses of Sulfur Tetrafluoride in Organic Synthesis

Anatolii Ivanovich Burmakov, Boris Vasil'evich Kunshenko, Lyubov' Antonovna Alekseeva and Lev Moiseevich Yagupolskii

Polytechnical Institute, Shevchenko St. 1, 270044, Odessa, USSR and Institute of Organic Chemistry, Murmanskaya St. 5, 252660, Kiev, USSR

Contents

6.1   Introduction

Since the publication in 1959–1960 of the pioneering works [1,2] which showed that by using sulfur tetrafluoride it is possible to effect the selective substitution of carbonyl oxygen in aldehydes, ketones, carboxylic acids, and of the hydroxy group in alcohols by fluorine, interest in this fluorinating agent has been constantly growing. Similar reactions with sulfur tetrafluoride have been reported for the compounds in which oxygen is double-bonded with phosphorus, arsenic [3,4], germanium [5], antimony [6], and iodine [7,8], as well as for dicarbonyl compounds [9]. The reactions of SF_4 with compounds containing double and triple carbon–nitrogen bonds gave the representatives of a new class of compounds–iminosulfur difluoride derivatives [10, 11].

Several reviews on the reactions of organic compounds with SF_4 have been published [12–15], but they cover literature only until 1973. Recently a lot of new data considerably expanding our knowledge on the synthetic possibilities of SF_4 have appeared. This was also stimulated by the development of convenient methods for its synthesis [16–22].

Dialkylaminosulfur trifluorides formed in the reaction of SF_4 with silylated amines were found to be the SF_4 analogues capable of substituting the hydroxy groups and carbonyl oxygen by fluorine [23,24].

The reactions of SF_4 with polyfunctional organic compounds have been studied, and methods have been developed for the synthesis of aliphatic, aromatic, and heterocyclic fluorine-containing alcohols, ketones, carboxylic acids, etc.

The efficiency of SF_4 as an agent for the substitution of oxygen by fluorine was found to sharply increase when the reactions of oxygen-containing organic compounds with SF_4 are carried out in anhydrous hydrogen fluoride.

More and more data are appearing indicating that in a number of reactions SF_4 may have an oxidating activity. The use of fluorinating systems on the basis of SF_4, such as SF_4–HF–$Cl_2(Br_2)$ and SF_4–HF–S_2Cl_2, allows substitution of hydrogen by fluorine, and halofluorination of various classes of organic compounds.

In the present review, an attempt has been made to summarize the literature data referring chiefly to recent years, on the new uses of sulfur tetrafluoride in organic synthesis.

6.2   Reactions of Organic Compounds with Sulfur Tetrafluoride in Anhydrous Hydrogen Fluoride

Sulfur tetrafluoride is widely used to introduce fluorine into organic compounds. However substitution of carbonyl oxygen by fluorine using sulfur tetrafluoride in many cases fails to yields good results. A very important method of transforming the aromatic carboxylic acids to trifluoromethyl-containing compounds possesses essential limitations. Thus, for example, the yields of benzotrifluoride and its derivatives from benzoic acid and its analogues having electron-donating substituents (CH_3, OCH_3) do not exceed 8–15% [25]. Even under severe conditions (at 300 °C and above), polycarboxylic acids fail to give vicinal poly(trifluoromethyl)benzenes [26]. The yields of fluorinated products are also low in the reactions of SF_4 with carboxylic acids and ketones containing several strong electron-accepting groups in the molecule.

An important achievement in sulfur tetrafluoride chemistry, substantially expanding its preparative utility, was the modification of its reactions by adding significant amounts (10 to 30 mol per mol of the carbonyl compound) of anhydrous HF. This raised the yields of fluorinated products, and in many cases allowed one to alter the reaction route.

6.2.1 Aromatic Carboxylic Acids

The use of excess hydrogen fluoride in the reactions of aromatic carboxylic acids with SF_4 raises the yields of trifluoromethyl derivatives steeply both with electron-donating and electron-accepting substituents, and enables the reactions to take place under much milder conditions (Table 1).

Even in very severe conditions (300 °C and more), the reactions of SF_4 with sterically hindered aromatic and heterocyclic polycarboxylic acids give only poly(trifluoromethyl)benzoyl fluorides [26,31], while in HF they afford the respective poly(trifluoromethyl)benzenes and poly(trifluoromethyl)furanes [32,36] (Table 2).

The sterically hindered benzene-, naphthalene-, and furane polycarboxylic acids with two carboxyl groups screened on both sides are transformed by SF_4 to cyclic compounds containing the CF_2–O–CF_2 fragment. The reaction proceeds via anhydride formation.

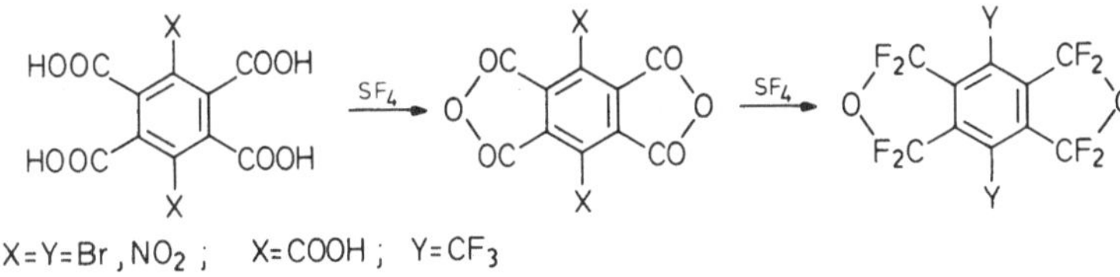

$$X = Y = Br, NO_2 ; \quad X = COOH ; \quad Y = CF_3$$

Substitution of carbonyl oxygen by fluorine in anhydrides proceeds with difficulty. The use of HF in these reactions allows one to obtain fluorinated products in high yields (Table 3).

6.2.2 Esters of Carboxylic Acids

6.2.2.1 Reactions Forming Trifluoromethyl Derivatives

The reactions of esters of carboxylic acids with SF_4 give trifluoromethyl derivatives and alkyl fluorides. These reactions require much more severe

Table 1. Reactions of benzene- and naphthalenecarboxylic acids with SF_4

Acid	Products	Without HF				In HF (20–30 mol per mol of acid) [27]		
		T(°C)	Time (h)	Yield (%)	Ref.	T(°C)	Time (h)	Yield (%)
Benzoic	$C_6H_5CF_3$	120	6	22	2	100	10	93
Benzoic	$C_6H_5CF_3$	160	6[a]	16	25	—	—	—
4-Methylbenzoic	$4\text{-}CH_3C_6H_4CF_3$	160	6[a]	12	25	120–140	10	80
4-Methoxybenzoic	$4\text{-}CH_3OC_6H_4CF_3$	160	6[a]	8	25	120–140	10	55
4-Chlorobenzoic	$4\text{-}ClC_6H_4CF_3$	160	6[a]	24	25	90–100	10	70
1-Naphthoic	$1\text{-}CF_3C_{10}H_7$	160 to 250	10[a]	43	28	50	10	60
4-Nitrobenzoic	$4\text{-}NO_2C_6H_4CF_3$	169	6[a]	66	2	80	10	92
4-Methylnaphthoic-1	$4\text{-}CH_3C_{10}H_6CF_3\text{-}1$	80	8	0[b]	29	75	10	70
4-Nitronaphthoic-1	$4\text{-}NO_2C_{10}H_6CF_3\text{-}1$	160 to 230	20	53	29	40	10	83
Naphthalene-1,4-dicarboxylic	$1,4\text{-}(CF_3)_2C_{10}H_6$	200	12	55	29	50	10	78
Naphthalene-2,6-dicarboxylic	$2,6\text{-}(CF_3)_2C_{10}H_6$	190–200	12	68	30	80	10	90

[a] In benzene
[b] $4\text{-}CH_3C_{10}H_6COF\text{-}1$ (96%)

Table 2. Reactions of benzene- and furanepolycarboxylic acids with SF$_4$ in HF

Acid	Ratio of acid:SF$_4$:HF (mol)	T(°C)	Time (h)	Products	Yield (%)	Ref.
1,2,3-Benzenetricarboxylic	1:15:100	320	10	1,2,3-Tris(trifluoromethyl)benzene	71	32
1-Trifluoromethyl-2,6-benzenedicarboxylic	1:7:0	280	18	1,2,3-Tris(trifluoromethyl)benzene	25	32
1-Trifluoromethyl-2,6-benzenedicarboxylic	1:7:20	300	12	1,2,3-Tris(trifluoromethyl)benzene	92	32
1,2,3,5-Benzenetetracarboxylic	1:20:50	310	10	1,2,3,5-Tetrakis(trifluoromethyl)-benzene	37	32
1-Trifluoromethyl-2,4,6-benzenetricarboxylic	1:20:100	310	10	1,2,3,5-Tetrakis(trifluoromethyl)-benzene	55	32
5-Nitrobenzene-1,2,3-benzenetricarboxylic	1:10:20	250	50	1,2,3-Tris(trifluoromethyl)-5-nitrobenzene	42	33
2,3,4-Tris(trifluoromethyl)benzoic	1:20:0	320	10	1,2,3,4-Tetrakis(trifluorooethyl)-benzene	8	32
2,3,4-Tris(trifluoromethyl)benzoic	1:20:100	320	10	1,2,3,4-Tetrakis(trifluoromethyl)-benzene	81	32
1-Trifluoromethyl-2,3,5,6-benzenetetracarboxylic	1:30:50	300	100	1,2,3,4,5-Pentakis(trifluoromethyl)-benzene	53	34
2,3,4-Furanetricarboxylic	1:10:20	150	10	2,3,4-Tris(trifluoromethyl)furane	76	35
3,4,5-Tris(trifluoromethyl)-2-furanecarboxylic	1:4:20	150	10	2,3,4,5-Tetrakis(trifluoromethyl)-furane	74	36

Table 3. Reactions of benzene-, naphthalene-, and furanepolycarboxylic acids containing two sterically hindered carboxylic groups with SF_4 in HF

Acid	Ratio of acid : SF_4 : HF (mol)	T(°C)	Time (h)	Products	Yield (%)	Ref.
1,2,3,4-Benzenetetracarboxylic	1:20:0	200	14	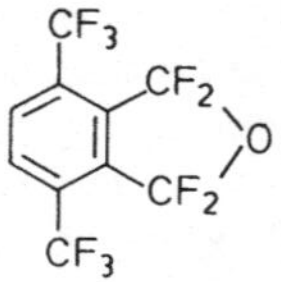	76	31
3,6-Bis(trifluoromethyl)phthalic	1:10:0	20	12	3,6-Bis(trifluoromethyl)phthalic anhydride	92	37
3,6-Bis(trifluoromethyl)phthalic	1:10:0	200	18	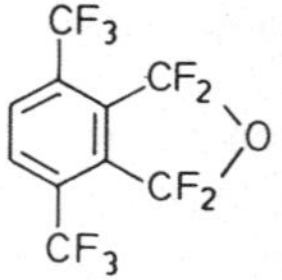	74	37
3,6-Dibromobenzenetetracarboxylic	1:10:0	20	12	3,6-Dibromo-1,2:4,5-benzene-tetracarboxylic anhydride	92	37
3,6-Dibromobenzenetetracarboxylic	1:10:0	240–250	30	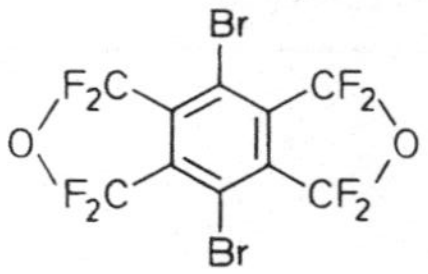	84	37

Acid	Ratio	Temp.	Time (h)	Product	Yield	Ref.
3,6-Dinitrobenzenetetracarboxylic	1:20:50	160	15		41	33
1,8-Naphthalenedicarboxylic	1:10:0	20	12	1,8-Naphthalenedicarboxylic anhydride	97	37
1,8-Naphthalenedicarboxylic	1:10:0	220	10		63	37
4,5-Dinitronaphthalene-1,8-di-carboxylic	1:6:0	260	12	4,5-Dinitronaphthalene-1,8-di-carboxylic anhydride	100	29
4,5-Dinitronaphthalene-1,8-di-carboxylic	1:6:20	180	10		80	29
3,6-Dinitronaphthalene-1,8-di-carboxylic	1:6:0	240	10		35	29

Table 3. (*Continued*)

Acid	Ratio of acid:SF_4:HF (mol)	T(°C)	Time (h)	Products	Yield (%)	Ref.
3,6-Dinitronaphthalene-1,8-dicarboxylic	1:6:20	150	10	same	100	29
1,4,5,8-Naphthalenetetracarboxylic	1:6:0	260	12	1,8:4,5-Naphthalenetetra-carboxylic anhydride	100	29
1,4,5,8-Naphthalenetetracarboxylic	1:6:20	240	10	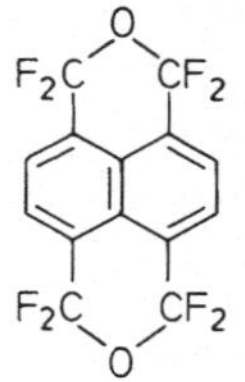 (naphthalene bis-dioxolane structure with CF_2, F_2C, O groups)	80	29
Furanetetracarboxylic	1:10:10	190	25	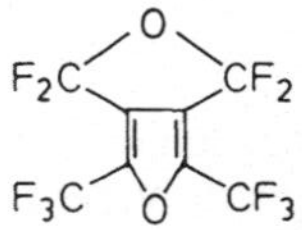 (furan structure with CF_2, CF_3, F_2C, F_3C, O groups)	54	35

conditions than those of carboxylic acids with SF_4. For example, methyl benzoate being heated with SF_4 to 300 °C forms benzotrifluoride with only a 15% yield [2]. Using an excess of anhydrous HF in the reaction of sulfur tetrafluoride with esters of carboxylic acids allows one to decrease the reaction temperature. Thus methyl benzoate, methyl 4-chlorobenzoate, and methyl 4-nitrobenzoate react with SF_4 in the presence of HF to give the respective derivatives of benzotrifluoride at 75 to 80 °C [38].

$$4\text{-X--}C_6H_4COOMe + SF_4 \xrightarrow{\text{HF}} 4\text{-X--}C_6H_4CF_3$$

$$X = H \ (45\%), \ Cl \ (70\%), \ NO_2 \ (20\%)$$

In the reaction of alkyl-substituted malonates with SF_4 in HF under mild conditions, the transformation of one of the ethoxycarbonyl groups to trifluoromethyl occurs [39].

$$EtOCOCR(Bu)COOEt + SF_4 \xrightarrow[25\,°C]{\text{HF}} EtOCOCR(Bu)CF_3$$

$$R = H, \ Me$$

6.2.2.2 Reactions Forming α, α-Difluoroethers

The exchange of the carbonyl oxygen of an ester group for two fluorine atoms first effected by Sheppard and his co-workers [40–42] was used for the synthesis of aryl perfluoroalkyl ethers.

$$ROCOF + SF_4 \xrightarrow[100 \text{ to } 175\,°C]{\text{HF}} ROCF_3$$

$$ArOCOX + SF_4 \xrightarrow[170 \text{ to } 220\,°C]{\text{HF}} ArOCF_2X$$

$$X = F, \ CF_3, \ C_2F_5$$

The reaction is carried out in the presence of small amounts of HF (0.2–1 mol per mol of the starting ester) and gives satisfactory results provided that there are electron-accepting substituents. Aryl trifluoroacetates containing electron-donating substituents fail to give the aryl pentafluoroethyl ethers, leading only to tar products [43]. On the other hand, accumulation of strong electron-accepting groups in the aromatic ring leads to decreased yields of pentafluoroethoxy-benzenes. Thus 2,4-dinitrophenyl pentafluoroethyl ether is obtained only with a 19% yield. The use of greater amounts of HF (up to 10 mol per mol) in the reaction of aryl trifluoroacetates with SF_4 allows these reactions to take place at 20 °C instead of 175 to 220 °C. In this case the yields of aryl pentafluoroethyl ethers increase by a factor of 2 or 3 [43]. Table 4 gives the comparative data on the effects of substituents and excess of anhydrous HF in the reactions of SF_4 with phenyl trifluoroacetates.

Table 4. Pentafluoroethoxybenzenes $XC_6H_4OC_2F_5$ formed in the reaction of $XC_6H_4OCOCF_3$ with SF_4 [43]

X	Method	Yield (%)	X	Method	Yield (%)
H	A	61.0	4-CH$_3$	A	0
2-Cl	A	35.6	4-CH$_3$	B	45.6
2-Cl	B	71.4	2-CH$_3$O	A	0
3-Cl	A	45.7	2-CH$_3$O	B	35.6
4-Cl	A	46.3	2-NO$_2$	A	29.2
4-Cl	B	73.8	3-NO$_2$	A	69.0
4-F	A	36.7	4-NO$_2$	A	82.0
4-F	B	68.6	2,4-(NO$_2$)$_2$	A	19.3
2-CH$_3$	A	0	2,4-(NO$_2$)$_2$	B	55.8
2-CH$_3$	B	39.5	2-NO$_2$-4-Me	A	25.4
3-CH$_3$	A	0			
3-CH$_3$	B	62.5			

Method A. The reagents (ester: SF_4 : HF $= 1:2:1$, mol) are heated to 100 to 175 °C for 6 h.

Method B. The reagents (ester: SF_4 : HF $= 1:2:10$, mol) are kept at 20 °C for 15 h.

Using an excess of HF in the reactions of aryl trifluoroacetates with SF_4 allows easy synthesis of vicinal poly(pentafluoroethoxy)benzenes from sterically hindered trifluoroacetates of polyatomic phenols. Thus 1,2,3-tris(trifluoroacetoxy)benzene heated with SF_4 at 150 °C in the presence of small additions of HF (mol per mol of ester) forms chiefly 2,6-bis(pentafluoroethoxy)phenyl trifluoroacetate, the yield of 1,2,3-tris(pentafluoroethoxy)benzene being only 11% [44]. In an excess of HF (10 mol per mol of ester), the latter is formed with a 60% yield already at 100 °C [45]. In the same manner, SF_4 reacts with 1,2,3,4-tetrakis(trifluoroacetoxy)benzene [45].

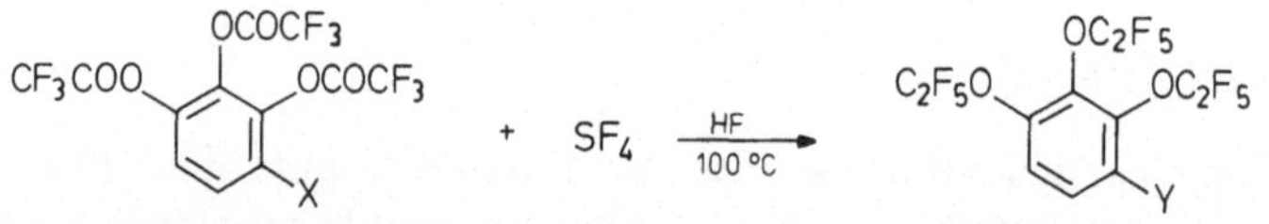

X=Y=H (60%); X=OCOCF$_3$, Y=OC$_2$F$_5$ (79%)

Likewise, 2,3,6-tris(pentafluoroethoxy)-4-nitrophenyl trifluoroacetate gives 2,3,5,6-tetrakis(pentafluoroethoxy)-4-nitrobenzene with a 66% yield, and 3,4,5,6-tetrakis(pentafluoroethoxy)phenylene 1,2-bis(trifluoroacetate) forms hexakis(pentafluoroethoxy)benzene.

Sulfur tetrafluoride in HF is widely used to prepare various perfluoroalkyl ethers, especially those containing strong electron-accepting substituents [47].

$$(CF_3)_3COCOC(CF_3)_3 + SF_4 \xrightarrow[250\,°C]{HF} (CF_3)_3COCF_2OC(CF_3)_3$$

Patent [48] describes difluoroformals containing nitro groups instead of perfluoroalkyl ones.

$$(CF(NO_2)_2CH_2O)_2C=O + SF_4 \xrightarrow{HF} (CF(NO_2)_2CH_2O)_2CF_2$$

In the reaction of SF_4 with esters of fluorine-containing glycols and perfluorodicarboxylic acids, the carbonyl oxygen of ester groups is substituted by fluorine atoms, and free carbonyl groups are only transformed to fluoroformyl ones [63].

$$(HOCO(CF_2)_3COO)_2R_F + SF_4 \xrightarrow[150\,°C]{HF} (FCO(CF_2)_3CF_2O)_2R_F$$

The reactions of poly- and perfluoroacyloxy- derivatives of methyl benzoate with SF_4 in HF at 75 to 80 °C also produce polyfluoro- or perfluoroalkyl ethers, with the methoxycarbonyl group transformed to the trifluoromethyl one [49].

$$R_FCOOC_6H_4COOMe + SF_4 \xrightarrow{HF} R_FCF_2OC_6H_4CF_3$$

$$R_F = C_3F_7,\ H(CF_2)_4,\ C_4F_9,\ H(CF_2)_6,\ C_6F_{13}$$

6.2.3 Ketones

With ketones, as with carboxylic acids, the use of large amounts of HF in the reactions of SF_4 yields good results.

In the absence of HF, acetone reacts with SF_4 at 110 °C to form 2,2-difluoropropane with a 60% yield [2]. When this reaction is carried out in HF at 20 °C (6 to 8 mol of HF per mol of ketone), 2,2-difluoropropane is obtained in a quantitative yield. In a similar way SF_4 reacts with ethyl methyl ketone and diethyl ketone [50].

$$RCOR' + SF_4 \xrightarrow[20\,°C,\ 10\,h]{HF} RCF_2R'$$
$$98–100\%$$

$$R = R' = Me;\ R = R' = Et;\ R = Me,\ R' = Et$$

Accumulation of electron-accepting substituents in the ketone molecule deactivates the carbonyl group in the reaction with SF_4, and may be transformed to the difluoromethylene one only in rigid conditions in HF.

In this way perfluoroanthraquinone was converted to perfluoro-9,10-dihydroanthracene [51].

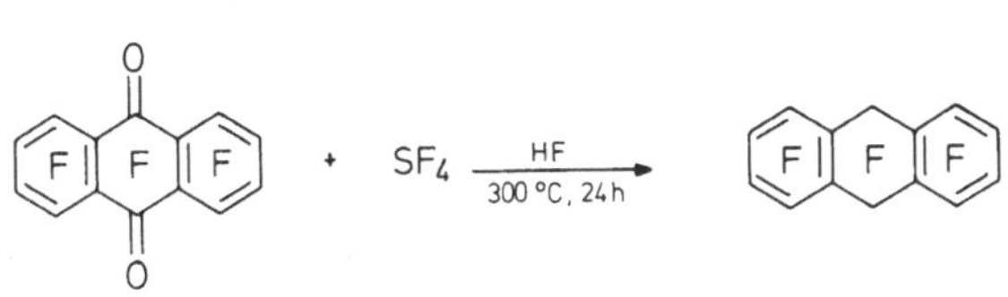

The reaction of SF_4 with 5,6-dinitroacenaphthenequinone at 290 °C leads to 5,6-dinitro-1,1,2,2-tetrafluoroacenaphthene [52] with a 70% yield. The same reaction carried out in HF gave tetrafluoroacenaphthene with a quantitative yield already at 150 °C [29].

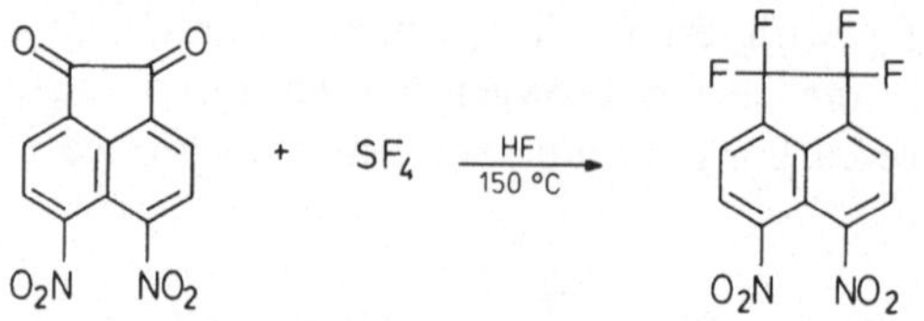

To substitute the carbonyl oxygen by two fluorine atoms in 4-bromotetra-fluorophenyl oxaperfluoroalkyl ketones, these compounds must be heated with SF_4 and HF to 180 °C [53].

$$4\text{-}BrC_6F_4COR_FOR_F + SF_4 \xrightarrow[180\,°C,\,18\,h]{HF} 4\text{-}BrC_6F_4CF_2R_FOR_F$$

$$R_FO\text{-}R_F = C_3F_7O[CF(CF_3)CF_2O]_nCFCF_3,\ C_2F_5O(CF_2CF_2O)_nCF\text{-},$$
$$CF_3O(CF_2O)_nCF_2\text{-},\ C_3F_7O(CF_3)CF\text{-}$$

An excess of HF should also be used in the reactions of sterically hindered ketones with SF_4. Thus if cyclobutanone [54] is transformed by SF_4 to difluorocyclobutane already at 20 °C, and cyclopentanone is easily converted to difluorocyclopentane [55], 1,1,3,3-tetramethyl-2,2,4,4-tetrafluorocyclobutane may only be obtained from the respective diketone on heating with SF_4 and HF to 160 °C [2].

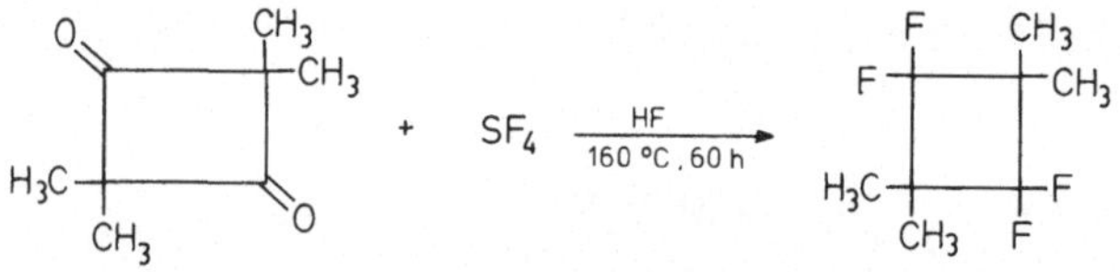

The HF addition is also favourable for the reactions of some ketosteroids with SF_4. The yields of difluorosteroids, though not very high, are still much higher in this case than in the reactions catalysed by even such efficient catalyst as boron trifluoride [56–58].

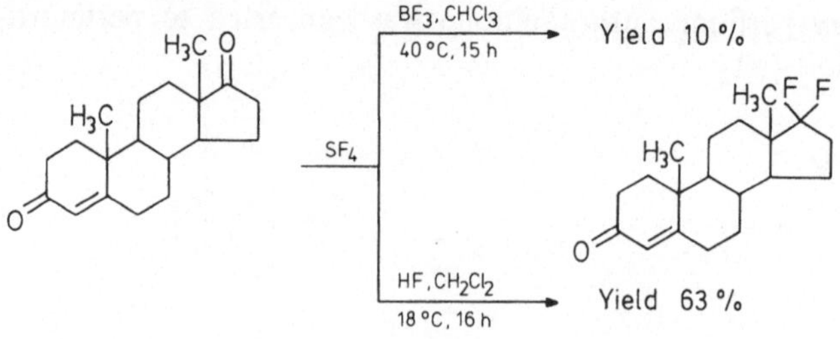

The examples of organic reactions of SF_4 in anhydrous HF considered in this Section allow to regard this system as a new fluorinating agent for introduction of fluorine atoms into organic molecules.

Sulfur tetrafluoride in anhydrous HF forms an extremely reactive ion F_3S^+ [59], which adds to the carbonyl group and facilitates further attack of the resulting cation by the fluoride ion.

$$\begin{array}{c} \diagdown \\ \diagup \end{array}C=O \quad + \quad F_3S^+ \quad \longrightarrow \quad \begin{array}{c} \diagdown \\ \diagup \end{array}\overset{+}{\underset{F^-}{C}}-OSF_3 \quad \underset{-SOF_2}{\longrightarrow} \quad \begin{array}{c} \diagdown \\ \diagup \end{array}CF_2$$

The reactions involving hydrogen fluoride possibly proceed via the initial addition of HF at the carbonyl group [47].

$$\begin{array}{c} \diagdown \\ \diagup \end{array}C=O \quad + \quad HF \quad \rightleftharpoons \quad \underset{F}{\overset{|}{-C}}-OH$$

$$A$$

Fluorocarbinol A further reacts with SF_4 or F_3S^+. Hydrogen fluoride present in the liquid phase promotes a shift of the equilibrium towards formation of carbinol A and acts as a solvent.

Combination of these factors results in the fact that the reactions of carbonyl compounds with SF_4 in HF proceed under much milder conditions and with greater yields of fluorinated products.

6.3 Use of Sulfur Tetrafluoride for the Synthesis of Fluorine-Containing Organic Compounds with Functional Groups

The previous Section showed sulfur tetrafluoride, especially in anhydrous HF, to be an efficient agent for substitution of oxygen in the hydroxy- and carbonyl compounds by fluorine.

At the same time SF_4 may be effectively used to obtain fluorine-containing alcohols, ketones, and carboxylic acids. For that purpose, the reactions of SF_4 with polyfunctional compounds are used. The selective substitution of oxygen by fluorine in one or several functional groups may be carried out using the appropriate reaction conditions, reagent ratio, steric factors, reactivity differences, different additions and solvents, and the distinctions of the reactions of polyfunctional organic compounds with SF_4.

6.3.1 Effect of Reagent Ratio and Conditions in the Reactions of Polyfunctional Oxygen-Containing Compounds With SF_4

The influence of these factors in the reactions of SF_4 with dicarboxylic acids may be used for the synthesis of various carboxylic acids containing trifluoromethyl groups. This is especially dramatically manifested in the aromatic series (Table 5).

The reactions of SF_4 with aliphatic dicarboxylic acids containing 2 or 3 carbons between the carboxylic groups, have their own distinctions (Table 6). Along with hexafluoroalkanes, the reaction gives no trifluoroalkanoyl fluorides as in the case of adipic and sebacic acids, but the isomeric or cyclic derivatives of 2,2,5,5-tetrafluorotetrahydrofurane and 2,2,6,6-tetrafluorotetrahydropyrane in a 30–70% yield. Besides, small amounts of linear $\alpha,\alpha,\alpha',\alpha'$-tetrafluoroalkyl ethers are formed in this reaction (Table 6) [61].

Cyclic tetrafluoroethers $(\underset{\underset{\displaystyle O}{\boxed{}}}{CF_2(A)CF_2})$ may be used for the synthesis of trifluoroalkanoic acids, since upon heating with anhydrous HF they are transformed to trifluoroalkanoyl fluoride isomers [61].

$$HOOC(A)COOH + SF_4 \rightarrow CF_3(A)CF_3 + \underset{\underset{\displaystyle O}{\boxed{}}}{CF_2(A)CF_2} + [CF_3(A)CF_2]_2O$$

$$\underset{\underset{\displaystyle O}{\boxed{}}}{CF_2(A)CF_2} \underset{-H^+}{\overset{H^+}{\rightleftharpoons}} [^+CF_2(A)CF_2OH] \overset{F^-}{\underset{-HF}{\longrightarrow}} CF_3(A)COF$$

Thus 2,2,5,5-tetrafluorotetrahydrofurane is converted to 3,3,3-trifluorobutyryl fluoride, and 2,2,6,6-tetrafluorotetrahydropyrane—to 4,4,4-trifluoropentanoyl fluoride [61].

$$\underset{\underset{\displaystyle O}{\boxed{}}}{CF_2(CH_2)_nCF_2} + HF \xrightarrow[100 \text{ to } 170\,°C]{} CF_3(CH_2)_nCOF$$

$$n = 2(52\%),\ 3(96\%)$$

The reactions of polyketones with SF_4 proceed stepwise. Therefore even in an excess of SF_4, 1,3-dibenzoylperfluoropropane may be converted to the octafluoro- and decafluoro-derivative [2].

$$PhCO(CF_2)_3COPh + 3SF_4 \xrightarrow[150 \text{ to } 220\,°C]{} \begin{cases} \xrightarrow[20\ h]{} PhCO(CF_2)_4Ph \\ \\ \xrightarrow[51\ h]{} Ph(CF_2)_5Ph \end{cases}$$

The reaction of SF_4 with acenaphthenequinone gives, depending on the amount of SF_4 used, 1,1-difluoroacenaphthenone or 1,1,2,2-tetrafluoroacenaphthene as the major reaction products [62].

Table 5. Reactions of aromatic dicarboxylic acids with SF_4

Acid HOOC(A)COOH	Ratio of SF_4:acid (mol)	T(°C)	Time (h)	Yield (%)			Ref.
				$CF_3(A)CF_3$	$CF_3(A)COF$	FOC(A)COF	
1,2-Benzenedicarboxylic	3:1	20	6	—	—	98	37
1,2-Benzenedicarboxylic	5:1	120	6	43	23	—	2
1,4-Benzenedicarboxylic	3:1	150	8	—	74	—	25
1,4-Benzenedicarboxylic	5:1	120	6	76	3	—	2
1-Trifluoromethylbenzene-2,6-dicarboxylic	6:1	250	30	17	60	—	32
Naphthalene-1,2-dicarboxylic	2:1	100	6	—	—	95	30
Naphthalene-1,2-dicarboxylic	3:1	100	6	—	90	—	30
Naphthalene-2,3-dicarboxylic	2:1	100	6	—	—	93	30
Naphthalene-2,3-dicarboxylic	3:1	100	6	—	92	—	30
Naphthalene-2,3-dicarboxylic	5:1	200	12	42	—	—	30
Naphthalene-2,6-dicarboxylic	2:1	100	6	—	—	98	30
Naphthalene-2,6-dicarboxylic	3:1	100	10	—	95	—	30
Furan-3,4-dicarboxylic	4:1[a]	100–115	12	70	—	—	60
Furan-3,4-dicarboxylic	2.8:1[b]	100	10	—	79	—	60

[a] With 2 moles of HF
[b] With 10 moles of HF

Table 6. Reactions of alkanedicarboxylic acids with SF_4

Acid HOOC(A)COOH	T(°C)	Time (h)	Yield (mol. %)			Ref.
			$CF_3(A)CF_3$	$CF_2(A)CF_2$ $\overset{\displaystyle}{\underset{O}{\rule{1cm}{0.4pt}}}$	$(CF_3(A)CF_2)_2O$	
HOOCCH$_2$CH$_2$COOH	60	3	11	35	12	61
HOOC(CH$_2$)$_3$COOH	5	48	34	10	16	61
HOOC(CH$_2$)$_4$COOH	150	8	19	—	—	2
HOOC(CH$_2$)$_8$COOH	120	6	27	—	—	2
HOOC(CH$_2$)$_8$COOH	120	6	87	—	6	2
HOOCCH$_2$CHClCOOH	55	3	15	24	6	61
HOOCCHClCHClCOOH	200	6	20	52	—	61
HOOCCHBrCHBrCOOH	180	6	30	37	—	61

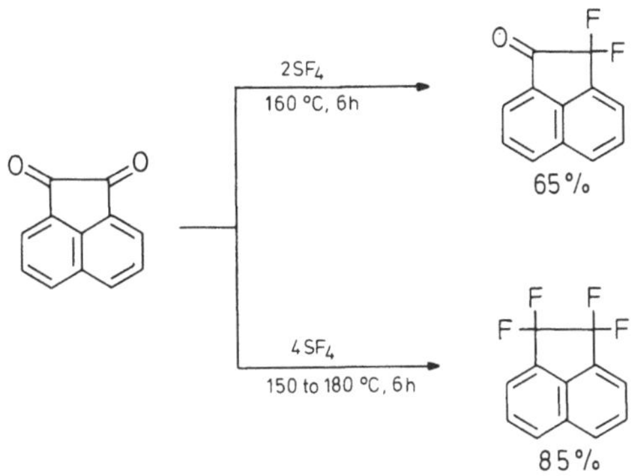

2- and 4-pentafluoroethoxyphenols were obtained by the reaction of tri-fluoroacetates with the calculated amount of SF_4, with subsequent hydrolysis of the unchanged trifluoroacetoxy group [46].

$$C_6H_4(OCOCF_3)_2 + SF_4 \xrightarrow[100\,°C,\ 2\,h]{0.8\ HF} C_6H_4(OC_2F_5)OCOCF_3$$

$$35\%$$

$$\longrightarrow C_6H_4(OC_2F_5)OH$$

6.3.2 Steric Effects in the Reaction of SF_4 with Polyfunctional Oxygen-Containing Organic Compounds

The steric effect on the route of the reactions of polyfunctional compounds with sulfur tetrafluoride may be most effectively used to obtain the trifluoromethyla-ted aromatic carboxylic and dicarboxylic acids.

The reactions of SF_4 with vicinal benzenepolycarboxylic acids with sterically hindered carboxylic groups lead, even in rigid conditions, to trifluoromethyl- or poly(trifluoromethyl)benzoyl and -phthaloyl fluorides and difluorides in high-yield (Table 7). For the sake of comparison it should be noted that 4-nitro-phthalic acid reacts with SF_4 in the same conditions as the 3-nitro-isomer, giving the product of substitution of both carboxyl groups by the trifluoromethyl ones [26].

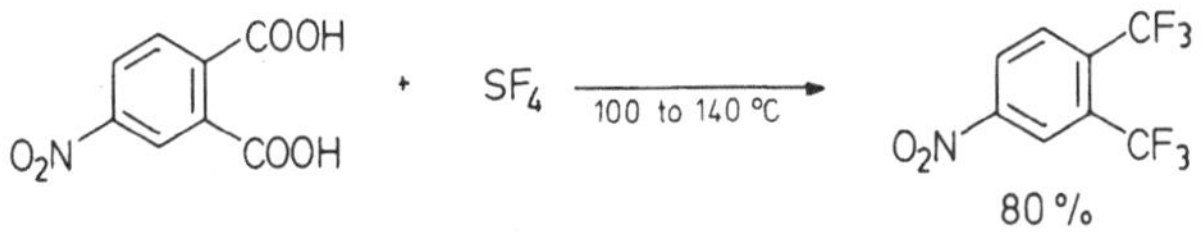

The sterically non-hindered 1,2,4,5-benzenetetracarboxylic acid reacts with SF_4 even in milder conditions than in the reaction of SF_4 with 1,2,3,5-benzenetetracarboxylic acid, forming 1,2,4,5-tetrakis(trifluoromethyl)benzene with a 70% yield [2].

Table 7. Reactions of sterically hindered benzenepolycarboxylic acids with SF_4

Acid	Ratio of SF_4 : acid (mol)	T(°C)	Time (h)	Product	Yield (%)	Ref.
3-Nitrobenzene-1,2-dicarboxylic	4.5:1	100–140	8	3-Nitro-6-trifluoromethyl-benzoyl fluoride	85	26
1,2,3-Benzenetricarboxylic	8:1	150–175	15	2,6-Bis(trifluoromethyl)-benzoyl fluoride	84	26
1,2,3-Benzenetricarboxylic	4:1	120–150	11	3-Trifluoromethylphthaloyl difluoride	69	63
1,2,3,4-Benzenetetracarboxylic	6:1	120–170	10	3,6-Bis(trifluoromethyl)-phthaloyl difluoride	76	63
1,2,3,5-Benzenetetracarboxylic	10:1	150–200	14	2,4,6-Tris(trifluoromethyl)-benzoyl fluoride	79	31
3-Trifluoromethyl-1,2,4-benzenetricarboxylic	10:1	150–220	19	2,3,6-Tris(trifluoromethyl)-benzoyl fluoride	76	37
2,5,6-Tris(trifluoromethyl)-1,3-benzenedicarboxylic	8:1	290	15	2,3,5,6-Tetrakis(trifluoro-methyl)benzoyl fluoride	82	64

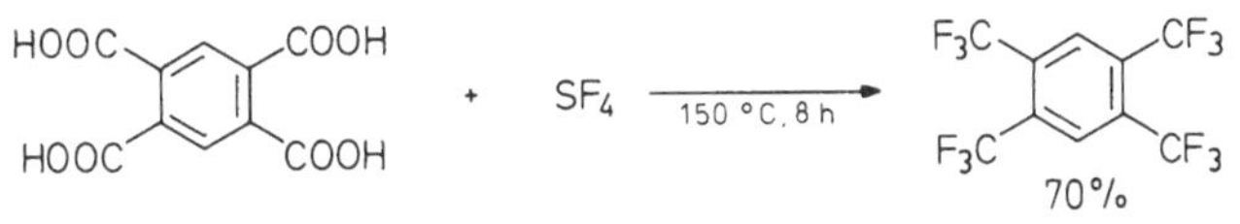

Benzenepolycarboxylic acids having at least four vicinal carboxyl groups possess a strong steric hindrance for the transformation of middle COOH group to the trifluoromethyl ones. Under the mild conditions as seen from Table 7, this leads to 3,6-bis(trifluoromethyl)phthaloyl difluoride in a high yield. Upon heating with SF_4 in more stringent conditions, the trifluoromethyl derivatives of 1,1,3,3-tetrafluorophthalane are formed [31,65].

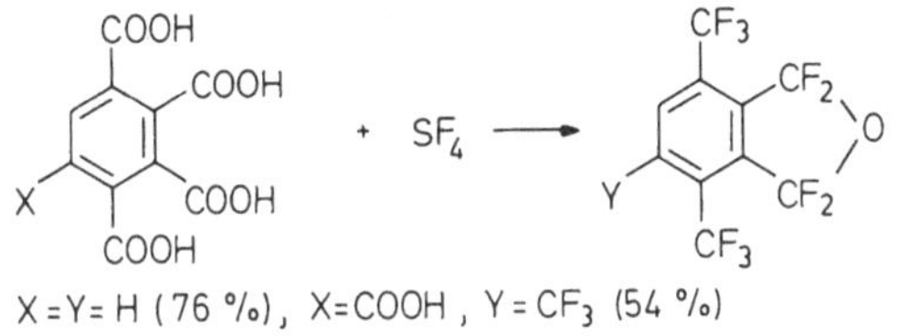

X = Y = H (76 %), X = COOH , Y = CF$_3$ (54 %)

Benzenehexacarboxylic acid reacts with SF_4, also giving a mixture of cyclic compounds containing the CF_2OCF_2 group [37,65].

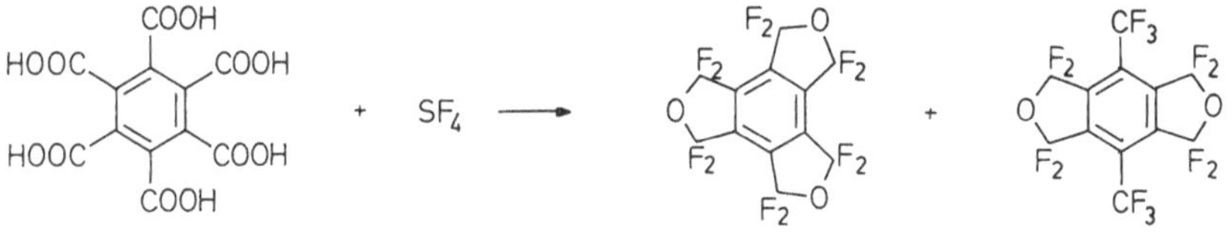

The difluoromethylene groups of the tetrafluorophthalane cycle are extremely stable against alkaline solutions, but are more easily hydrolysed in acid solutions than the trifluoromethyl ones. Hence treatment of trifluoromethyl-containing tetrafluorophthalanes with concentrated H_2SO_4 affords the respective poly(trifluoromethyl)phthalic acids in high yields [37,65].

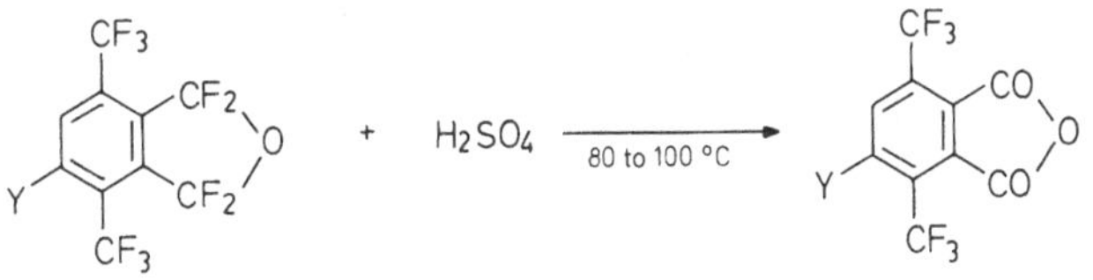

The reaction of naphthalene-1,8-dicarboxylic acid and its derivatives with SF_4 give no bis(trifluoromethyl)naphthalenes but the derivatives of 1,1,3,3-tetrafluoro-1H-naphtho[1,8-c,d]-pyrane [29,37].

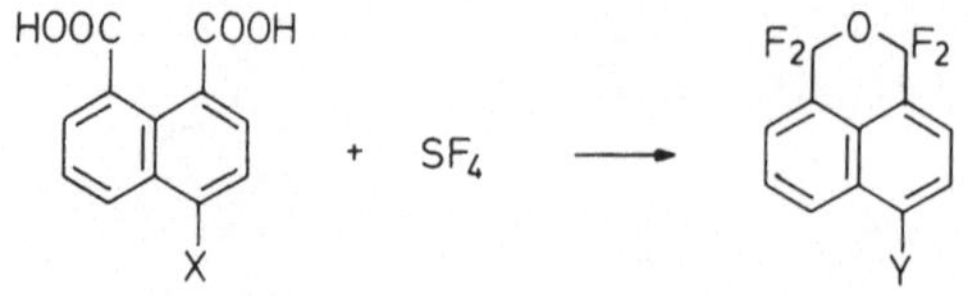

X=Y=H (63%), X=Y=NO$_2$ (85%), X=COOH, Y=CF$_3$ (70%)

The hydrolysis of the tetrafluoropyrane cycle by H$_2$SO$_4$ proceeds under mild conditions, similarly to the hydrolysis of the tetrafluorophthalane cycle, which allowed to obtain 4-trifluoromethylnaphthoic acid in a high yield from trifluoromethylnaphthopyrane [29].

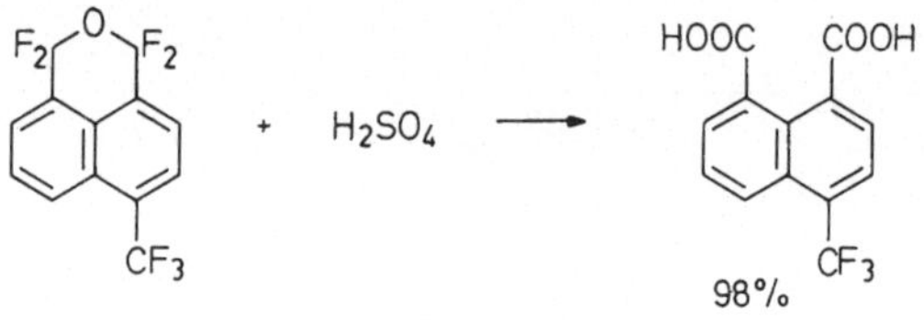

98%

As in the case of aromatic carboxylic acids, in the reactions of vicinal poly(trifluoroacetoxy)benzenes with SF$_4$, the steric hindrance was used to synthesize various poly(pentafluoroethoxy)phenols. Thus in the reaction of 1,2,3-tris(trifluoroacetoxy)benzene with SF$_4$, the middle CF$_3$COO group remains intact, and after hydrolysis, 2,6-bis(pentafluoroethoxy)phenol is formed with a high yield [44]. In a similar way SF$_4$ reacts with 1,2,3,4-tetrakis(trifluoroacetoxy)benzene [45].

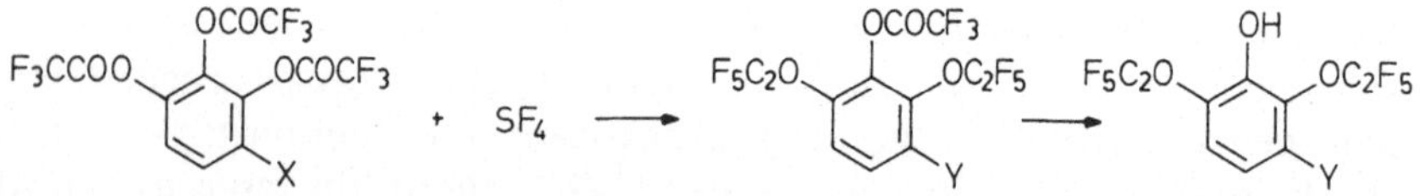

X=Y=H (89 %), X=OCOCF$_3$ Y=OC$_2$F$_5$ (79 %)

The reaction of hexakis(trifluoroacetoxy)benzene with SF$_4$ forms a mixture consisting mainly of 1,2-dioxytetrafluoroethylidenetetrakis(pentafluoroethoxy)benzene. The perfluorodioxyalkylidene groups are less stable to acid hydrolysis than the pentafluoroethoxy ones, and upon treatment of this cyclic product with oleum, tetrakis(pentafluoroethoxy)pyrocatechol is formed [45].

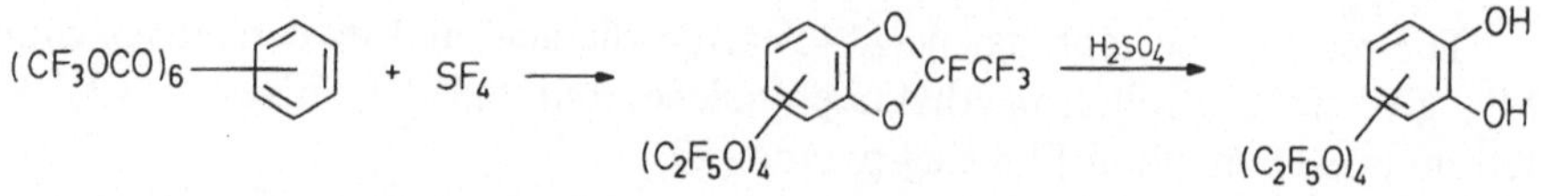

6.3.3 Differences in the Reactivities of Oxygen-Containing Functional Groups Towards SF$_4$

The reactivity of some functional groups towards SF$_4$ strongly depends on whether they constitute a part of the aliphatic, aromatic, or heterocyclic compounds. For example, phenol hydroxyl is not substituted by fluorine in the reaction with SF$_4$. This allows one to use the reaction of aromatic hydroxy-carboxylic acids with SF$_4$ to obtain various trifluoromethyl-containing phenols.

2- And 4-trifluoromethylphenols are unstable when heated in the presence of HF. But in the reaction of hydroxybenzoic acids with SF$_4$ in HF, the transformation of the carboxyl group to the trifluoromethyl one proceeds in such mild conditions that no resinification of the products occurs. An exception to this is salicylic acid which reacts with SF$_4$ and HF on cooling, and even under these conditions gives tar products. In the synthesis of 2-trifluoromethylphenol, the hydroxy group needs protection by acetylation.

Table 8 summarises the conditions of the reactions of hydroxybenzoic acids with SF$_4$ and product yields.

A larger-scale reaction of 3-hydroxybenzoic acid with SF$_4$ in HF (up to 1 kg of the acid) was reported [67] to give a lot of tar products. Benzene added to the reaction mixture as a solvent prevents resinification of the products and allows one to obtain trifluoromethylphenols in large quantities.

By contrast with hydroxybenzoic acids, the hydroxy group in tropolone and its derivatives is easily substituted by fluorine, the carbonyl group remaining intact [2,68].

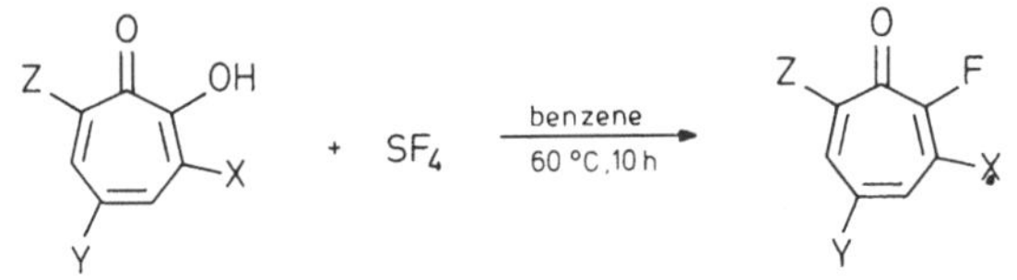

X=Y=Z=H (28 %), X=Y=Z=Br (57 %)

The hydroxy and carbonyl groups adjacent to heteroatoms are stable against SF$_4$. Thus the reaction of 6-hydroxynicotinic acid with SF$_4$ gives only 5-trifluoromethyl-2-hydroxypyridine [69].

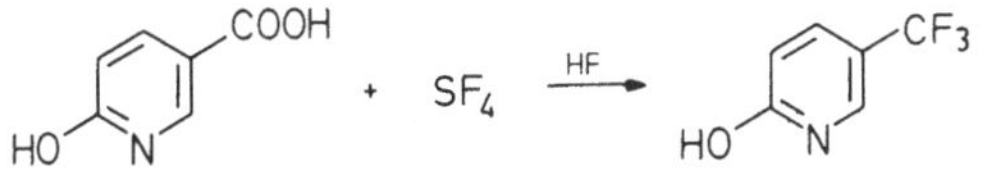

The reactions of 5-formyl- and 5-carboxyuracyl with SF$_4$ give the corresponding difluoromethyl and trifluoromethyl derivatives [70,71].

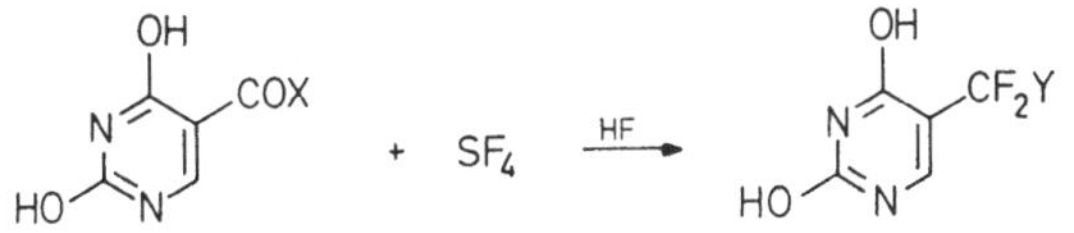

X=Y=H (60 %), X=OH   Y=F (77 %)

Table 8. Reactions of hydroxybenzoic acids with SF_4 [66]

Acid	Ratio of acid : SF_4 : HF (mol)	$T(°C)$	Time (h)	Product	Yield (%)
5-Bromo-2-hydroxybenzoic	1:4:25	25	15	5-Bromo-2-hydroxybenzotri-fluoride	45
5-Nitro-2-hydroxybenzoic	1:5:24	25	12	5-Nitro-2-hydroxybenzotri-fluoride	73
3-Hydroxybenzoic	1:4:25	20	15	3-Hydroxybenzotrifluoride	75
4-Hydroxybenzoic	1:4:25	10	90	4-Hydroxybenzotrifluoride	90
3,5-Dinitro-4-hydroxybenzoic	1:5:25	10	80	3,5-Dinitro-4-hydroxybenzo-trifluoride	92

In the same manner SF_4 reacts with 1-methyl-5-formyluracyl [72]. The 4,6-dihydroxy-1,2,5-triazine-3-carboxylic acid is transformed by SF_4 to 4,6-dihydroxy-3-trifluoromethyl-1,2,5-triazine [73].

The selective fluorination of polyketones with SF_4 was demonstrated by Shirota and his co-workers [74]. The oxo groups at 2- and 5-positions of tetrone are easily converted to the difluoromethylene ones, the other keto groups remain intact [74].

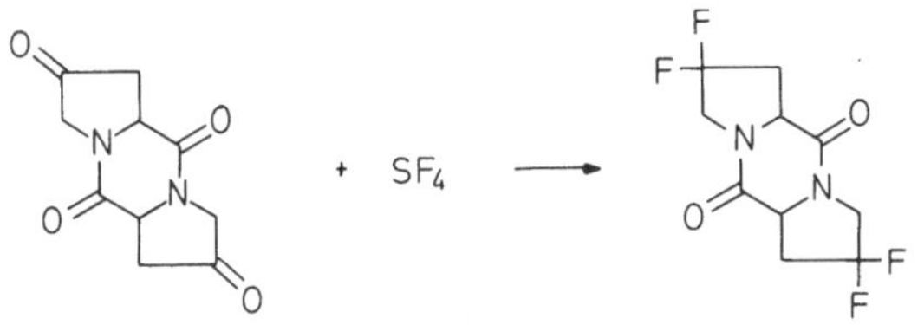

5-Carboxy-2-pyrone also reacts with SF_4 with preservation of the oxo group [69].

The oxygen-containing functional groups conjugated with the unsaturated bonds seem to be less active than the non-conjugated ones. This is vividly seen in the reactions of unsaturated dicarboxylic acids with SF_4 affording the products of substitution of the trifluoromethyl group only for one of the two carboxyl groups. Thus the reaction of SF_4 with itaconic and 1-methyl-1-cyclobutene-2,3-dicarboxylic acids may give the products in which the carboxyl group adjacent to the carbon atom of the double bond forms only the fluoroformyl bond [2].

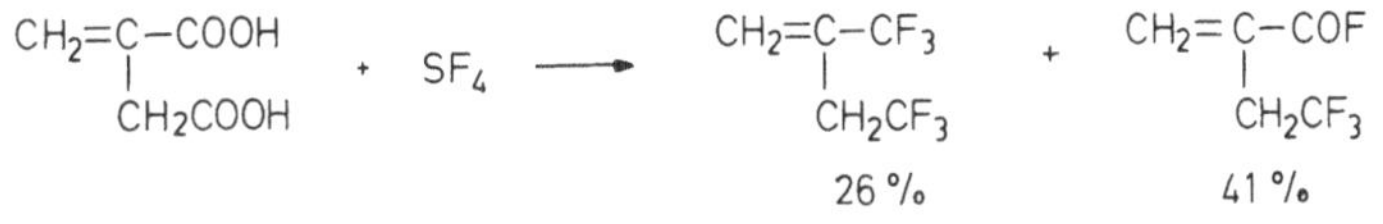

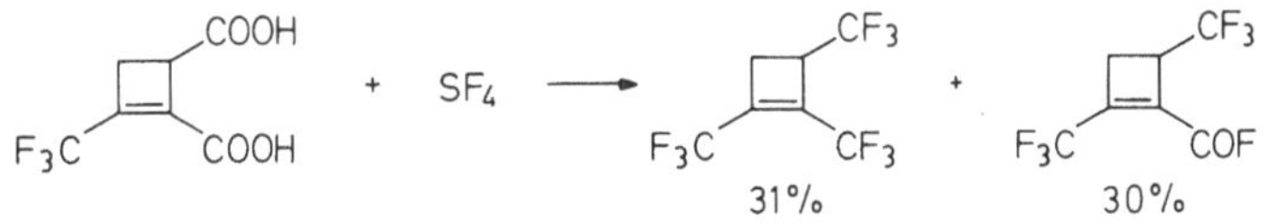

The same reasons account for the low reactivity of the conjugated 3-oxo group in α,β-unsaturated steroids, whereas in the saturated analogues it is substituted easier by the difluoromethylene group in the reaction with SF_4 than other groups.

The reactivity of the keto groups at different carbon atoms of the steroid frame decreases in the series: 3-keto > 6-keto > 17-keto > 20-keto-11-ketosteroid > 20-keto-11-deoxysteroid > conjugated 3-keto > 11-keto [56,57]. The aldehyde group of steroids is very easily transformed to the difluoromethyl one. Table 9 lists the results of the selective reactions of di- and triketosteroids and aldosteroids.

Table 9. Reactions of polyoxosteroids and aldosteroids with SF_4 in HF–dichloromethane

Steroid	Reaction conditions	Product	Yield (%)	Ref.
5α-Androst-1-ene-3,11,17-trione	18°C, 16 h	17,17-Difluoro-5α-androst-1-ene-3,11-dione	66	56
Androst-4-ene-3,17-dione	18°C, 16 h[a]	17,17-Difluoroandrost-4-ene-3-one	63	56
5α-Androstane-3,17-dione	40°C, 15 h[a]	3,3-Difluoro-5α-androstan-17-one	37	56
		3,3,17,17-Tetrafluoro-5α-androstane	11	
11-Ketoprogesterone	18°C, 16 h	20,20-Difluoro-11-ketoprogesterone	37	56
Progesterone	18°C, 16 h	20,20-Difluoroprogesterone	36	56
5α-Pregnane-3,20-dione	40°C, 15 h[a]	3,3-Difluoro-5α-pregnane-20-one	33	57
		3,3,20,20-Tetrafluoro-5α-pregnane	5	
5β-Pregnane-3,20-dione	40°C, 15 h[a]	3,3-Difluoro-5β-pregnane-20-one	29	57
		3,3,20,20-Tetrafluoro-5β-pregnane	7	
5β-Pregnane-11,20-dione	18°C, 16 h	20,20-Difluoro-5β-pregnan-11-one	50	57
5α-Fluoropregnan-3β-ol-6,20-dione acetate	20°C, 10 h	5α,6,6-Trifluoropregnan-3β-ol-20-one acetate	58	58
5α-Fluoropregnan-3β-ol-6,20-dione acetate	20°C, 16 h	5α-6,6,20,20-Pentafluoropregnan-3β-ol acetate	29	58
5β-Pregnan-3α-ol-11,20-dione acetate	18°C, 16 h	20,20-Difluoro-5β-pregnan-3α-ol-11-one acetate	58	56
5α-Fluoropregnane-3β,17α-diol-6,20-dione diacetate	20°C, 10 h	5α,6,6-Trifluoropregnane-3β,17α-diol-20-one diacetate	61	58

[a] In HF–CHCl₃

6.3.4 Selective Reactions of SF₄ with Aliphatic Polyhydroxy, Hydroxycarbonyl, and Dicarbonyl Compounds

The methods of selective substitution of oxygen in one or several functional groups by fluorine atoms in the aliphatic polyatomic alcohols, hydroxyketones, and diketones, hydroxy and oxo acids and their esters are based on the recently found peculiarities of the behaviour of these compounds with sulfur tetrafluoride.

6.3.4.1 Polyatomic Alcohols

A specific feature of the reactions of vicinal diatomic alcohols with SF₄ is the formation of fluoroalkyl fluorosulfites. In this case only one of the hydroxy groups is substituted by the fluorine atom. Apparently, at first the intermediate cyclic difluorosulfuranes of type B are formed, which react with the fluoride ion to give fluorosulfites. The latter are stable against sulfur tetrafluoride but easily react with water or HF, giving the respective alcohols [75].

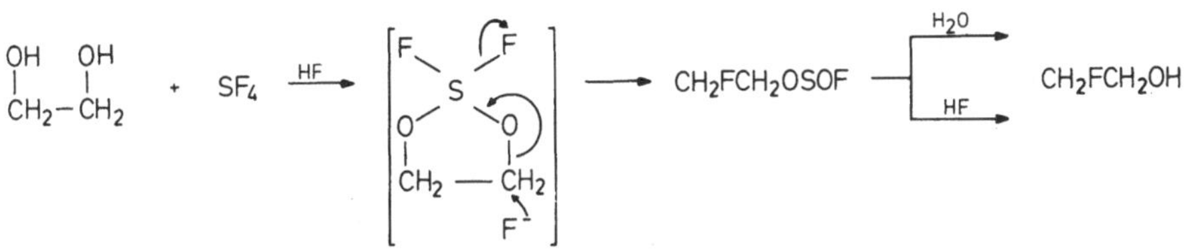

The reaction is of general character and may serve as the preparative method to obtain various monofluoro-substituted alcohols.

In the case of asymmetric glycols, the regioselectivity of the process depends on the electronic nature of groups in the substrate. As the fluoride ion attack at the intermediate cyclic difluorosulfuranes B seems to be preceded by the protonation of one of oxygen atoms, the electron-donating groups promote substitution of fluorine for the adjacent hydroxyl group.

A reverse effect is produced by the electron-accepting groups. As a result, d,l-propanediol-1,2 is transformed by SF₄ with subsequent hydrolysis only to the primary alcohol—d,l-2-fluoropropanol, and d,l-3,3,3-trifluoropropanediol-1,2 gives d,l-1,3,3,3-tetrafluoro-2-propanol in a quantitative yield [89].

$$CH_3CH-CH_2 + SF_4 \xrightarrow{NaF} CH_3CHFCH_2OSOF \xrightarrow{H_2O} CH_3CHFCH_2OH$$

$$\underset{OH\ OH}{} \qquad 60\%$$

$$CF_3CH-CH_2 + SF_4 \xrightarrow{\text{NaF}} CF_3CH(OSOF)CH_2F \xrightarrow{\text{H}_2\text{O}} CF_3CHOHCH_2F$$
$$\overset{|}{OH}\ \overset{|}{OH} \qquad\qquad\qquad 98\%$$

The reactions of SF_4 with compounds having vicinal hydroxy groups may be exemplified by the reactions of SF_4 with esters of tartaric acids [90–92]. Heating of dimethyl (+)- and (−)-tartrates with SF_4 leads to a mixture of erythro-2-fluoro-1,2-bis(methoxycarbonyl)ethyl fluorosulfite 1, dimethyl (−)(2S:3S)-2-fluoro-3-hydroxysuccinate 2, and dimethyl meso-2,3-difluoro-succinate 3, the components being the same upon variation of the reaction temperature from 50 to 180 °C. The reaction proceeds with complete reversal of configuration at one of carbon atoms, and with its complete preservation at the second carbon atom [91,92].

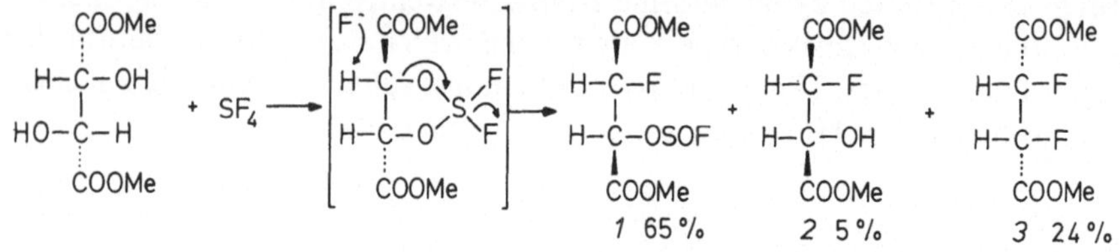

The authors of [92] found that the hydroxyester 2 is formed upon treatment of fluorosulfite 1 with HF evolving in the reaction of dimethyl tartrate with SF_4. The subsequent substitution of the hydroxy group by fluorine under the action of SF_4 in the presence of HF leads to the difluoroester 3. To obtain mono-fluoromaloate 2, the reaction with SF_4 is carried out at 20 °C in the presence of an HF acceptor—sodium fluoride. Fluorosulfite 1 is obtained here in a quantitative yield and after hydrolysis gives the hydroxyester 2. The product of substitution of fluorine for both hydroxy groups—dimethyl difluorosuccinate 3 may also be obtained in a quantitative yield in the reaction of dimethyl tartrate with SF_4 in excess HF.

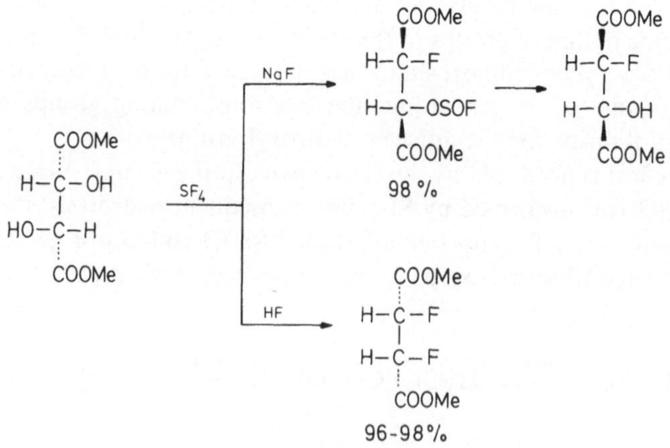

The intermediate formation of difluorosulfuranes and fluorosulfites seems to be also characteristic for the reactions of 1,3- and 1,4-diols. However these compounds were not isolated. The reactions of 1,3-propanediol and 1,3-butanediol with SF_4 gave the respective fluorine-containing alcohols.

$$HOCH_2CH_2CH_2OH + SF_4 \rightarrow CH_2FCH_2CH_2OH$$

$$CH_3CHOHCH_2CH_2OH + SF_4 \rightarrow CH_3CHFCH_2CH_2OH$$

Treatment of 1,4-butanediol with SF_4 yields, along with 4-fluoro-1-butanol, an appreciable amount of tetrahydrofuran formed by dehydration.

$$HOCH_2(CH_2)_2CH_2OH \; + \; SF_4 \longrightarrow CH_2F(CH_2)_2CH_2OH \; + \; \begin{matrix} CH_2-CH_2 \\ | \qquad | \\ CH_2 \quad CH_2 \\ \diagdown \; O \; \diagup \end{matrix}$$

Accumulation of electron-accepting groups in the glycol molecule increases the stability of carbon–oxygen bonds to such an extent that these bonds do not undergo cleavage under the action of the fluoride ion. Thus perfluoropinacone reacts with SF_4 to give a quantitative yield of perfluorinated spirosulfurane 4 [93].

$$\begin{matrix} (CF_3)_2C-OH \\ | \\ (CF_3)_2C-OH \end{matrix} \; + \; SF_4 \xrightarrow{20\,°C} \begin{matrix} (CF_3)_2C-O\diagdown \quad \diagup O-C(CF_3)_2 \\ | \qquad\qquad S \qquad | \\ (CF_3)_2C-O\diagup \quad \diagdown O-C(CF_3)_2 \end{matrix}$$
$$4 \quad 100\%$$

Accumulation of the hydroxy groups in the molecules of polyatomic alcohols leads to a sharp decrease of the yield of fluorine-containing compounds due to the strong resinification and carbonisation of the products. This may be avoided by carrying out the reactions in anhydrous HF. In this case the reaction of glycerol with SF_4 shows the regularities found for the reactions of SF_4 with diatomic alcohols. Thus the reaction of SF_4 with glycerol in HF at low temperatures gave 3-fluoro-1,2-propylene sulfite in a high yields, which seems to be formed as a result of hydrolysis of the intermediate difluorosulfurane C [80].

$$\begin{matrix} CH_2OH \\ | \\ CHOH \\ | \\ CH_2OH \end{matrix} \; + \; SF_4 \xrightarrow[-40\,°C]{HF} \left[\begin{matrix} CH_2-O\diagdown \\ | \qquad\quad SF_2 \\ CH-O\diagup \\ | \\ CH_2OSF_3 \end{matrix} \xrightarrow[-F^-]{\substack{HF \\ -SOF_2^-}} \begin{matrix} CH_2-O\diagdown \\ | \qquad\quad SF_2 \\ CH-O\diagup \\ | \\ CH_2F \end{matrix} \right] \longrightarrow \begin{matrix} CH_2-O\diagdown \\ | \qquad\quad S=O \\ CH-O\diagup \\ | \\ CH_2F \end{matrix}$$
$$C \qquad\qquad D \quad 82\%$$

In more rigid conditions, at 20 °C, difluorosulfurane C undergoes the F^- attack to form fluorosulfite E which is transformed in HF to 1,3-difluoro-2-propanol.

$$C + HF \longrightarrow CH_2FCH(OSOF)CH_2F \xrightarrow{HF} CH_2FCHOHCH_2F$$
$$E$$

6.3.4.2 Aliphatic Hydroxy Ketones

The aliphatic 2-hydroxy ketones react with SF_4 in HF, readily forming trifluoroalkanes [94]. In certain conditions it is possible to carry out the selective substitution of carbonyl oxygen or the hydroxy group in hydroxy ketones by fluorine. In the absence of HF, hydroxy ketones react with SF_4 in mild conditions, giving an equimolar mixture of trifluoroalkanes and difluoroalkyl fluorosulfites. The reactions of SF_4 with hydroxy ketones, unlike those with glycols, are intermolecular. This is indicated by the formation of an equimolar mixture of 2-fluoroketones and fluorosulfites containing the carbonyl group. With the increased reaction time, there occurs the substitution of carbonyl oxygen by fluorine to form fluorosulfites containing the CF_2 group and the respective fluoroalkanes.

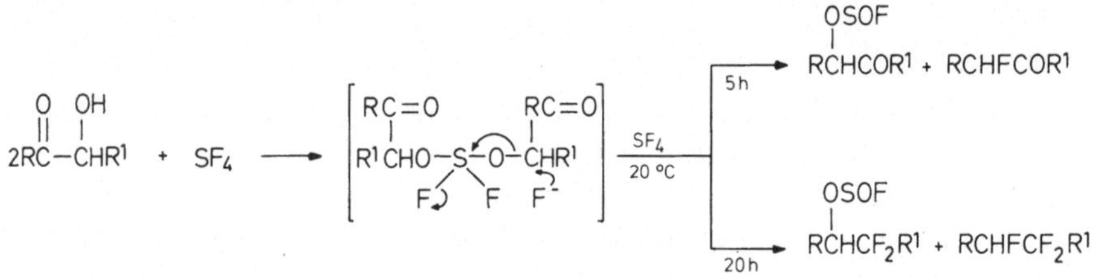

R=R¹=Et , R=R¹=Pr , R=Me R¹=H

The hydrolysis of fluorosulfites $RCH(OSOF)CF_2R'$ affords α,α-difluoro-alkylcarbinols $RCHOHCF_2R'$. Despite the fact that the yields of these compounds do not exceed 50%, the reaction of SF_4 with 2-hydroxy ketones may serve as a convenient preparative synthesis of alcohols with the difluoromethylene group at the 2-position.

The reactions of SF_4 with 2-hydroxy ketones follow a different route if diethyl ether is used as a solvent. Only 2-fluoroketones are formed with high yields [95].

$$RCHOHCOR' + SF_4 \xrightarrow{\text{ether}} RCHFCOR'$$

$$R = R' = Et, Pr$$

The role of diethyl ether is, apparently, to bind the liberated HF which serves as a catalyst in the substitution of carbonyl oxygen by two fluorine atoms.

3-Hydroxy ketones with an active methylene group are easily dehydrated by SF_4. This reaction gives a mixture of unsaturated ketone and trifluoroalkane [94].

$$RCHOHCH_2COR' + SF_4 \rightarrow RCH=CHCOR' + RCHFCH_2CF_2R'$$

6.3.4.3 Aliphatic Hydroxy Carboxylic Acids

In 2-hydroxy carboxylic acids, the presence of the electron-accepting carboxylic group vicinal to hydroxyl significantly increases the stability of intermediate alkoxytrifluorosulfuranes F formed in the reactions with SF_4. Due to this, the elimination of thionyl fluoride and substitution of the OSF_3 group by fluorine are difficult. The carbon atom of the carbonyl group in 2-hydroxy acids is much more positive than in 2-hydroxy ketones. Hence there occurs not the intermolecular reaction but the intramolecular fluorine migration from the OSF_3 group to the carbonyl carbon and formation first of difluorosulfuranes G, and then—trifluoroalkyl fluorosulfites [83].

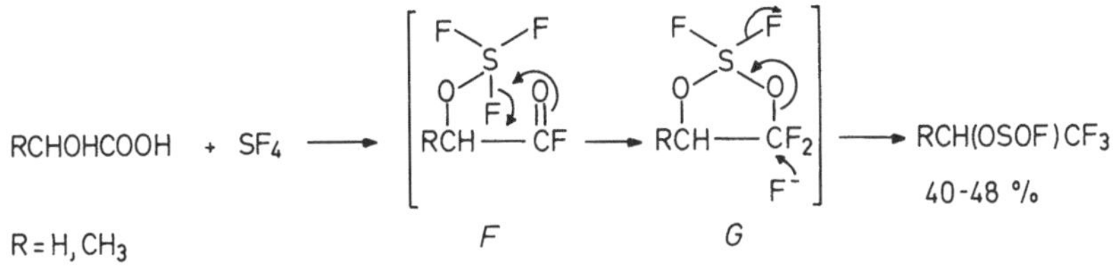

$R = H, CH_3$

The reactions of malic and tartaric acids with SF_4 lead to formation of the cyclic derivatives of 2,2,5,5-tetrafluorotetrahydrofuran [83,84]. The yield of linear and cyclic fluorosulfites in the reaction of d,l-malic acid with SF_4 reaches 75% [83].

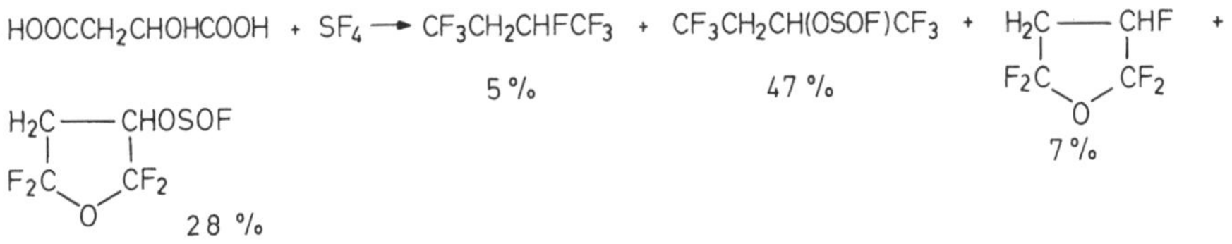

The cyclic and linear fluorosulfites are the main products in the reaction of $(+)$-(L)-tartric acid with SF_4 [84].

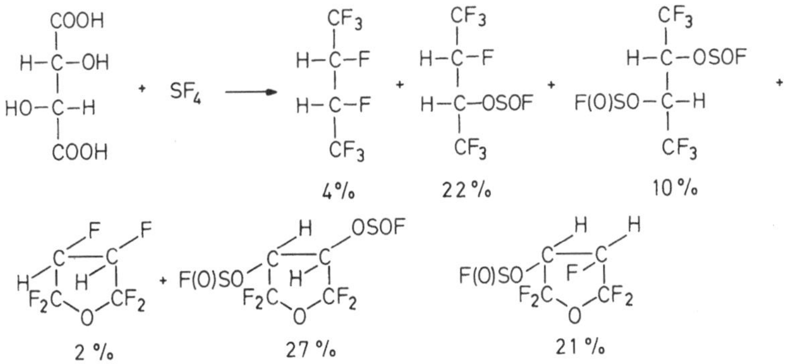

The hydrolysis of fluorosulfites obtained by treatment of 2-hydroxy carboxylic acids with SF_4 leads to the fluorinated aliphatic alcohols [83,84].

CF_3CH_2OH , $CH_3CHOHCF_3$, $CF_3CHOHCH_2CF_3$, $CF_3CHOHCHOHCF_3$,

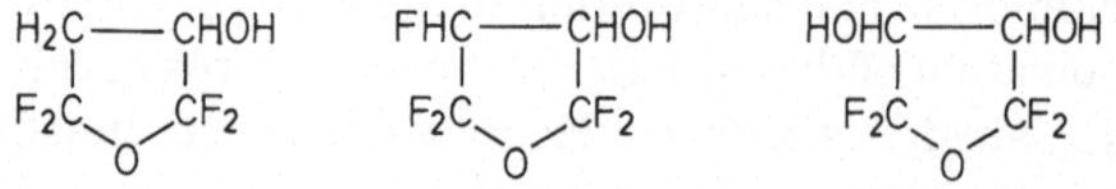

$$\begin{array}{ccc} H_2C\!\!-\!\!\!-\!\!CHOH & FHC\!\!-\!\!\!-\!\!CHOH & HOHC\!\!-\!\!\!-\!\!CHOH \\ \mid \quad\quad \mid & \mid \quad\quad \mid & \mid \quad\quad \mid \\ F_2C \quad CF_2 & F_2C \quad CF_2 & F_2C \quad CF_2 \\ {}_O & {}_O & {}_O \end{array}$$

In the case of 3-hydroxy carboxylic acids, the formation of fluorosulfites in the reaction with SF_4 was only observed in the presence of electron-accepting substituents adjacent to the hydroxy group [85].

$$CF_3CHOHCH_2COOH + SF_4 \rightarrow CF_3CH(OSOF)CH_2CF_3$$
$$100\%$$

$$\xrightarrow{\ H_2O\ } CF_3CHOHCH_2CF_3$$
$$100\%$$

In the absence of such substituents the reactions lead to fluorinated ethers and esters [85].

$$HOCH_2CH_2COOH + SF_4 \rightarrow CF_3CH_2CH_2OCH_2CH_2CF_3$$
$$+ CF_3CH_2CH_2OC(O)CH_2CH_2F$$

Introduction of one more electron-accepting substituent at the 3-position of 3-hydroxy acid results in a strong electron density shift from oxygen of the hydroxy group. This suppresses formation of alkoxytrifluorosulfurane. Heating of 4,4,4-trifluoro-3-trifluoromethyl-3-hydroxybutanoic acid with SF_4 to 50 °C yields only hydroxyacyl fluoride 5 [85].

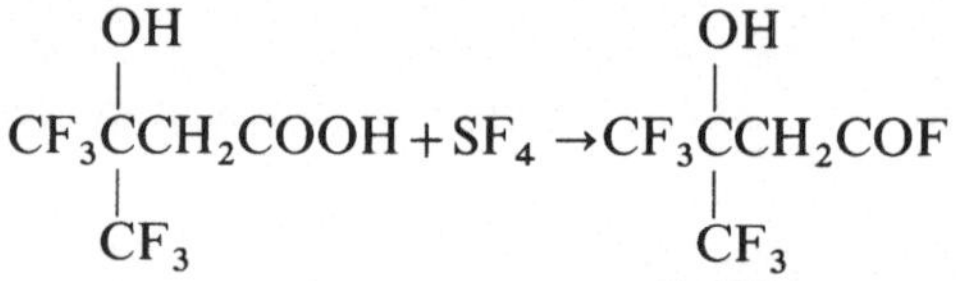

$$\begin{array}{ccc} \quad OH & & \quad OH \\ \quad\mid & & \quad\mid \\ CF_3CCH_2COOH + SF_4 \rightarrow & & CF_3CCH_2COF \\ \quad\mid & & \quad\mid \\ \quad CF_3 & & \quad CF_3 \\ & & 5,\ 98\% \end{array}$$

In more rigid conditions, at 150 to 170 °C and in the presence of HF, the products are fluoroalkanol 6, oxetane 7, and a small amount of tris(trifluoromethyl)ethylene [85].

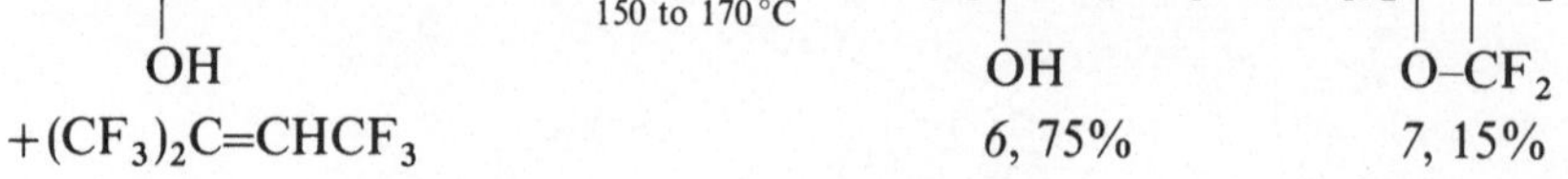

$$(CF_3)_2CCH_2COOH + SF_4 \xrightarrow[\text{150 to 170 °C}]{HF} (CF_3)_2CCH_2CF_3 + (CF_3)_2C\!\!-\!\!CH_2$$
$$\begin{array}{ccc} \mid & \mid & \mid\ \ \mid \\ OH & OH & O\!\!-\!\!CF_2 \end{array}$$

$$+ (CF_3)_2C = CHCF_3 \qquad\qquad 6,\ 75\% \qquad 7,\ 15\%$$
$$10\%$$

Alkanol 6 is obtained from hydroxyacyl fluoride 5 by substitution of carbonyl oxygen by two fluorine atoms.

The hydroxy group also seems to react partially in rigid conditions, forming the intermediate alkoxytrifluorosulfurane H. The OSF_3 group, like in the case of

4,4,4-trifluoro-3-hydroxybutanoic acid, reacts with the carbonyl group of the acyl fluoride. But here there is no formation of fluorosulfite, and the removal of SOF_2 gives oxetane 7.

$$(CF_3)_2CCH_2COF \ \underset{OH}{|} \ + \ SF_4 \ \xrightarrow{HF} \ \left[(CF_3)_2C \overset{CH_2}{\underset{O}{\diagup}} \overset{}{\underset{S}{\diagup}} O{=}CF \ \underset{F}{\overset{F}{\diagdown}} \right] \ \xrightarrow{-SOF_2} \ (CF_3)_2C{-}CH_2 \ \underset{O-CF_2}{|} $$

7

Tris(trifluoromethyl)ethylene is formed as a result of dehydration of alkanol 6 by SF_4.

Thus the reactions of accessible 2- and 3-hydroxy carboxylic acids with SF_4, demanding no preliminary protection of hydroxy group, are of the general character, and afford fluorine-containing alcohols after hydrolysis of fluorosulfites. This offers a route to various aliphatic and heterocyclic alcohols and glycols containing three or more fluorines in a molecule.

From the 4-hydroxy carboxylic acids, no products of the selective substitution of carboxyl groups by the trifluoromethyl one may be obtained. Lactones formed from 4-hydroxy carboxylic acids react with SF_4 in the presence of HF traces with cycle cleavage to give, depending on conditions, 4-fluoroacyl fluorides, or 1,1,1,4-tetra-fluoroalkanes [86].

$$CH_2CH_2CH_2C{=}O + SF_4 \ \xrightarrow{\quad}$$

$$\xrightarrow[120\,°C,\,8\,h]{} CH_2FCH_2CH_2COF \quad 80\%$$

$$\xrightarrow[160\,°C,\,12\,h]{} CH_2FCH_2CH_2CF_3 \quad 90\%$$

6.3.4.4 Esters of Hydroxy Carboxylic Acids

The previous Section showed the reaction of 2- and some 3-hydroxy carboxylic acids with SF_4 to be a route to fluorine-containing alcohols. The reactions of esters of 2- and 3-hydroxy carboxylic acids with SF_4 proceed in a different way and afford alkyl fluoroalkanoates.

An ester group is stable against SF_4, and only in rigid conditions or in the presence of excess HF is transformed to the trifluoromethyl one [38]. 2- And 3-hydroxyalkanoates react with SF_4 under mild conditions and in the presence of sodium fluoride, giving an equimolar mixture of fluoroalkanoate and alkoxycarbonylalkyl fluorosulfite, with a high yield [87].

$$\underset{\underset{COOMe}{|}}{\overset{\overset{R}{|}}{\underset{(CH_2)_n}{\overset{CHOH}{|}}}} + \ SF_4 \ \xrightarrow[20\,°C]{NaF} \ \left[\underset{\underset{COOMe}{|}}{\overset{\overset{R}{|}}{\underset{(CH_2)_n}{\overset{CH-O-S-O-CH}{|}}}} \underset{\underset{COOMe}{|}}{\overset{\overset{R}{|}}{\underset{(CH_2)_n}{|}}} (F^- \right] \ \longrightarrow \ \underset{\underset{COOMe}{|}}{\overset{\overset{R}{|}}{\underset{(CH_2)_n}{\overset{CHOSOF}{|}}}} + \ \underset{\underset{COOMe}{|}}{\overset{\overset{R}{|}}{\underset{(CH_2)_n}{\overset{CHF}{|}}}}$$

R = Me n = 0 ; R = COOMe n = 1

In this way only the esters of hydroxy acids react having no strong electron-accepting groups adjacent to the hydroxy group.

For alkyl d,l-3,3,3-trifluorolactates the route of the reaction with SF_4 is quite different to the reaction of alkyl lactates. In this case, the carbonyl oxygen is substituted by two fluorines, and the hydroxy group is transformed to the fluorosulfite one. The hydrolysis of the alkoxyalkyl fluorosulfites gave 1,1-difluoro-2-hydroxyalkyl ethers 8, 9 [87].

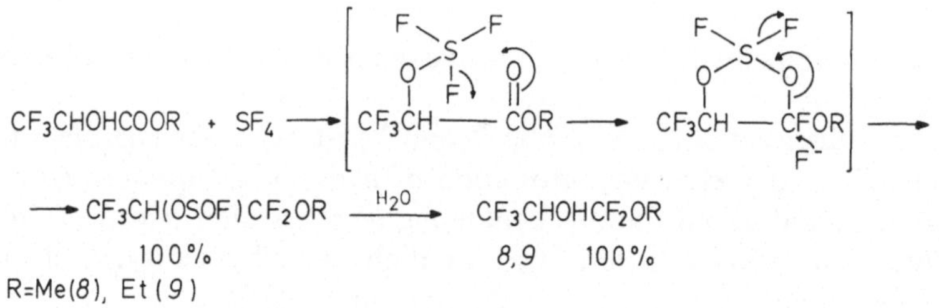

The same route is followed by the reaction of SF_4 with dimethyl *erythro*-3-fluoromaleate. As shown above, dimethyl maloate reacts with SF_4, giving only the products of intramolecular interaction. However the presence of at least one fluorine atom adjacent to the hydroxy group completely changes the reaction route. Here, as in the reaction of 3,3,3-trifluoro-2-hydroxypropanoates with SF_4, there occurs an intramolecular reaction leading to fluorine-containing hydroxy-alkyl ethers.

$$MeOCOCHOHCHFCOOMe + SF_4 \xrightarrow{NaF}$$

$$MeOCOCH(OSOF)CHFCF_2OMe \xrightarrow{NaHCO_3}$$

$$MeOCOCHOHCHFCF_2OMe \xrightarrow{SF_4,\ NaF}$$

$$MeOCF_2CH(OSOF)CHFCF_2OMe \xrightarrow{NaHCO_3}$$

$$MeOCF_2CHOHCHFCF_2OMe$$

Thus as a result of the intermittent reactions of dimethyl fluoromaloate and its derivatives with SF_4 and hydrolysis of fluorosulfites formed in this reaction, the carbonyl oxygens in both ester groups are substituted by fluorines, the hydroxy group remaining preserved.

6.3.4.5 Aliphatic Hydroxyamino Acids

An efficient method of substitution of the hydroxy group by fluorine in hydroxyamino acids has been worked out by Kollonitsch and his co-workers

[88–92]. The method involves the reaction of SF_4 with hydroxyamino acids at low temperatures (-78 to $-40\,°C$) in anhydrous HF. The authors of [89] explain the influence of HF on the process of fluorodeoxylation by the fact that it promotes the dissociation of SF_4, which leads to formation of a more electrophilic species F_3S^+ (scheme 1) and plays an important role in ionisation of alkoxytrifluorosulfurane (scheme 3). In the process of substitution of the hydroxy group by fluorine, HF obviously promotes the dissociation of alkoxytrifluorosulfurane, leading to the formation of an easier-leaving group $O\overset{+}{S}F_2$ (scheme 4). Besides, HF may serve as the fluoride ion source in the reactions of substitution of the OH group by fluorine.

$$SF_4 + HF \rightarrow {}^+SF_3 + HF_2^- \tag{1}$$

$$R–OH + {}^+SF_3 \rightarrow ROSF_3 + H^+ \tag{2}$$

$$R–OSF_3 \rightarrow RF + OSF_2 \tag{3}$$

$$R–OSF_3 + HF \rightarrow R–OSF_2^+ + HF_2^- \xrightarrow[\substack{-SOF_2 \\ -HF}]{} R^+ + F^- \rightarrow RF \tag{4}$$

Fluorodeoxylation of the hydroxyamino acids was studied in the case of serine and its derivatives, D- and L-threonine, and allo-threonine, as well as β-hydroxy-D,L-phenylalanine. D-Serine is transformed to 3-fluoro-D-alanine [90].

$$HOCH_2CH(NH_2)COOH + SF_4 \xrightarrow[-78\,°C]{HF} CH_2FCH(NH_2)COOH$$

$$51\%$$

In a similar way reacts the deuterated analogue of D-serine. In the same conditions β-hydroxy-D,L-phenylalanine forms β-fluoro-D,L-phenylalanine.

$$PhCHOHCH(NH_2)COOH + SF_4 \xrightarrow[-78\,°C]{HF} PhCHFCH(NH_2)COOH$$

$$65\%$$

Two possible routes of the reactions of hydroxyamino acids with SF_4 in HF at $-78\,°C$ have been suggested [90]. Fluorodeoxylation of L-threonine and its epimer—L-allo-threonine was shown to form the mixtures in both cases where the products with reversed configuration predominate.

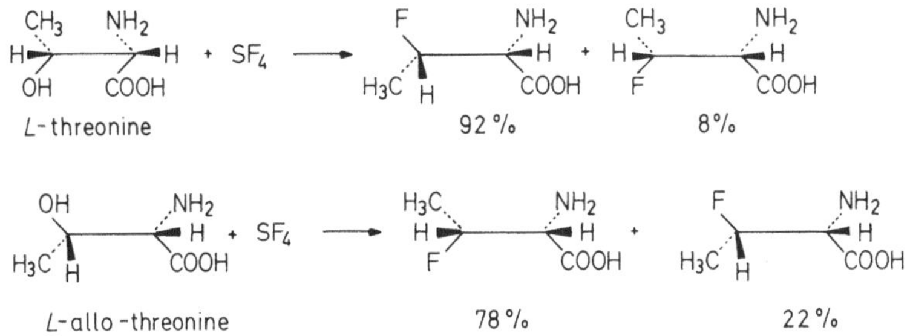

Similar results have been obtained for D-serine. On the basis of these data, the authors conclude that in the threonine–*allo*-threonine system the reactions follow the S_N2 mechanism.

On the other hand, fluorodeoxylation of 2-methylserine gives, along with the expected 2-(fluoromethyl)alanine, 40% of 1-amino-cyclopropanecarboxylic acid.

$$
\underset{\underset{NH_2}{|}}{\overset{\overset{CH_3}{|}}{HOCH_2CCOOH}} + SF_4 \xrightarrow{HF} \underset{\underset{NH_2}{|}}{\overset{\overset{CH_3}{|}}{FH_2CCCOOH}} + \quad \begin{array}{c} H_2C \\ | \\ H_2C \end{array}\!\!C\!\!\begin{array}{c} NH_2 \\ \\ COOH \end{array}
$$

Formation of the latter is explained by the fact that the reactions proceed by the S_N1 mechanism via the carbonium ion. The reactive primary carbonium ion is incorporated at the C–H bond of the CH_3 group.

The authors did not manage to carry out deoxylation of diastereomeric 3-hydroxyaspartic acids, whereas the reactions of their esters readily proceed at 0 °C.

$$MeOCOCHOHCH(NH_2)COOMe + SF_4 \xrightarrow[0\,°C]{HF}$$

$$MeOCOCHFCH(NH_2)COOMe$$

Fluorodeoxylation of hydroxyamino acids by SF_4 may be improved by carrying out the reaction in HF and in the presence of BF_3 or $AlCl_3$ which markedly facilitate substitution of the OH group by fluorine. This method affords high yields of α-fluoromethylamino acids from the respective hydroxy-derivatives [91,92].

$$
\underset{\underset{NH_2}{|}}{\overset{\overset{CH_2OH}{|}}{RCH_2CCOOH}} + SF_4 \xrightarrow{HF,\ BF_3\ or\ AlCl_3} \underset{\underset{NH_2}{|}}{\overset{\overset{CH_2F}{|}}{RCH_2CCOOH}}
$$

$$R = C_6H_5,\ HOC_6H_4,\ (HO)_2C_6H_3,\ NH_2CH_2CH_2$$

6.3.4.6 Aliphatic Dialdehydes and Diketones

Glyoxal [93], diacetyl [94], and perfluorodiacetyl [95] easily react with SF_4 in mild conditions to form fluorine-containing spiro-sulfuranes.

$$
\underset{}{\overset{\overset{O\ \ O}{||\ ||}}{RC-CR}} + SF_4 \longrightarrow \begin{array}{c} RFC-O \\ | \\ RFC-O \end{array}\!\!S\!\!\begin{array}{c} O-CFR \\ | \\ O-CFR \end{array}
$$

$$R=H\ (50\ \%),\ CH_3\ (35\%),\ CF_3\ (100\%)$$

Asymmetric 1,2-diketones react with SF_4 to give mixtures of tetrafluoroalkanes and fluoroolefins [94].

$$RCOCOCH_2R' + SF_4 \rightarrow RCF_2CF_2CH_2R' + RCF_2CF=CHR'$$

$$\ 60\text{–}65\%\ 25\text{–}30\%$$

6.3.4.7 Aliphatic Oxo Carboxylic Acids and Their Esters

Treatment of 2-oxo carboxylic acids with SF_4 leads to decarboxylation and formation of trifluoroalkanes containing one carbon less than the molecule of the starting acid [96].

$$RCOCOOH + SF_4 \rightarrow RCF_3$$

$R = Me, Ph$

The reaction of 2-oxoglutaric acid with SF_4 [96] gives the same products as the reaction of SF_4 with succinic acid [61], but the yield of 2,2,5,5-tetrafluoro-tetrahydrofuran is twice as high.

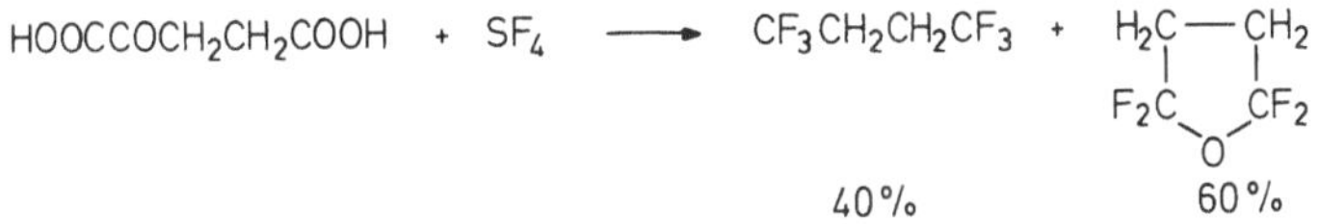

$$HOOCCOCH_2CH_2COOH + SF_4 \longrightarrow CF_3CH_2CH_2CF_3 + \underset{40\%}{} \quad \underset{60\%}{}$$

The selective substitution of the oxo group by two fluorine atoms with preservation of the carboxyl group is only feasible in some 3-oxo dicarboxylic and 4-oxo carboxylic acids. Thus 3-oxo-glutaric acid reacts with SF_4 at 100 °C to give 3,3,5,5-tetrafluoropentanoyl fluoride, or in HF–1,1,1,3,3,5,5,5-octafluoro-pentane [96].

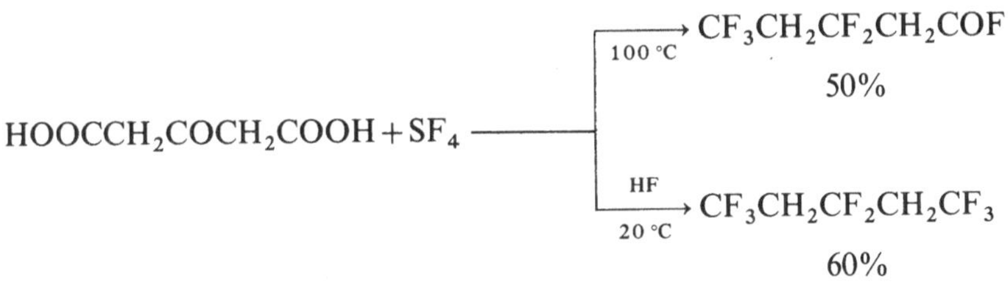

$$HOOCCH_2COCH_2COOH + SF_4 \longrightarrow$$

$$\xrightarrow[100\,°C]{} CF_3CH_2CF_2CH_2COF \quad 50\%$$

$$\xrightarrow[20\,°C]{HF} CF_3CH_2CF_2CH_2CF_3 \quad 60\%$$

The varying reactivity of carbonyl groups seems to be due to the presence of enol forms in the starting acid. Enolization accounts for the fact that , in contrast to 2-oxoglutaric acid—which as shown above releases CO in the reaction with SF_4— the oxomalonic acid gives, along with decarbonylation products, compounds where the oxo group is substituted by two fluorines [96].

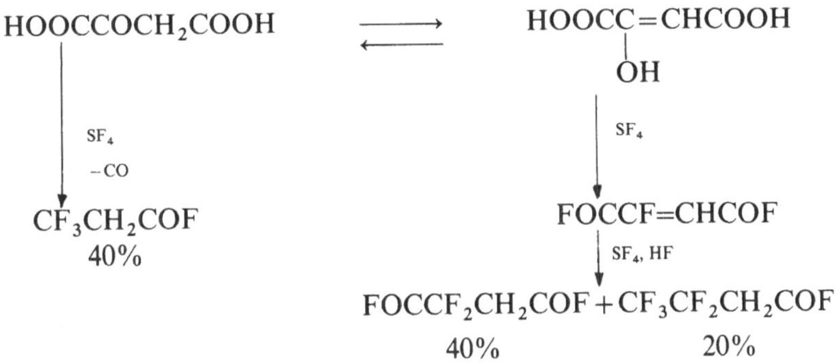

$$HOOCCOCH_2COOH \quad \underset{\longleftarrow}{\longrightarrow} \quad \underset{\overset{|}{OH}}{HOOCC=CHCOOH}$$

$$\downarrow \underset{-CO}{SF_4} \qquad\qquad\qquad \downarrow SF_4$$

$$\underset{40\%}{CF_3CH_2COF} \qquad\qquad FOCCF=CHCOF$$

$$\qquad\qquad\qquad\qquad \downarrow SF_4, HF$$

$$\underset{40\%}{FOCCF_2CH_2COF} + \underset{20\%}{CF_3CF_2CH_2COF}$$

Under more severe conditions, a mixture of 1,1,1,3,3,3-hexafluoropropane and 1,1,1,2,2,4,4,4-octafluorobutane is formed [96].

The reaction of SF_4 with levulic acid gives, depending on conditions, 4-fluoro-4-valerolactone, 4,4-difluoropentanoyl fluoride, and 1,1,1,4,4-pentafluoropentane in high yields [96].

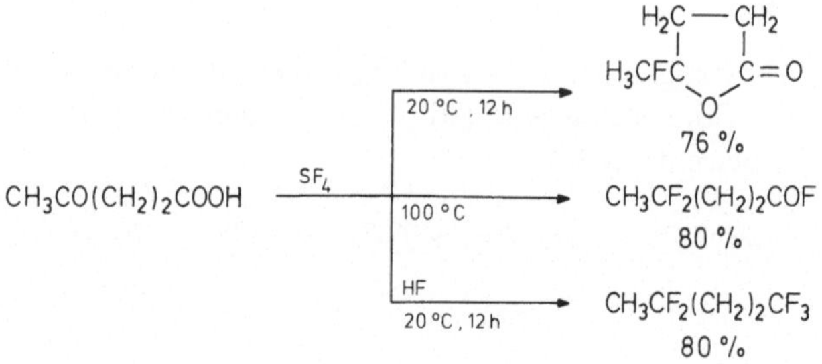

$$CH_3CO(CH_2)_2COOH \xrightarrow{SF_4} \begin{cases} \xrightarrow{20\ °C\ ,\ 12\ h} & \begin{array}{c} H_2C - CH_2 \\ | \quad\quad | \\ H_3CFC \quad C=O \\ \diagdown O \diagup \end{array} \quad 76\ \% \\ \\ \xrightarrow{100\ °C} & CH_3CF_2(CH_2)_2COF \quad 80\ \% \\ \\ \xrightarrow[20\ °C\ ,\ 12\ h]{HF} & CH_3CF_2(CH_2)_2CF_3 \quad 80\ \% \end{cases}$$

Thus, only in some cases do the reactions of oxo carboxylic acids with SF_4 afford difluorocarboxylic acids with satisfactory yields. For the synthesis of various difluorocarboxylic acids containing fluorine at the 2-, 3-, or 4-position, the most universal method is the reaction of alkyl oxocarboxylates with SF_4 in HF [97–99].

Table 10 lists the results of the reactions of esters of oxo mono- and dicarboxylic acids.

This method is inapplicable only for esters of oxo 2-polyfluorocarboxylic acids. In this case, the reaction with SF_4 proceeds with substitution of carbonyl of the ester group by two fluorines and preservation of the oxo group. The reaction gives alkoxypentafluoroacetones—the representatives of a new type of ketones [99,100].

$$CF_3COCOOR + SF_4 \rightarrow CF_3COCF_2OR$$

$$R = CH_3,\ C_2H_5$$

In the same way SF_4 reacts with 2-oxo carboxylic acids having other perfluoro- and polyfluoroalkyl groups, such as C_2F_5, $H(CF_2)_4$ [99].

A similar reaction is observed upon heating of diethyl mesoxalate with SF_4 in HF, in more rigid conditions than those given in Table 10. Along with diethyl difluoromalonate, the reaction gives ethyl 2,2,3,3-tetrafluoro-3-ethoxy-propanoate [98].

$$EtOOCCOCOOEt + SF_4 \xrightarrow[80\ °C,\ 14\ h]{HF} EtOOCCF_2COOEt$$
$$50\%$$

$$+ EtOCF_2CF_2COOEt$$
$$50\%$$

Table 10. Reactions of ester oxocarboxylates and oxodicarboxylates with SF_4

Ester	Ratio of ester:SF_4:HF (mol)	T(°C)	Time (h)	Product	Yield (%)	Ref.
$CH_3COCOOEt$	1:1:10	20	12	CH_3CF_2COOEt	78	97
CH_3COCH_2COOEt	1:1:10	20	12	$CH_3CF_2CH_2COOEt$	85	97
$CH_3COCH(C_2H_5)COOEt$	1:1:10	20	15	$CH_3CF_2CH(C_2H_5)COOEt$	80	98
$CH_3COC(C_2H_5)_2COOEt$	1:1:10	20	15	$CH_3CF_2C(C_2H_5)_2COOEt$	78	98
$CH_3COCH(C_3H_7)COOEt$	1:1:10	20	15	$CH_3CF_2CH(C_3H_7)COOEt$	76	98
$C_2H_5COCH_2COOEt$	1:1:10	20	15	$C_2H_5CF_2CH_2COOEt$	82	97
CH_2FCOCH_2COOEt	1:1:10	20	10	$CH_2FCF_2CH_2COOEt$	95	99
CHF_2COCH_2COOEt	1:1:10	20	10	$CHF_2CF_2CH_2COOEt$	98	99
CF_3COCH_2COOEt	1:1:10	20	10	$CF_3CF_2CH_2COOEt$	75	99
$CF_3COCHFCOOEt$	1:1:10	50	10	$CF_3CF_2CHFCOOEt$	50	99
				$CF_3CF(OEt)CH_2CF_3$	25	99
$CH_3CO(CH_2)_2COOEt$	1:1:10	20	12	$CH_3CF_2(CH_2)_2COOEt$	76	97
$CH_3CO(CH_2)_2COOEt$	1:1.5	95	11	$CH_3CF_2(CH_2)_2COOEt$	16	2
$EtOCOCOCOOEt$	1:1:10	20	12	$EtOCOCF_2COOEt$	80	97
$EtOCOCOCH_2COOEt$	1:1:10	20	12	$EtOCOCF_2CH_2COOEt$	76	97
$(CH_2COOEt)_2CO$	1:1:10	20	12	$(CH_2COOEt)_2CF_2$	85	97
$(CH_2COOEt)_2CO$	1:2	80	6	$(CH_2COOEt)_2CF_2$	29	2
$(CH_2COOEt)_2CO$	1:2	80	6	$(CH_2COOEt)_2CF_2$	50	101
$(CH_2CH_2COOEt)_2CO$	1:1:5	18–65	20	$(CH_2CH_2COOEt)_2CF_2$	66	102

6.3.4.8   Reactions of SF$_4$ with the Tautomeric Forms of 1,3-Diketones and 3-Oxocarboxylates

Aliphatic 1,3-diketones react with SF$_4$ in the presence of HF, readily forming tetrafluoroalkanes [115].

$$RCOCHR^1COR^2 + SF_4 \xrightarrow{\ HF\ } RCF_2CHR^1CF_2R^2$$
$$80\%$$

$$R = R^2 = Me, \ R^1 = H; \ R=R^1 = R^2 = Me; \ R=Me, \ R^1 = H, \ R^2 = Et$$

Without added HF, the reactions of 1,3-diketones with SF$_4$ yield mixtures of products containing, apart from tetrafluoroalkanes and difluoroketones, compounds with double and triple carbon–carbon bonds [113].

$$CH_3COCH_2COCH_3 + SF_4 \rightarrow CH_3COCH_2CF_2CH_3 + CH_3CF_2CH_2CF_2CH_3$$
$$70\% \qquad\qquad\qquad 8\%$$

$$+ \ CH_3COCH{=}CFCH_3 + CH_3COC{\equiv}CCH_3$$
$$15\% \qquad\qquad 7\%$$

The reactions of 1,3-diketones with SF$_4$ proceed with participation of enol forms. Initially the reaction apparently gives the unsaturated trifluorosulfurane J whose electronic conjugation promotes elimination of SOF$_2$ and F$^-$, leading to the intermediate carbocation K. Cation K may be stabilised either by addition of the fluoride ion, or by proton elimination. As a result, fluoro-alkenones and alkynones are formed [103].

$$CH_3\underset{\underset{\displaystyle OH}{|}}{C}{=}CHCOOCH_3 + SF_4 \ \rightarrow \ \left[CH_3\underset{\underset{\displaystyle OSF_3}{|}}{C}{=}CHCOCH_3 \ \rightarrow \ CH_3\overset{+}{C}{=}CHCOCH_3 \right]$$
$$\qquad\qquad\qquad\qquad\qquad\qquad J \qquad\qquad\qquad\qquad K$$

$$K\underset{\underset{\displaystyle -H^+}{\Large\longrightarrow}}{\overset{\overset{\displaystyle +F^-}{\Large\longrightarrow}}{\Big|}}\ \begin{matrix} CH_3CF{=}CHCOCH_3 \\[4pt] CH_3C{\equiv}CCOCH_3 \end{matrix}$$

In the studies of the reactions of 1,3-diketosteroids with SF$_4$ in benzene, Kaufmann and his co-workers [104] obtained the products of substitution of enol hydroxyl by fluorine and the phenyl group, which indicates the existence of the intermediate vinyl cations.

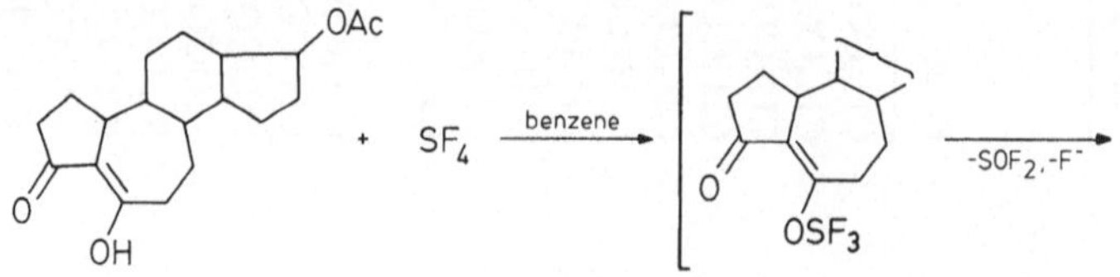

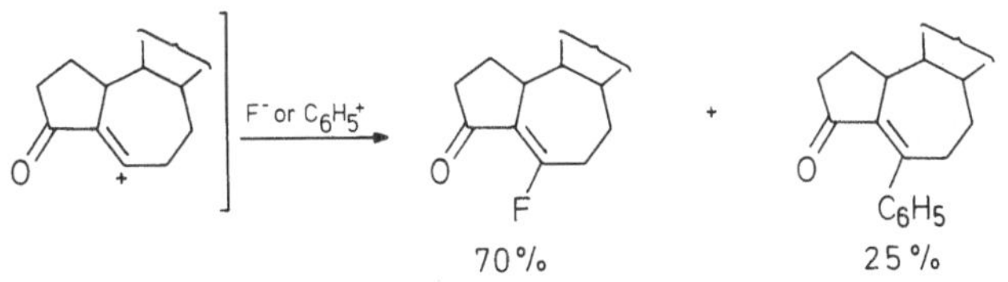

70%   25%

4,4-Difluoro-2-pentanone is obtained as a result of addition of HF to 4-fluoro-3-penten-2-one. Transformation of difluoroketone to 2,2,4,4-tetrafluoropentane proceeds only in the presence of HF as a catalyst. Therefore, carrying out the reactions of 1,3-diketones with SF_4 in the presence of sodium fluoride as an HF acceptor and in mild conditions, one obtains difluoroketones as the main product [103].

$$RCOCHR^1COR^2 + SF_4 \xrightarrow{NaF} RCOCHR^1CF_2R^2$$

75–80%

$$R = R^2 = Me, R^1 = H; R = R^1 = R^2 = Me$$

In diethyl ether the reaction of 1,3-diketones with SF_4 proceeds only via the enol form, giving unsaturated monofluoroketones with high yields [82].

$$RCOCHR^1COR^2 + SF_4 \xrightarrow{ether} RCF=CR^1COR^2$$

80%

$$R = R^2 = Me, R^1 = H; R = R^1 = R^2 = Me$$

The reactions of fluorine-containing 1,3-diketones differ from the similar reactions of their non-fluorinated analogues. The reactions of SF_4 with 1,1,1-trifluoroalkan-2,4-diones proceed with the C–C bond cleavage to form 2,3,3,3-tetrafluoropropene and the respective acyl fluorides [103].

$$CF_3COCH_2COR + SF_4 \rightarrow CF_3CF=CH_2 + RCOF$$

3,3-Dihydrohexafluoro-2,4-pentandione reacts with SF_4 in HF, giving only 3-hydroperfluoro-2-pentene [103].

$$CF_3COCH_2COCF_3 + SF_4 \xrightarrow{HF} CF_3CF=CHCF_2CF_3$$

80%

The reactions with SF_4 via enol forms are also specific to acetoacetate and its derivatives. Without added HF, the reaction of acetoacetate and ethyl 3-oxopentanoate with SF_4 leads to a mixture of difluoroalkanoates, fluoroalkenoates, and alkynoates [97,105].

$$RCOCH_2COOEt + SF_4 \xrightarrow{100\,°C,\ 8\,h} RCF_2CH_2COOEt + RCF=CHCOOEt$$

35%   25%

$$+ RC≡CCOOEt$$

40%

In the enol forms of α-alkylacetoacetates without the mobile α-hydrogen atom, proton eliminates from the methyl group of the intermediate alkoxytrifluorosulfuranes, which results in formation of compounds with cumulated double carbon–carbon bonds [97].

$$CH_3COCHRCOOEt + SF_4 \rightarrow CH_3CF_2CHRCOOEt + CH_3CF=CRCOOEt$$
$$35\% \qquad\qquad 10\%$$

$$+ CH_2=C=CRCOOEt$$
$$40\%$$

R=Pr, i-Pr

Thus the reactions of 3-oxocarboxylates with SF_4 may present a convenient preparative method for the synthesis of unavailable alkynoates and 2,3-alkadienoates.

6.4 Fluorination and Halofluorination of Organic Compounds with SF_4 in HF in the Presence of Halogens or Sulfur Monochloride

Until recently the reactions of SF_4 with organic compounds proceeding with substitution of hydrogen by fluorine had been considered to be anomalous.

Anthrone reacts with SF_4 in anhydrous HF, forming 10,10-difluoroanthrone [106].

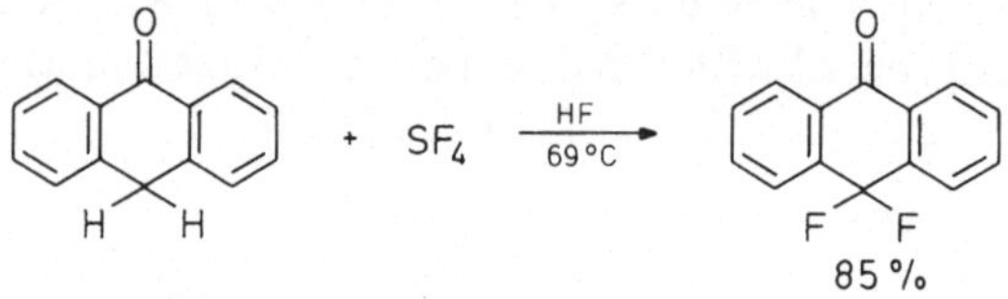

In the reaction of d,l-lactide with SF_4 in HF, along with substitution of two carbonyl oxygens by fluorines, there occurs substitution of the tertiary hydrogen atom by fluorine to form two stereoisomeric 2,5-dimethyl-2-hydroperfluorodioxanes [86].

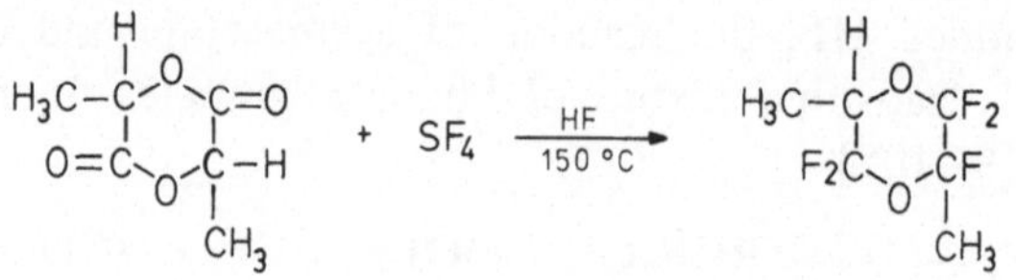

The chlorine-containing cyclic carbonates react with SF_4 in anhydrous HF, giving the products of substitution of carbonyl oxygen, chlorine, or hydrogen by fluorine [107].

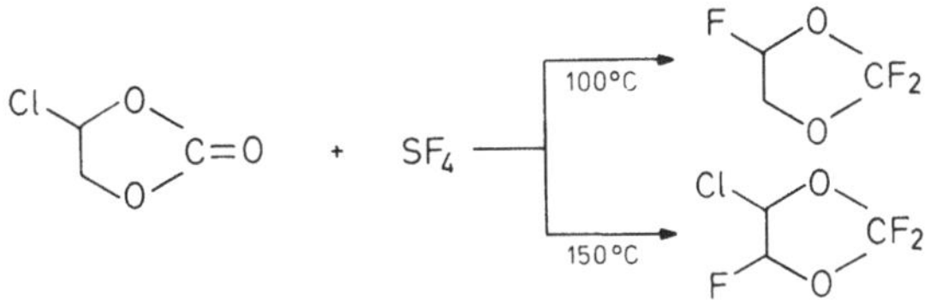

The reaction of 2-methylcyclohexanone with SF_4 yields, along with 2-methyl-1,1-difluorocyclohexane, 10% of 2-methyl-1,1,2-trifluorocyclohexane [108].

Adamantane easily reacts with SF_4 to form a mixture of fluoroadamantanes containing one to four fluorines at molecular nodes [109].

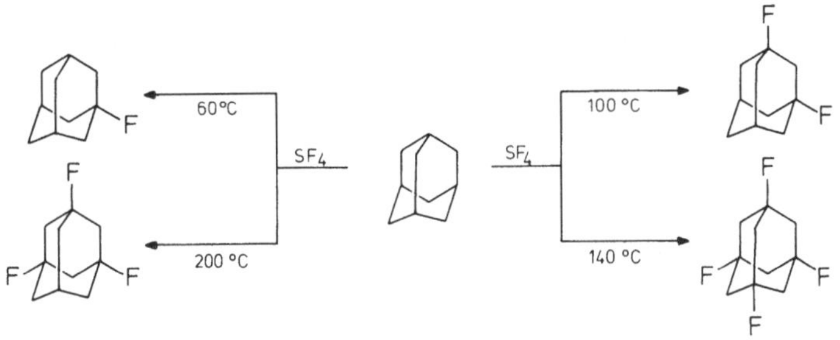

Introduction of the fourth fluorine atom into the molecule is only feasible in anhydrous HF.

Adamantane-carboxylic acids and their derivatives react with SF_4 in anhydrous HF like adamantane. Apart from the transformation of the carboxyl group to the trifluoromethyl one, the reaction involves substitution of two or three hydrogens in the node positions of adamantane by fluorine. Adamantane-1-carboxylic acid is transformed at 140 °C to 1-trifluoromethyl-3,5-difluoro-adamantane with a high yield. Raising the reaction temperature to 210 °C leads to substitution of fluorine for all the tertiary hydrogens in the adamantane molecule [110–112].

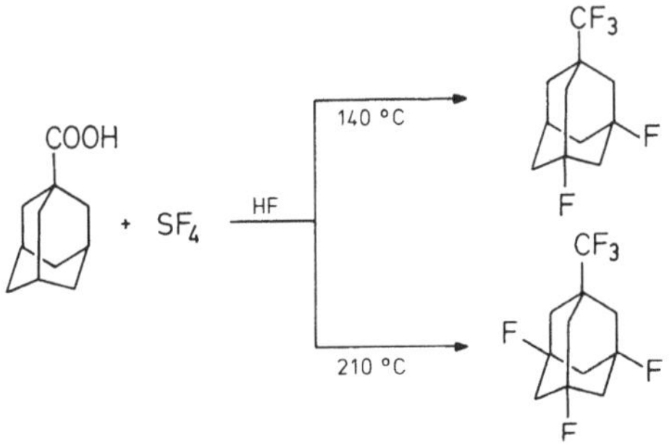

Substitution of hydrogen or oxygen atoms to form trifluoromethyl-containing tertiary amines proceeds in the reaction of N,N-dialkylformamides with SF_4, in the presence of KF [113].

$$HCONR_2 + SF_4 \xrightarrow{KF} CF_3-NR_2$$

In all of the above examples, the fluorination processes were conducted using the technical-grade sulfur tetrafluoride obtained from sulfur dichloride and sodium fluoride, and always containing the chlorine or sulfur chloride impurities.

Recently, it has been suggested [114] that even insignificant admixtures of chlorine or sulfur chlorides should promote substitution of hydrogen in some organic compounds by fluorine in the reaction with SF_4. To verify this suggestion, sulfur tetrafluoride was purified twice by keeping it over the amalgamated copper turnings and mercury. In the reaction of adamantane with SF_4 purified from chlorine and sulfur chlorides, no hydrogen substitution by fluorine occurs and unchanged adamantane is returned. Addition of the catalytic amounts of chlorine or sulfur chlorides to the reaction mixture leads to formation of fluoroadamantanes with high yield.

Under the same conditions (in the presence of HF and the catalytic amounts of chlorine or sulfur chlorides) SF_4 reacts with 2,2'-dichlorodiethyl ether and 1,4-dioxan. The reaction gives linear or cyclic ethers containing three to five fluorines [114,115].

$$CH_2ClCH_2OCH_2CH_2Cl + SF_4 \xrightarrow{HF}$$

$$\xrightarrow{55\,°C} CF_2ClCHFOCH_2CH_2Cl \quad 16\%$$

$$\xrightarrow{110\,°C} CF_3CHFOCH_2CH_2Cl \quad 58\%$$

$$\xrightarrow{200\,°C} CF_3CHFOCH_2CH_2F \quad 63\%$$

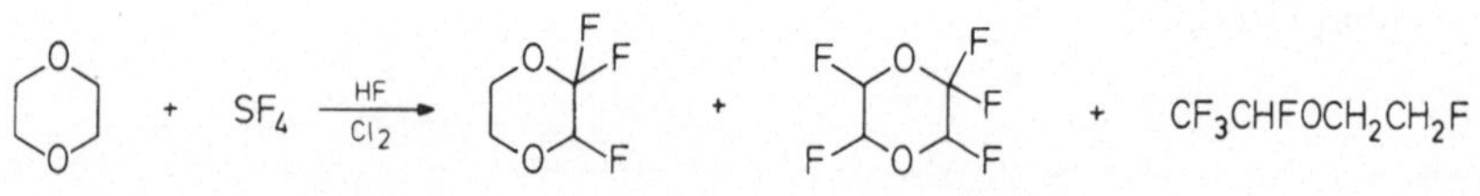

Formation of 1,2,2,2,2'-pentafluorodiethyl ether in fluorination of dioxan occurs as a result of cleavage of difluorodioxan under the action of HF and subsequent reaction of the hydroxy ether formed with SF_4.

The reaction of 2,2'-dimethoxydiethyl ether (diglyme) with SF_4–HF in the presence of chlorine at 160 °C proceeds not only with hydrogen substitution by fluorine but also with cyclisation, giving 1,1,2-trifluoridioxan and methyl fluoride with a good yield [115].

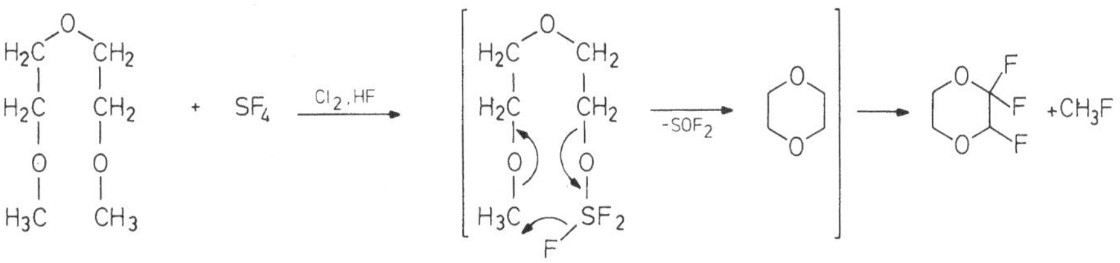

The above reactions of organic compounds with SF_4 in HF, in the presence of catalytic amounts of chlorine or sulfur chlorides involve not the direct hydrogen substitution by fluorine but chlorination with subsequent substitution of chlorine by fluorine. The presence of SF_4 in the reaction mixture promotes electrophilic chlorination of organic compounds as a result of polarisation of the halogen molecule. The chlorination leads to formation of hydrogen chloride which readily reacts with SF_4 at $-78\,°C$ [116].

$$SF_4 + 4HCl \rightarrow SCl_4 + 4HF$$

$$SCl_4 \rightarrow SCl_2 + Cl_2$$

$$2SCl_2 \rightarrow S_2Cl_2 + Cl_2$$

This sequence of reactions results in regeneration of chlorine and allows one to obtain fluorine-containing compounds with high yields, using the catalytic amounts of the chlorinating agent.

6.4.1 Reactions of Organic Compounds with the SF_4–HF–Hlg$_2$ System

The reactions of organic compounds with SF_4 in HF in the presence of stoichiometric (but not catalytic) amounts of chlorine involve not only fluorinations but also chlorofluorinations, giving high yields of products.

The conjugated chlorofluorination is known to take place in the reactions of unsaturated compounds with molecular chlorine in anhydrous HF [117]. But the yields of chlorofluoroalkanes do not exceed 20%. In general, raising the yields of halofluorination products implies raising the concentration of halogenonium cations in the reaction mixture and withdrawal of the halide anions. This problem is solved by various methods: as a source of halogenonium cation, hexachloromelamine or N-chloroamides are used [118]; the halofluorination reactions are carried out in the presence of pyridine to raise the fluoride ion concentration [119]; silver nitrate is used to bind the chloride ion and remove it from the reaction [119].

The conjugated chlorofluorination is largely facilitated by adding SF_4 to the HF–Cl_2 system. Being a Lewis acid, sulfur tetrafluoride promotes polarisation of the chlorine molecule, raising the Cl^+ concentration in the reaction medium. On the other hand, SF_4 transforms to chloride ion [116,120] to molecular chlorine, removing them from the reaction sphere, which completely rules out addition of the chlorine molecule at the double bond. Finally, SF_4 reacts with anhydrous

HF, giving the F_3S^+ and HF_2^- ions [59]. This not only promotes formation of unsaturated compounds from chloro- or fluoroalkanes under the action of F_3S^+, but also leads to a high concentration of fluoride ions for the conjugated halofluorination.

Aliphatic hydrocarbons react with SF_4 in anhydrous HF in the presence of chlorine (with the ratio of $SF_4:HF:Cl_2 = 1:5:2$) to form vicinal chlorofluoro-alkanes [120]. Butane is transformed under the action of SF_4–HF–Cl_2 already at 20 °C to a mixture of diastereomeric 2-chloro-3-fluorobutanes. The initial stage of the reaction is chlorination of butane to 2-chlorobutane which is further transformed to 2-butene. As a result of the electrophilic attack of the latter by the Cl^+ cation, carbocation L is formed, which is stabilised by adding the fluoride ion to form two diastereomers of 2-chloro-3-fluorobutane.

$$CH_3CH_2CH_2CH_3 + SF_4 \xrightarrow[20\,°C]{HF} [CH_3CHClCH_2CH_3$$

$$\xrightarrow[-HCl]{} CH_3CH=CHCH_3$$

$$\xrightarrow{Cl^+} CH_3CHCl\overset{+}{C}HCH_3] \xrightarrow{F^-} CH_3CHClCHFCH_3$$
$$L$$

In more rigid conditions, 2-chloro-3-fluorobutanes undergo elimination of hydrogen chloride and addition of stoichiometric amounts of "ClF" to form 2-chloro-3,3-difluorobutane.

$$CH_3CHClCHFCH_3 + SF_4 \xrightarrow[70\,°C]{HF,\ Cl_2} [CH_3CH=CFCH_3]$$

$$\xrightarrow{Cl^+,\ F^-} CH_3CHClCF_2CH_3$$

In a similar way the SF_4–HF–Cl_2 system reacts with pentane, giving stereo-isomeric 2,3-dichloro-4-fluoropentanes with high yield.

Vicinal chlorofluoroalkanes are also formed by the reaction of SF_4–HF–Cl_2 with difluoroalkanes and aliphatic ketones [120].

$$\left.\begin{array}{l} RCH_2COCH_2R' \\ RCH_2CF_2CH_2R' \end{array}\right\} \xrightarrow{SF_4,\ Cl_2,\ HF} RCHClCF_2CH_2R'$$

$$R = R' = H;\ R = Me,\ R' = H;\ R = R' = Me$$

Difluoroalkanes formed at the initial stage of the reaction of ketones with SF_4–HF–Cl_2 eliminate HF to form fluoroalkenes. Further addition of stoichio-metric equivalents of chlorine monofluoride at the double bond of fluoroalkenes leads to the formation of vicinal chlorodifluoroalkanes.

$$RCH_2COCH_2R' + SF_4 \xrightarrow{HF} [RCH_2CF_2CH_2R' \xrightarrow[-HF]{}$$

$$RCH=CFCH_2R'] \xrightarrow{SF_4,\ Cl_2,\ HF} RCHClCF_2CH_2R'$$

The mechanism of the reactions of SF_4–HF–Cl_2 with alkanes, difluoroalkanes, and aliphatic ketones involving the initial formation of unsaturated compounds with subsequent addition of stoichiometric equivalents of "ClF", is confirmed by the reactions of this system with olefins.

In the reaction of trichloroethylene with SF_4–HF–Cl_2, there occurs the addition of stoichiometric equivalents of "ClF". The conjugated chlorofluorination of trichloroethylene proceeds according to the Markownikov rule, as confirmed by the ionic mechanism of the reaction. Variation of the reaction temperature from 20 to 150 °C does not affect the halofluorination process. A substantial effect on it is produced by the reagent ratio in the SF_4–HF–Cl_2 system. Variation of SF_4 concentration from 0.25 to 1.5 mol per mol of the starting olefin increases the yield of the conjugated halofluorination products, allows one to conduct the process under milder conditions and reduces the reaction time.

The stereospecificity of the conjugated halofluorination reactions of unsaturated compounds with SF_4–HF–Br_2 has been studied in the case of Z- and E-1,2-dichloroethylenes. This reaction is strictly *anti*-stereospecific. In this case, Z-1,2-dichloroethylene is transformed to *threo*-1-bromo-2-fluoro-1,2-dichloroethane, and the E-isomer gives *erythro*-1-bromo-2-fluoro-1,2-dichloroethane in about quantitative yields. The absence of a regio-adduct and formation of *threo*- and *erythro*-diastereomers in quantitative yields indicates that the reaction proceeds only via the bromonium ions M, N which are more stable than the chloronium ones and are not transformed to the latter.

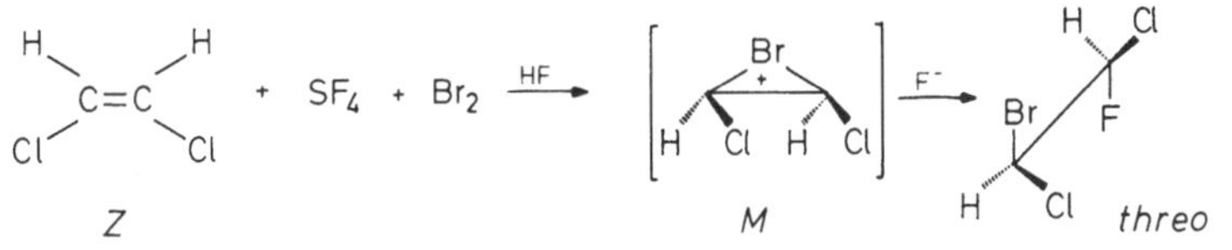

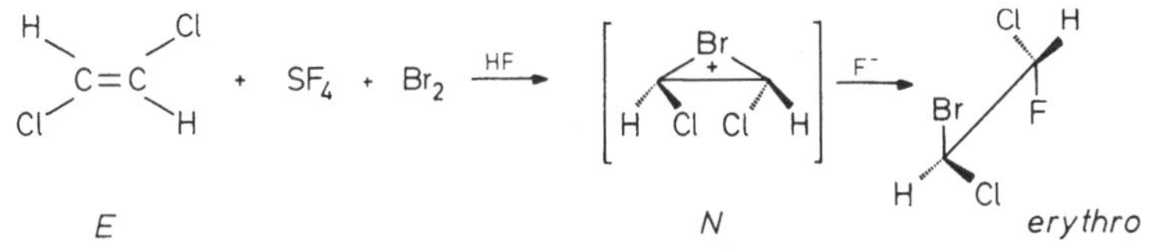

The reactions of the ethers with SF_4 in the presence of HF and an excess of molecular chlorine lead to formation of chlorofluoroethers containing the CCl_3–CHF group. The reactions of diethyl and 2,2'-dichlorodiethyl ethers with SF_4–HF–Cl_2 afford respectively 2,2,2-trichloro-2-fluorodiethyl ether *10* and 2,2,2,2'-tetrachloro-1-fluorodiethyl ether *11* with high yields [120].

$$CH_3CH_2OCH_2CH_3 + SF_4 + Cl_2 \xrightarrow[80\,°C]{HF} CCl_3CHFOCH_2CH_3$$

10

$$CH_2ClCH_2OCH_2CH_2Cl + SF_4 + Cl_2 \xrightarrow{HF} CCl_3CHFOCH_2CH_2Cl$$

11

At the reaction temperature raised to 100 to 140 °C, diethyl and 2,2'-dichloro-diethyl ether react with SF_4–HF–Cl_2 with cleavage of the C–O bond. The chlorofluorinated ethers *10* and *11* formed under mild conditions apparently undergo ether bond cleavage under the action of SF_4–HF at the elevated temperature to form trifluorosulfurane *0*, which eliminates thionyl fluoride to form 1,2-difluorotrichloroethane [120].

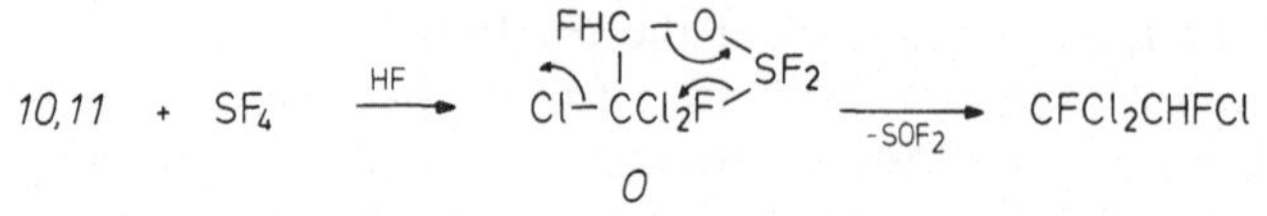

A similar transfer of chlorine in trifluorosulfuranes takes place in the reactions of chloral [121] and 2-trichloromethylhexafluoro-2-propanol [122] with SF_4.

The reaction of dioxan with SF_4–HF in excess chlorine at 120 °C yields chlorodifluorodioxan *13*, along with trifluorodioxan *12*. Raising the reaction temperature to 150 °C leads to the predominant formation of trifluorodioxan *12*. This results from the chlorine substitution by fluorine in compound *13* under the action of SF_4–HF. At a lower concentration of SF_4 and HF in the reaction mixture, the main product is chlorodifluorodioxan *13* [120].

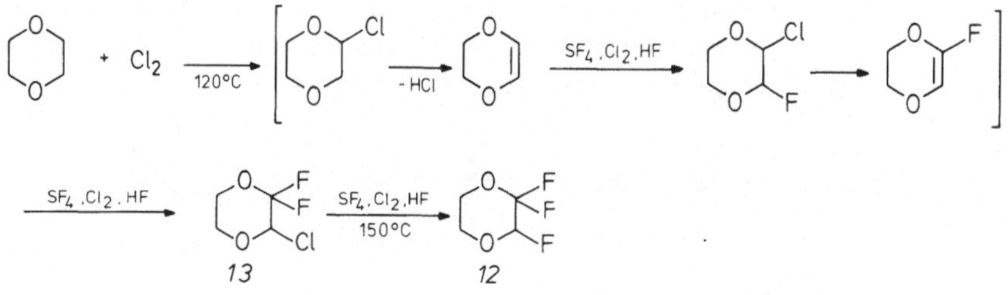

The reaction of 2,3-dichlorodioxan with SF_4–HF leads to trifluorodioxan *12* even without adding chlorine, as it is formed in situ by the reaction of HCl with SF_4 [116], HCl is formed as a result of chlorine substitution by fluorine in 2,3-dichlorodioxan under the action of HF.

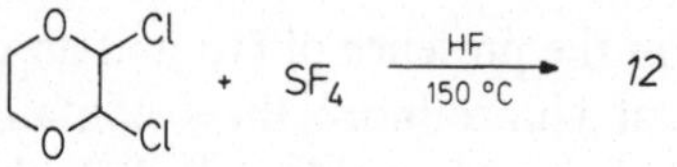

The reaction of tetrahydrofuran with SF_4–HF–Cl_2 under mild conditions and according to the same scheme as for dioxan gives 3,3-dichloro-2-fluorotetrahydrofuran *14* with a good yield [120].

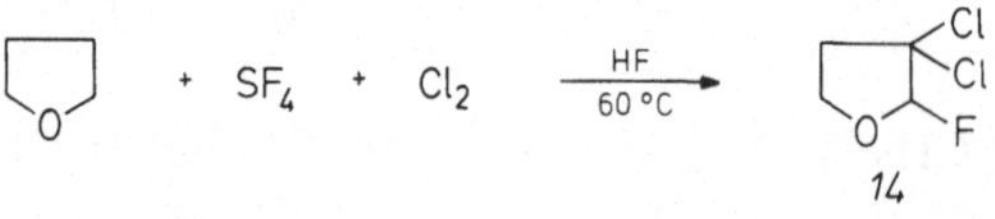

In more rigid conditions this reaction leads to cycle fission to form a mixture of chlorofluorobutanes *15–17*.

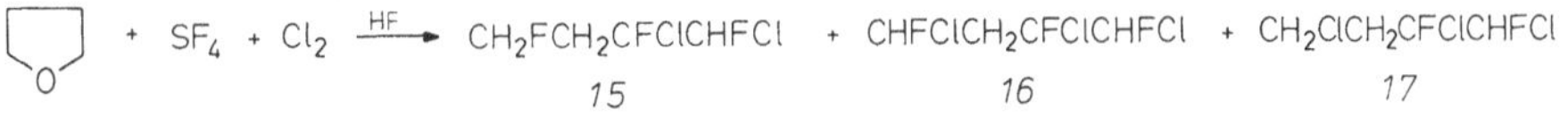

Dichlorofluorotetrahydrofuran *14* supposedly undergoes cleavage at 130 °C under the action SF_4–HF to form trifluorosulfurane *P* which eliminates thionyl fluoride to form chlorofluorobutane *15*.

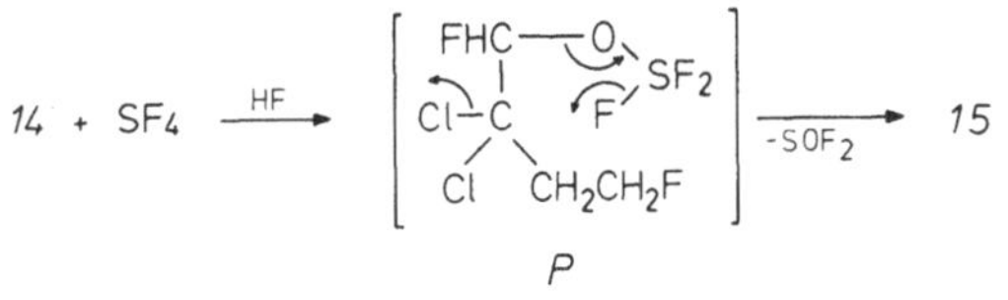

Chlorofluoroalkane *16* is formed as a result of chlorination of compounds *14* at the free α-position and subsequent cycle cleavage of compound *18* under the action of SF_4–HF [120].

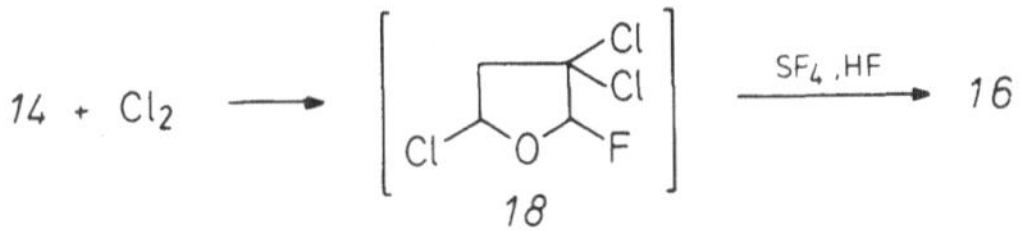

Cleavage of tetrahydrofuran cycle of compound *14* by HCl in the presence of SF_4 leads to the formation of chlorofluorobutane *17*.

Substituted furans also undergo the conjugated chlorofluorination reactions with SF_4–HF–Cl_2 (Br_2). In this case, addition of the stoichiometric equivalents of "ClF" of "BrF" occurs at the 2,5-positions of the furan cycle, leading to the respective dihydrofurans in high yields. The reaction of 2,5-dibromofuran with SF_4–HF–Br_2 affords 2,2,5,5-tetrafluoro-2,5-dihydrofuran in a 80% yield. The reaction proceeds via the bromofluorination stage with subsequent substitution of bromine in allyl positions by fluorine [123].

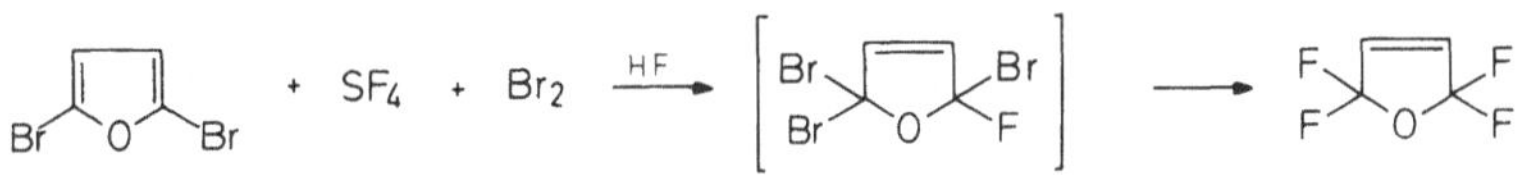

5-Nitrofurfurol reacts with SF_4–HF–Cl_2 with conjugated chlorofluorination of the furan cycle, as well as substitution of chlorine, the nitro group, and oxygen of the aldehyde group by fluorine to form 2,5,5-trifluoro-2-(difluoromethyl)-2,5-dihydrofuran *19*. The latter is also formed in the reaction of SF_4–HF–Cl_2 with 2-(difluoromethyl)-5-nitrofuran [124].

The furancarboxylic acids as rule react with SF_4, forming the trifluoromethyl derivatives [35,60,125,126]. However 2-trifluoromethylfuran may not be obtained from furan-2-carboxylic acid. Under mild conditions the reaction yields furan-2-carbonyl fluoride, which, upon further heating of the reaction mixture, is completely resinified [60]. The reaction of furan-2-carboxylic acid with SF_4–HF–Cl_2 leads to dihydrofurans 20 and 21 containing chlorine in the nucleus or side chain [127].

The chlorine- and fluorine-containing furans 20 and 21 are also obtained from 5-chlorofuran-2-carboxylic acid. The initial stage of the reaction involves chlorofluorination of the furan ring to form dihydrofuran 22, and subsequently carbocation Q is formed, which is in agreement with the literature data [128]. The presence of the chlorine atom adjacent to the positively charged carbon atom in carbocation Q promotes the transformation of the latter to the chloronium cation R as a result of coordination of chlorine p-electrons with the carbocationic centre. The fluoride ion attacks ion R at either of two positions: at the carbon atom of the difluoromethylene group to form compound 20, or at the chlorine-bonded carbon of heterocycle, which leads to migration of chlorine to the side chain to form compound 21.

5-Bromofuran-2-carboxylic acid reacts with SF_4–HF–Br_2 in the similar way [129].

The presence of electron-accepting substituents in the furan ring suppresses the conjugated halofluorination, indicating the electrophilic mechanism of the reaction.

Dihydrofurans containing chlorine in heterocycle 23 and in side chain 24 are formed from furan-2,5-dicarboxylic acid under the action of SF_4–HF–Cl_2 only at 150 °C.

2-Trifluoromethylfuran-3,5-dicarboxylic acid and 2-trifluoromethylfuran-3,4-dicarboxylic acid react with SF_4–HF–Cl_2 to give the respective 2,3,5-tris(trifluoromethyl)furan and 2,3,4-tris(trifluoromethyl)furan. Further chlorofluorination of these compounds does not take place. Substitution of the trifluoromethyl group in 2-trifluoromethylfuran-3,4-dicarboxylic acid by the methyl group leads to the electrophilic addition of stoichiometric equivalents of "ClF" [130].

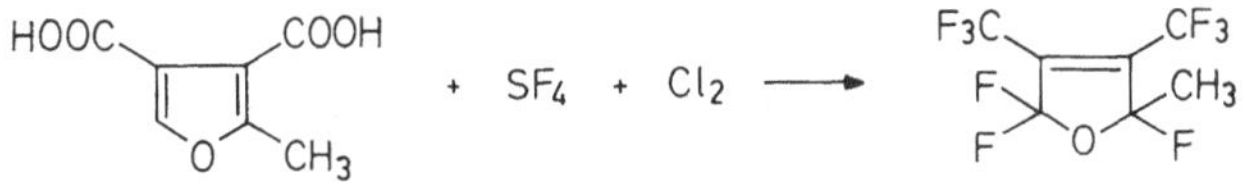

The reaction of 5-nitrofuran-2-carboxylic acid with SF_4–HF–Cl_2 proceeds in a similar way as the reaction of that system with 5-nitrofurfurol [124].

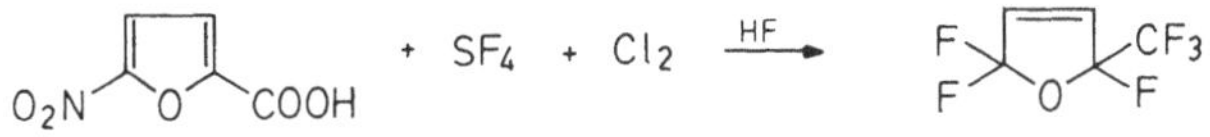

Formation of fluorine-containing dihydrofurans via the stage of conjugated halofluorination is confirmed by the reactions of SF_4–HF–Cl_2 with 2-methoxycarbonyl- and 2-ethoxycarbonylfurans [131]. The reaction of 2-methoxycarbonylfuran gives, depending on the conditions, fluorine-containing dihydrofurans 25–27 corresponding to the products formed at different stages of the reaction of furan derivatives with SF_4–HF–Cl_2.

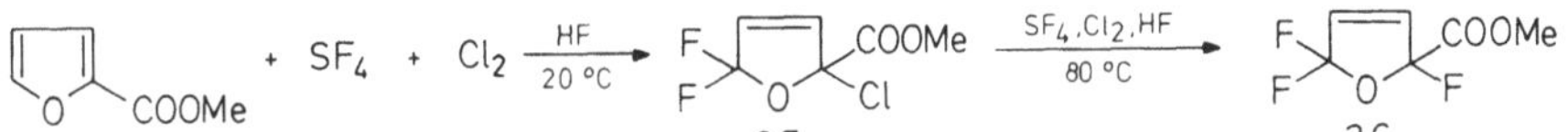

Formation of compound 27 is a first instance of substitution of carbonyl oxygen in an ester group by two fluorines in the reaction with SF_4–HF–Cl_2.

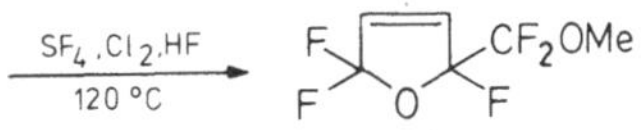

6.4.2 Reactions of Organic Compounds with the SF_4–HF–S_2Cl_2 System

It has been shown recently that sulfur tetrafluoride may be used to obtain fluorine-containing sulfides, disulfides, and sulfene chlorides with good yields.

The reaction of propene with SF_4–HF–S_2Cl_2 under mild conditions gives 2,2'-difluorodipropyl disulfide 28 whose structure corresponds to the addition of stoichiometric equivalents of sulfur monofluoride to two propene molecules. Sulfur tetrafluoride, being a Lewis acid, polarises the S_2Cl_2 molecule, which corresponds to the addition of the ClS_2^+ ion to propene, to form the carbocation

S. The latter reacts with the fluoride ion, giving fluoroalkyldisulfenyl chloride 29 which reacts in a similar way with the next propene molecule to form the disulfide 28 [50].

$$CH_3CH=CH_2 + SF_4 + S_2Cl_2$$

$$\xrightarrow{HF} [CH_3\overset{+}{C}HCH_2SSCl \xrightarrow{F^-} CH_3CHFCH_2SSCl]$$
$$ S 29$$

$$29 + SF_4 + S_2Cl_2 + CH_3CH=CH_2 \rightarrow CH_3CHFCH_2SSCH_2CHFCH_3$$
$$ 28$$

The addition of stoichiometric equivalents of "S_2F_2" to propene proceeds according to the Markownikov rule. Substitution of the methyl group of propene by the trifluoromethyl one completely changes orientation of the addition. 3,3,3-Trifluoropropene reacts with SF_4–HF–S_2Cl_2, forming a mixture of two stereoisomeric sulfides 30 [132].

$$CF_3CH=CH_2 + SF_4 + S_2Cl_2 \xrightarrow{HF} (CH_2FCH)_2S$$
$$ | $$
$$ F_3C$$
$$ 30$$

Formation of sulfides but not disulfides in this reaction, as in the reaction of propene, is explained in the following way. 3,3,3-Trifluoropropene is initially transformed to thiosulfene chloride 31 which undergoes cleavage by chlorine to give sulfene chloride 32. The reaction of the latter with the next 3,3,3-trifluoropropene molecule leads to sulfides 30 [132].

$$CF_3CH=CH_2 + SF_4 + S_2Cl_2 \xrightarrow{HF} [^+CH_2CH(CF_3)SSCl$$

$$\xrightarrow{F^-} CH_2FCH(CF_3)SSCl \xrightarrow[-SCl_2]{Cl_2} CH_2FCH(CF_3)SCl]$$
$$ 32 31$$

$$\xrightarrow{CF_3CH=CH_2, SF_4, HF} 30$$

Perhaloolefins react with SF_4–HF–S_2Sl_2 under more severe conditions than 3,3,3-trifluoropropene. In this case the reaction stops at the stage of formation of sulfene chlorides. 1,2-Dichlorodifluoroethylene reacts with SF_4–HF–S_2Cl_2 at temperatures above 60 °C, giving 1,2-dichloroperfluoroethylsulfur chloride.

$$CFCl=CFCl + SF_4 + S_2Cl_2 \xrightarrow[60\,°C]{HF} CF_2ClCFClSSCl$$

$$\xrightarrow[-SCl_2]{Cl_2} CF_2ClCFClSCl$$
$$ 74\%$$

Chlorotrifluoroethylene reacts with SF_4–HF–S_2Cl_2 to form a mixture of regioisomeric chlorotetrafluoroethylsulfur monochlorides 33 and 34.

$$CF_2=CFCl + SF_4 + S_2Cl_2 \xrightarrow{\text{HF}} CF_3CFClSCl + CF_2ClCF_2SCl$$
$$\qquad\qquad\qquad\qquad\qquad\qquad\quad 33,\ 28\% \qquad\quad 34,\ 39\%$$

As shown above, aliphatic ketones and difluoroalkanes react with SF_4–HF–Cl_2, yielding vicinal chlorodifluoroalkanes. In the reaction of ketones and difluoroalkanes with SF_4–HF–S_2Cl_2, no chlorodifluoroalkanes are formed; depending on the temperature, the reaction gives polyfluoroalkyl sulfides and polyfluoroalkyl disulfides. The reaction products always contain molecular sulfur.

Acetone and 2,2-difluoropropane are transformed by SF_4–HF–S_2Cl_2, depending on conditions, to polyfluoroalkyl sulfides 35, 36, or disulfide 37 [50,133].

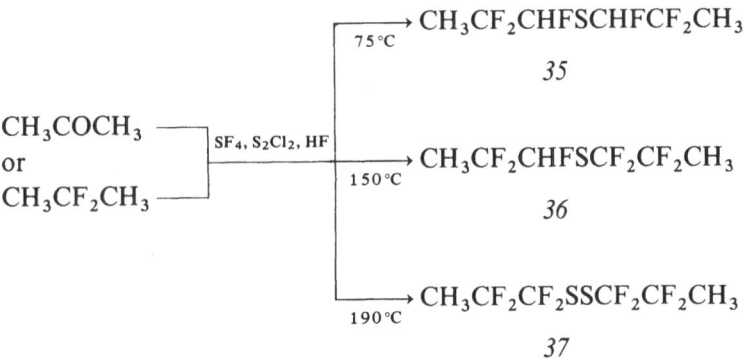

2,2-Difluoropropane formed at the first stage of the reaction of acetone with SF_4–HF reacts with the F_3S^+ ion with F^- elimination to form the carbocation T which releases the proton to give 2-fluoropropene. The latter reacts with SF_4–HF–S_2Cl_2 to yield 2,2,2′,2′-tetrafluorodipropyl sulfide. Its chlorination and subsequent substitution of chlorine by fluorine leads to sulfides 35.

$$CH_3COCH_3 + SF_4 \xrightarrow{\text{HF}} CH_3CF_2CH_3$$

$$\xrightarrow{+SF_3} [CH_3\overset{+}{C}FCH_3 \xrightarrow[-H^+]{} CH_3CF=CH_2$$
$$\qquad\qquad T$$

$$\xrightarrow{SF_4,\ S_2Cl_2,\ HF} (CH_3CF_2CH_2)_2S] \rightarrow 35$$

Ethyl methyl ketone is converted by SF_4–HF–S_2Cl_2 to the sulfide 38 and the disulfide 39. The same products are formed from 2,2-difluorobutane. The yield of disulfide *39* may be considerably improved by performing the reaction in the presence of sulfur [50].

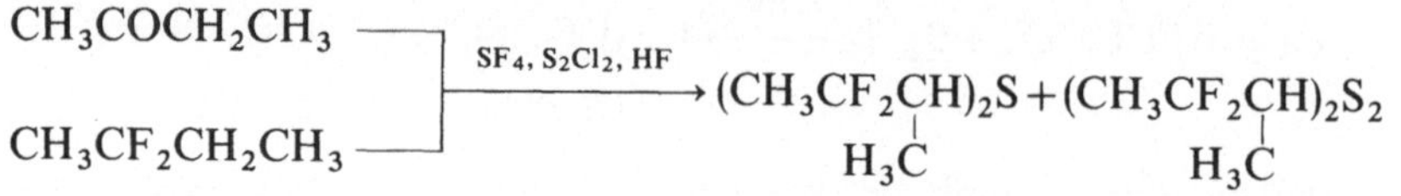

$$CH_3COCH_2CH_3$$
$$\xrightarrow{\text{SF}_4,\ \text{S}_2\text{Cl}_2,\ \text{HF}}$$
$$CH_3CF_2CH_2CH_3$$

$$(CH_3CF_2CH)_2S + (CH_3CF_2CH)_2S_2$$
$$\qquad\quad \overset{|}{H_3C} \qquad\qquad\qquad \overset{|}{H_3C}$$
$$\qquad\quad \textbf{38},\ 75\% \qquad\qquad\qquad \textbf{39},\ 15\%$$

High yields of sulfides and disulfides in the reactions of aliphatic ketones and difluoroalkanes with SF_4–HF–S_2Cl_2 are obtained even when using the catalytic amounts of S_2Cl_2. Hydrogen chloride liberated in the reaction reacts with SF_4, which not only leads to oxidation of the chloride ion but also establishes constant concentration of sulfur monochloride in the reaction mixture.

In the reaction of chlorofluoroolefins with SF_4–HF–S_2Cl_2, high yields of alkylsulfur monochlorides are obtained only when using an excess of sulfur monochloride.

The reactions of the furancarboxylic acids with SF_4–HF–S_2Cl_2 not only lead to substitution of the carboxyl groups by the trifluoromethyl ones, but also to the addition of two fluorine atoms at the 2,5-positions of the furan ring. Furan-2-carboxylic acid is transformed by SF_4–HF–S_2Cl_2 at 20 °C to a mixture of *cis*- and *trans*-2,5-difluoro-2-trifluoromethyl-2,5-dihydrofurans *40* and *41* with a high yield [127]. The reaction proceeds via the formation of compound *42*. The SSCl group in compound *42* is in the allyl position and is therefore easily substituted by fluorine.

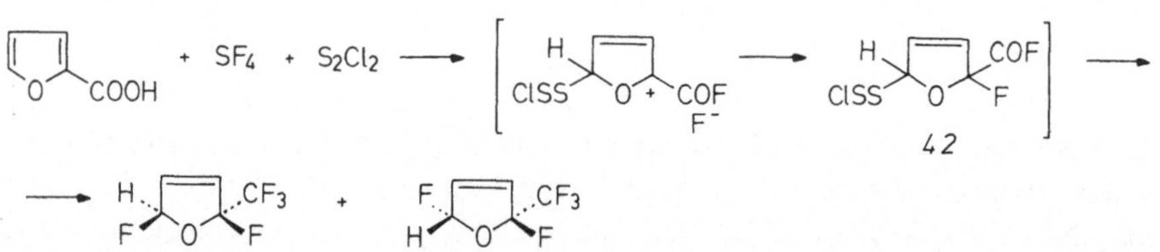

It is interesting to note that in the ^{19}F NMR spectra of compounds *40* and *41*, each type of fluorine atoms has two signals in the ratio of 5:4, having the same structure but different chemical shifts.

2,5-Dihydrofurans are known to have the "half-envelope" structure with an oxygen bridge extending out of the C_2–C_3–C_4–C_5 plane [134]. Therefore, different position of four substituents in the 2,5-positions of the furan ring relative to oxygen suggests possible existence of four diastereomers. This accounts for the formation of two *cis-40a,b* and two *trans-40a,b* in the reaction of furan-2-carboxylic acid with SF_4–HF–S_2Cl_2, and doubled signals in the ^{19}F NMR spectra of compounds *40* and *41*.

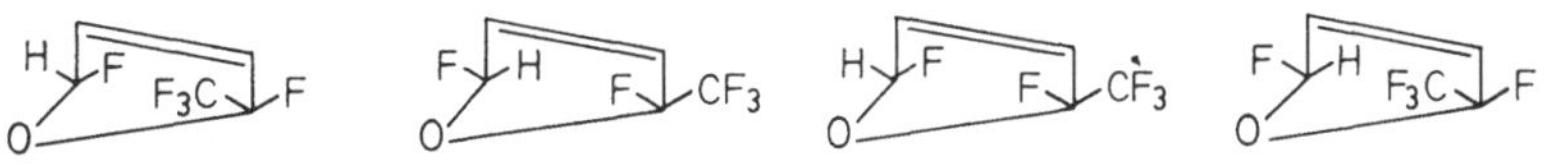

40 a   40 b   41 a   41 b

Increasing of the temperature of this reaction to 90 °C leads to the formation of 2,5,5-trifluoro-2-trifluoromethylfuran *43*. In similar conditions this product is also obtained from dihydrofurans *40* and *41*. The elevated reaction temperature seems to lead to HF elimination from compounds *40* and *41* by SF$_4$ in HF. Subsequently there occurs the addition of stoichiometric equivalents of "S$_2$FCl" at the 2,5-positions of the furan ring and substitution of the SSCl group by fluorine.

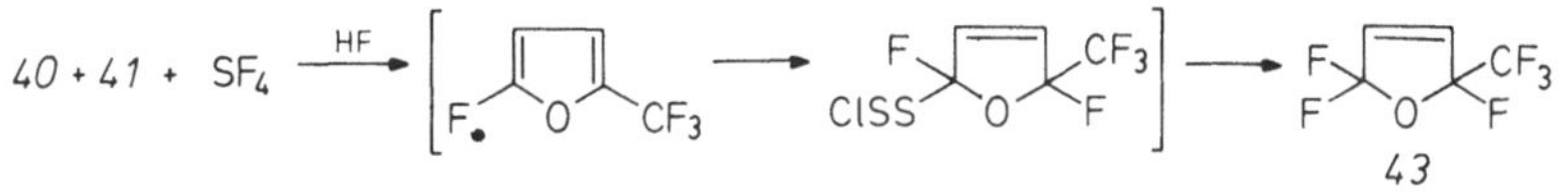

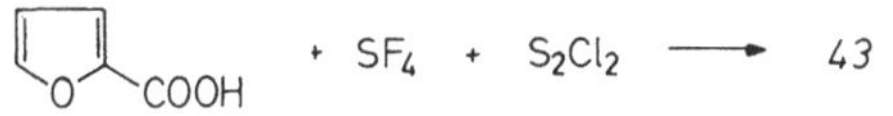

In the reaction of 5-chlorofuran-2-carboxylic acid with SF$_4$–HF–S$_2$Cl$_2$ dihydrofuran *43* is formed in about a quantitative yield even at 20 °C.

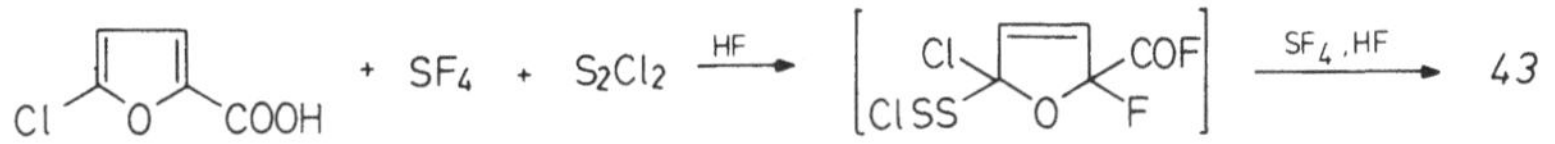

The reaction of furan-2,5-dicarboxylic acid with SF$_4$–HF–S$_2$Cl$_2$ leads to 2,5-difluoro-2,5-bis(trifluoromethyl)-2,5-dihydrofurans [127,131].

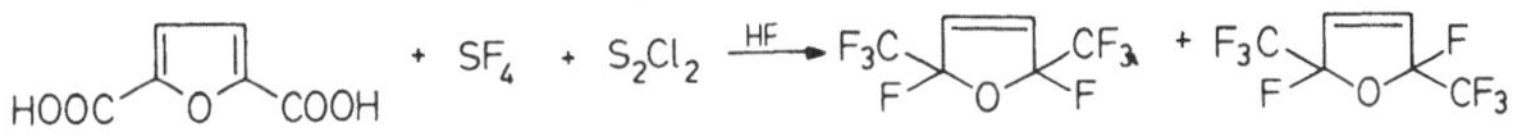

Thus the use of sulfur monochloride together with the SF$_4$–HF system is a method of introducing two fluorine atoms into the 2,5-positions of the furan ring. The addition of halogenating agents to the SF$_4$–HF system substantially expands the synthetic possibilities of sulfur tetrafluoride to afford high yields of chloro- and fluoro-alkanes, linear or cyclic ethers, 2,5-dihydrofurans and their derivatives, alkylsulfur monochlorides, dialkyl sulfides, and disulfides.

6.5 References

1. US Pat 2859245 (1958); (1959) Chem. Abs. 53:12236
2. Hasek WR, Engelhardt VA (1960) J. Amer. Chem. Soc. 82:543
3. Smith WC (1960) J. Amer. Chem. Soc. 82:6176
4. US Pat 2950306 (1960); (1962) Chem. Abs. 55:2569

250   A.I. Burmakov, B.V. Kunshenko, L.A. Alekseeva and L.M. Yagupolskii

5. Kondratenko NV, Syrova GP, Sheinker YuN, Popov VI, Yagupolskii LM (1971) Zh. Obshch. Khim. 41:2056
6. Yagupolskii LM, Kondratenko NV, Popov VI (1976) Zh. Obshch. Khim. 46:620
7. Yagupolskii LM, Lyalin VV, Orda VV (1968) Zh. Obshch. Khim. 38:2813
8. Lyalin VV, Syrova GP, Orda VV, Alekseeva LA, Yagupolskii LM (1970) Zh. Org. Khim. 6:1420
9. Harder RJ, Smith WC (1961) J. Amer. Chem. Soc. 83:3422
10. US Pat 2862029 (1959); (1959) Chem. Abs. 53:9152
11. Smith WC, Tallock CW (1960) J. Amer. Chem. Soc. 52:551
12. Yagupolskii LM, Burmakov AI, Alekseeva LA (1971) in: Reakzii i metody issledovaniya organicheskikh soedinenii, vol 22. Khimiya, Moskva, p 40
13. Boswell GA, Ripka WC (1973) Org. Reactions 21:1
14. Khardin AP, Gorbunov BN, Protopopov PA (1973) Khimiya chetyrekhftoristoi sery. Saratovskii universitet, Saratov, p 201
15. Martin DC (1967) An. New York Acad. Sci. 145:161
16. US Pat 2992073 (1961); (1961) Chem. Abs. 55:27813
17. Tallock CW, Fawcett FS, Smith WC, Gofman DD (1960) J. Amer. Chem. Soc. 82:539
18. Fawcett FS, Tallock CW (1963) Inorg. Synth. 7:119
19. Franz R (1980) J. Fluor. Chem. 15:423
20. US Pat 3399063 (1958); (1958) Chem. Abs. 53:78864
21. Naumann D, Padma DK (1973) Z. anorg. allg. Chem. 401:53
22. Becher W, Massone J (1974) Chem. Ztg. 98:117
23. Markovskii LN, Pashinnik VE, Kirsanov AV: Synthesis 1973:787
24. Middleton JW (1975) J. Org. Chem. 40:574
25. Burmakov AI, Alekseeva LA, Yagupolskii LM (1973) Zh. Org. Khim. 8:153
26. Yagupolskii LM, Burmakov AI, Alekseeva LA (1969) Zh. Obshch. Khim. 39:2053
27. Kunshenko BV, Burmakov AI, Alekseeva LA, Lukmanov VG, Yagupolskii LM (1974) Zh. Org. Khim. 10:886
28. Kunshenko BV, Alekseeva LA, Yagupolskii LM (1972) Zh. Org. Khim. 8:830
29. Kunshenko BV, Alekseeva LA, Yagupolskii LM (1974) Zh. Org. Khim. 10:1698
30. Kunshenko BV, Alekseeva LA, Yagupolskii LM (1973) Zh. Org. Khim. 9:1954
31. Burmakov AI, Alekseeva·LA, Yagupolskii LM (1970) Zh. Org. Khim. 6:144
32. Lukmanov VG, Alekseeva LA, Burmakov AI, Yagupolskii LM (1973) Zh. Org. Khim. 9:1019
33. Lukmanov VG, Alekseeva LA, Yagupolskii LM (1977) Zh. Org. Khim. 13:2129
34. Lukmanov VG, Alekseeva LA, Yagupolskii LM (1974) Zh. Org. Khim. 10:2000
35. Lyalin VV, Grigorash RV, Alekseeva LA, Yagupolskii LM (1975) Zh. Org. Khim. 11:460
36. Grigorash RV, Lyalin VV, Alekseeva LA, Yagupolskii LM (1978) 14:2623
37. Yagupolskii LM, Burmakov AI, Alekseeva LA, Kunshenko BV (1973) Zh. Org. Khim. 9:689
38. Fialkov YuA, Moklyachuk LI, Kremlev MM, Yagupolskii LM (1980) Zh. Org. Khim. 16:1476
39. US Pat 4237276 (1980)
40. Sheppard WA (1961) J. Amer. Chem. Soc. 83:4860
41. Sheppard WA (1964) J. Org. Chem. 29:1
42. Aldrich PE, Sheppard WA (1964) J. Org. Chem. 29:11
43. Belous VM, Alekseeva LA, Yagupolskii LM (1975) Zh. Org. Khim. 11:1672

44. Alekseeva LA, Belous VM, Yagupolskii LM (1974) Zh. Org. Khim. 10:1053
45. Yagupolskii LM, Belous VM, Alekseeva LA (1976) Zh. Org. Khim. 12:1287
46. Belous VM, Litvinova KD, Alekseeva LA, Yagupolskii LM (1976) Zh. Org. Khim. 12:1798
47. Pasquale (1973) J. Org. Chem. 38:3025
48. US Pat 4210710 (1979)
49. US Pat 4201876 (1981)
50. Muratov NN, Mokhamed NM, Kunshenko BV, Alekseeva LA, Yagupolskii LM (1985) Zh. Org. Khim. 21:1420
51. Bardon J, Childs AC, Parsons LM: J. Chem. Soc. Chem. Commun. 1982:534
52. Yagupolskii LM, Kunshenko BV, Alekseeva LA (1968) Zh. Obshch. Khim. 38:2592
53. Gopal N, Snyder CE, Tamborski C (1979) J. Fluor. Chem. 14:511
54. Conlin RT, Frey NM: J. Chem. Soc. Faraday Trans. 1979:2556
55. Ishikawa N, Kitazume T, Takaoka A (1979) J. Synth. Org. Chem. Jap. 37:606
56. Martin DG, Kagan F (1962) J. Org. Chem. 27:3164
57. Tadanier J, Cole W (1961) J. Org. Chem. 26:2436
58. Boswell GA (1966) J. Org. Chem. 31:991
59. Azeem M, Brownstein M (1969) Can. J. Chem. 47:4159
60. Lyalin VV, Grigorash RV, Alekseeva LA, Ýagupolskii LM (1975) Zh. Org. Khim. 11:1086
61. Dmowski W, Kolinski A (1978) Polish J. Chem. 52:71
62. Kunshenko BV, Alekseeva LA, Yagupolskii LM (1970) Zh. Org. Khim. 6:1286
63. Burmakov AI, Alekseeva LA, Yagupolskii LM (1969) Zh. Org. Khim. 5:1892
64. Yagupolskii LM, Lukmanov VG, Boiko VN, Alekseeva LA (1977) Zh. Org. Khim. 13:2388
65. Burmakov AI, Alekseeva LA, Yagupolskii LM (1970) Zh. Org. Khim. 6:2498
66. Blakitnii AN, Zalesskaya IM, Kunshenko BV, Fialkov YuA, Yagupolskii LM (1977) Zh. Org. Khim. 13:2149
67. Alekseeva LA, Belous VM, Lozinskii MO, Shendrik VP, Yagupolskii LM (1983) Ukr. Khim. Zh. 49:74
68. US Pat 2894989 (1959); (1961) Chem. Abs. 55:420
69. US Pat 4249009 (1981)
70. Mertes MP, Saheb SE (1963) J. Pharm. Sci. 52:508
71. Mertes MP, Saheb SE (1963) J. Med. Chem. 6:619
72. Sakai TT, Santi DV (1973) J. Med. Chem. 16:1079
73. US Pat 3324126 (1967); (1968) Chem. Abs. 68:39646
74. Shirota FN, Nagasawa HT, Elberling JA (1977) J. Med. Chem. 20:1176
75. Burmakov AI, Khassanein SM, Kunshenko BV, Alekseeva LA, Yagupolskii LM (1986) Zh. Org. Khim. 22:1273
76. Kozlova AM, Sedova LN, Alekseeva LA, Yagupolskii LM (1973) Zh. Org. Khim. 9:1418
77. Bell HH, Hudlicky M (1980) J. Fluor. Chem. 15:191
78. Burmakov AI, Motnyak LA, Kunshenko BV, Alekseeva LA, Yagupolskii LM (1981) J. Fluor. Chem. 19:151
79. Kryukova LYu, Kryukov LN, Kolomietz AF, Sokolskii GA, Knunyants IL: Izv. Akad. Nauk SSSR. Ser. Khim. 1979:1913
80. Khasanein SM, Burmakov AI, Bloshchiza FA, Yagupolskii LM (1987) Zh. Org. Khim. 22:1111
81. Stepanov IV, Burmakov AI (1985) Zh. Org. Khim. 21:45

82. Stepanov IV, Burmakov AI, Alekseeva LA, Yagupolskii LM (1986) Zh. Org. Khim. 22:227
83. Burmakov AI, Motnyak LA, Kunshenko BV, Alekseeva LA, Yagupolskii LM (1980) Zh. Org. Khim. 16:1401
84. Motnyak LA, Burmakov AI, Kunshenko BV, Sass VP, Alekseeva LA, Yagupolskii LM (1981) Zh. Org. Khim. 17:728
85. Motnyak LA, Burmakov AI, Kunshenko BV, Neizvestnaya TA, Alekseeva LA, Yagupolskii LM (1983) 19:720
86. Muratov NN, Burmakov AI, Kunshenko BV, Alekseeva LA, Yagupolskii LM (1982) Zh. Org. Khim. 18:1403
87. Motnyak LA, Burmakov AI, Kunshenko BV, Neizvestnaya TA, Alekseeva LA, Yagupolskii LM (1984) Zh. Org. Khim. 20:1169
88. Kollonitsch J, Marburg S, Perkins L (1975) J. Org. Chem. 40:3808
89. Kollonitsch J (1978) Isr. J. Chem. 17:53
90. US Pat 4096180 (1978)
91. US Pat 4215221 (1980)
92. US Pat 4288601 (1981)
93. Stepanov IV, Burmakov AI, Kunshenko BV, Alekseeva LA, Yagupolskii LM (1982) Zh. Org. Khim. 22:1812
94. Burmakov AI, Stepanov IV, Kunshenko BV, Sedova LN, Alekseeva LA, Yagupolskii LM (1982) Zh. Org. Khim. 18:1168
95. Hodges KS, Schomburg D, Weise, Schmutzler R (1977) J. Amer. Chem. Soc. 99:6096
96. Bloshchiza FA, Burmakov AI, Kunshenko BV, Alekseeva LA, Yagupolskii LM (1985) Zh. Org. Khim. 21:1414
97. Bloshchiza FA, Burmakov AI, Kunshenko BV, Alekseeva LA, Bel'ferman AL, Pazderskii YuA, Yagupolskii LM (1981) Zh. Org. Khim. 17:1417
98. Bloshchiza FA, Burmakov AI, Kunshenko BV, Alekseeva LA, Yagupolskii LM (1982) Zh. Org. Khim. 18:782
99. Bloshchiza FA, Burmakov AI, Kunshenko BV, Alekseeva LA, Yagupolskii LM (1986) Zh. Org. Khim. 22:750
100. Bloshchiza FA, Burmakov AI, Kunshenko BV, Alekseeva LA, Yagupolskii LM (1983) Zh. Org. Khim. 19:1761
101. Kul'chzkii MM, Il'chenko AYa, Yagupolskii LM (1973) Zh. Org. Khim. 9:827
102. Lack RE, Ganter C, Roberts JD (1968) J. Amer. Chem. Soc. 90:7001
103. Stepanov IV, Burmakov AI, Kunshenko BV, Alekseeva LA, Yagupolskii LM (1983) Zh. Org. Khim. 19:273
104. Kaufmann H, Fuher H, Kavloda T (1981) Tetrahedron 37:225
105. Burmakov AI, Bloshchiza FA, Kunshenko BV, Alekseeva LA, Yagupolskii LM (1980) Zh. Org. Khim. 16:2617
106. Appleguist DE, Searle R (1964) J. Org. Chem. 29:987
107. US Pat 4431786 (1983)
108. Spasov SL, Griffith DL, Glazer ES, Nagarayan K, Roberts JD (1967) J. Amer. Chem. Soc. 89:88
109. Khardin AN, Popov AD, Protopopov PA (1977) Zh. Vses. Khim. O-va 22:116
110. Aleksandrov AM, Danilenko TI, Konovalov EV, Krasnoshchek AN, Medvedeva TP (1974) Zh. Org. Khim. 10:1548
111. Aleksandrov AM, Kukhar' VP, Danilenko GI, Krasnoshchek AP (1977) Zh. Org. Khim. 13:1629
112. Aleksandrov AM, Sorochinskii AE, Krasnoshchek AP (1979) Zh. Org. Khim. 15:336

113. Dmowski W, Kaminski M (1983) J. Fluor. Chem. 23:207
114. Kunshenko BV, Muratov NN, Burmakov AI, Alekseeva LA, Yagupolskii LM (1983) J. Fluor. Chem. 22:105
115. Muratov NN, Kunshenko BV, Burmakov AI, Alekseeva LA, Yagupolskii LM (1984) Zh. Org. Khim. 20:450
116. Padma DK (1977) Phosphorus and Sulphur 3:19
117. Knunyants IL, German LS: Izv. Akad. Nauk SSSR. Ser. Khim. 1966:1065
118. Pattison FLM, Peters DAV, Dean FH (1965) Can. J. Chem. 43:1689
119. Olah GA, Nojima M, Kerekes I: Synthesis 1973:780
120. Muratov NN, Omarov VO, Kunshenko BV, Burmakov AI, Alekseeva LA, Yagupolskii LM (1986) Zh. Org. Khim. 22:1806
121. Sargent PB, Krespan CG (1969) J. Amer. Chem. Soc. 91:415
122. Dear REA, Gilbert EE, Murray JJ (1971) Tetrahedron 27:3345
123. Kunshenko BV, Motnyak LA, Neizvestnaya TA, Il'nitzkii SO, Yagupolskii LM (1986) Zh. Org. Khim. 22:1791
124. Grigorash RV, Lyalin VV, Alekseeva LA, Yagupolskii LM (1978) Zh. Org. Khim. 14:844
125. Sherman WR, Freiffider M, Stone GR (1960) J. Org. Chem. 25:2048
126. Grigorash RV, Lyalin VV, Alekseeva LA, Yagupolskii LM: Khim. geterozikl. soed. 1977:1607
127. Kunshenko BV, Il'nitzkii SO, Motnyak LA, Lyalin VV, Yagupolskii LM (1987) Zh. Org. Khim. 23:833
128. Wielgat J, Domagaza Z (1982) J. Fluor. Chem. 20:785
129. Lyalin VV, Grigorash RV, Alekseeva LA, Yagupolskii LM (1981) Zh. Org. Khim. 17:1774
130. Lyalin VV, Grigorash RV, Alekseeva LA, Yagupolskii LM (1984) Zh. Org. Khim. 20:846
131. Kunshenko BV, Ilnitskii SO, Motnyak LA, Lyalin VV, Burmakov AI, Yagupolskii LM: Bull. Soc. Chim. Fr. 1986:974
132. Muratov NN, Mokhamed NM, Kunshenko BV, Alekseeva LA, Yagupolskii LM (1986) Zh. Org. Khim. 22:964
133. Kunshenko BV, Muratov NN, Burmakov AI, Alekseeva LA, Yagupolskii LM (1983) Zh. Org. Khim. 19:1342

7 Fluorination of Organic Compounds with Fluorosulfuranes

Leonid Nikolaevich Markovskii and **Valerii Efimovich Pashinnik**

Institute of Organic Chemistry, Murmanskaya St, 5, 252094 Kiev, USSR

Contents

7.1 Introduction

The sulfur tetrafluoride derivatives, such as alkyl- and aryltrifluorosulfuranes, dialkylaminotrifluoro- and bis(dialkylamino)difluorosulfuranes, and polyfluoro-alkoxytrifluorosulfuranes, are widely used in organic synthesis to substitute fluorine for oxygen in alcohols, aldehydes, ketones and acids, and for sulfur in thiocarbonyl compounds.

As opposed to sulfur tetrafluoride, most of the above fluorosulfuranes are liquids or crystalline substances, which allows one to conduct the reactions at atmospheric pressure and in glassware. Some derivatives of fluorosulfuranes, in particular, dialkylaminotrifluorosulfuranes, are more reactive than SF_4. All these factors provide high selectivity of fluorination, which is especially important in the case of polyfunctional compounds.

7.2 Fluorination with Alkyl- and Aryltrifluorosulfuranes

Alkyl- and aryltrifluorosulfuranes may be obtained by treatment of sulfur-containing compounds with fluorine, chlorine fluorides and iodine fluorides, and trifluoromethyl hypofluorite; by the electrochemical fluorination of sulfur-containing compounds; and by the addition of SF_4 to perfluoroolefins [1–11]. The most widely spread method for the preparation of these compounds is the general synthesis from dialkyl- and diaryl disulfides and silver difluoride suggested by Sheppard [12–14].

$$RSSR + 6AgF_2 \rightarrow 2RSF_3 + 6AgF$$

Thus treatment of diphenyl disulfide with AgF_2 gives phenyltrifluorosulfurane with a 60% yield, which is most frequently used for fluorination.

The reactions with aliphatic aldehydes and ketones are exothermal and are best controlled by using such solvents as dichloromethane or acetonitrile. The yields of the difluorides in these reactions do not exceed 50% [13].

$$AlkCHO + RSF_3 \rightarrow AlkCHF_2 + RSOF$$

$$AlkCOAlk' + RSF_3 \rightarrow AlkCF_2Alk' + RSOF$$

In mild conditions alkyl- and aryltrifluorosulfuranes also react with aromatic aldehydes. Ease of the reaction considerably depends on the electronic nature of the substituent in trifluorosulfurane. Thus the reaction of benzaldehyde with phenyltrifluorosulfurane at room temperature leads to benzal fluoride with a 80% yield [13]. A similar reaction with pentafluorophenyltrifluorosulfurane required heating to 100 °C, and the yield of benzal fluoride does not exceed 52% [14].

$$C_6H_5CHO + ArSF_3 \rightarrow C_6H_5CHF_2$$

$$Ar = C_6H_5 (80\%), C_6F_5 (52\%)$$

The aromatic ketones react with phenyltrifluorosulfurane in much more severe conditions, upon heating to 150 °C and in the presence of Lewis acids, for example, TiF_4 [13].

$$C_6H_5COC_6H_5 + PhSF_3 \xrightarrow{TiF_4} C_6H_5CF_2C_6H_5$$

The transformation of the carboxyl group to the trifluoromethyl one under the action of alkyl- and aryltrifluorosulfuranes occurs upon heating of the reagents to 120 to 150 °C in the presence of Lewis acids [13].

$$CH_3(CH_2)_5COOH + 2RSF_3 \rightarrow CH_3(CH_2)_5CF_3 + 2RSOF + HF$$

$$R = CF_3, C_6H_5$$

By contrast with similar reactions of SF_4 where such widespread Lewis acids as BF_3 and HF are used, these catalysts cannot be used for the reaction with phenyltrifluorosulfurane because of their high volatility, though good results have been obtained using a non-volatile Lewis acid—titanium tetrafluoride.

Dibasic acids such as acetylenedicarboxylic acid, also react with $PhSF_3$, with both carboxyl groups converted to the trifluoromethyl ones [15].

The relative reactivities of alkyl- and aryltrifluorosulfuranes and SF_4 in similar reactions have not been determined because of the different physical state of the reagents and the difficulty to compare the results. But the routes of substitution of oxygen by two fluorine atoms in the reactions of aldehydes and ketones with alkyl- and aryltrifluorosulfuranes are supposed to be the same as in the similar reactions of SF_4 [16].

7.3 Fluorination with Polyfluoroalkoxytrifluorosulfuranes

The subsitution of the hydroxy group in alcohols, and of the carbonyl group in aldehydes and ketones under the action of SF_4 is supposed to proceed via the stage of formation of alkoxytrifluorosulfuranes which rearrange with elimination of thionyl fluoride to form the respective fluoro-derivatives [16].

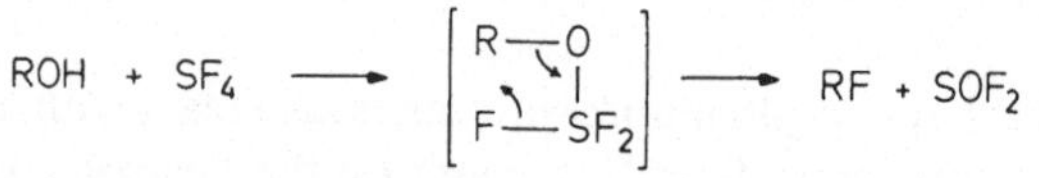

$$ROH + SF_4 \longrightarrow \begin{bmatrix} R-O \\ \diagdown \ | \\ F - SF_2 \end{bmatrix} \longrightarrow RF + SOF_2$$

Recent synthesis of stable polyfluoroalkoxytrifluorosulfuranes and studies of their thermolysis allowed to establish some tendencies of their thermal transformations leading to fluorinated products. α,α,ω-Trihydropolyfluoroalkoxytrifluorosulfuranes were obtained by the reaction of alcohols of the general formula $H(CF_2)_nCH_2OH$ (n = 2, 4, 6) and SF_4 in the presence of potassium or sodium fluorides [17].

$$H(CF_2)_nCH_2OH + SF_4 \xrightarrow[-70\,to\,-50\,°C]{KF \ or \ NaF} H(CF_2)_nCH_2OSF_3$$

The thermal stability of these compounds considerably depends on the size of polyfluoroalkyl substituent, increasing from sulfurane with n = 2 to sulfurane with n = 4 and 6. The transformations of polyfluoroalkoxytrifluorosulfuranes have been found to depend on the thermolysis conditions. Prolonged keeping of alkoxytrifluorosulfurane with n = 6 at 5 °C leads to symmetrization products, and heating of it to 180 °C during 30 min affords fluoroalkane with a quantitative yield [17].

$$H(CF_2)_6CH_2OSF_3 \begin{cases} \xrightarrow{5\,°C} SF_4 + [H(CF_2)_6CH_2O]_4S \\ \xrightarrow{180\,°C} SOF_2 + H(CF_2)_6CH_2F \end{cases}$$

Polyfluoroalkoxytrifluorosulfuranes may themselves behave as fluorinating agents. The reactions of 1,1,7-trihydrododecafluoroheptoxytrifluorosulfurane with carboxylic acids and their derivatives have been studied. The reaction route was shown to depend on the reaction temperature and the presence of Lewis acids in the reaction mixture. In the absence of Lewis acids, the reaction produces only acyl fluorides [18,19].

$$RCOOX + H(CF_2)_6CH_2OSF_3 \rightarrow RCOF + H(CF_2)_6CH_2OSOF + XF$$

$$R = CH_3, C_6H_5; \ X = H, Na, SiMe_3$$

In the presence of Lewis acids, the critical effect on the reaction route is produced by the temperature. At low temperatures, the reaction products are acyl fluorides, above $0\,°C$ the reaction gives polyfluoroalkyl esters of these acids [19].

$$\text{RCOOH} + \text{H(CF}_2)_6\text{CH}_2\text{OSF}_3 \underset{0\,°C}{\overset{-40\text{ to }-5\,°C}{\underset{\longrightarrow}{\overset{\longrightarrow}{\rule{0pt}{2.5em}}}}} \begin{array}{l} \text{RCOF} + \text{H(CF}_2)_6\text{CH}_2\text{OSOF} + \text{HF} \\[1em] \text{RCOOCH}_2(\text{CF}_2)_6\text{H} + \text{SOF}_2 + \text{HF} \end{array}$$

Polyfluoroalkoxytrifluorosulfuranes were shown to react with Lewis acids to form the polyfluoroalkoxydifluorosulfonium salt which is stable to $-5\,°C$, and above this temperature decomposes, liberating thionyl fluoride [19].

$$\text{H(CF}_2)_6\text{CH}_2\text{OSF}_3 + \text{A} \rightarrow [\text{H(CF}_2)_6\text{CH}_2\overset{+}{\text{O}}\text{SF}_2]^+\text{AF}^-$$

$$\underset{\overset{+}{\text{AF}^-}}{\text{H(CF}_2)_6\text{CH}_2\overset{+}{\text{O}}\text{S}\,\text{F}_2} \underset{\substack{-5\,°C \\ -\text{SOF}_2}}{\overset{\substack{\text{RCOOX} \\ -40\text{ to }-5\,°C}}{\begin{array}{l} \xrightarrow{\hspace{2em}} [\text{RCOOSF}_2\text{OCH}_2(\text{CF}_2)_6\text{H}] \\[1em] \rightarrow \text{RCOF} + \text{H(CF}_2)_6\text{CH}_2\text{OSOF} \\[1em] \xrightarrow{\hspace{2em}} [\text{H(CF}_2)_6\overset{+}{\text{C}}\text{H}_2 \quad \text{AF}^-] \end{array}}}$$

$$\xrightarrow{\text{RCOOX}} \text{RCOOCH}_2(\text{CF}_2)_6\text{H} + \text{XAF}$$

$\text{A} = \text{HF, BF}_3, \text{SbF}_5$

7.4 Fluorination with Dialkylaminotrifluoro- and Bis(dialkylamino)difluorosulfuranes

Much more accessible fluorinating agents are dialkylaminotrifluorosulfuranes (DAST). Methods for their preparation are based on the reaction of SF_4 with secondary amines, dialkylaminotrimethylsilanes, tetraalkyldiamides of methylphosphonic acid, tetraalkyldiamides of sulfinic acid, dialkylamides of alkylsulfinic acid, etc. [20–30,32]. The most widespread method for the synthesis of dialkylaminotrifluorosulfuranes has become the reaction of SF_4 with dialkylaminotrimethylsilanes in a solvent at low temperatures [26–28, 32].

$$\text{R}_2\text{NSiMe}_3 + \text{SF}_4 \xrightarrow[-70\text{ to}-50\,°C]{} \text{R}_2\text{NSF}_3 + \text{Me}_3\text{SiF}$$

$$\text{R} = \text{Me, Et; RR} = -(\text{CH}_2)_4-, \ -(\text{CH}_2)_5-, \ -(\text{CH}_2)_2\text{O}(\text{CH}_2)_2-$$

DAST are mobile liquids distillable in vacuum, easily hydrolyzable and they decompose slowly upon storage. The most stable compound among the DAST studied is morpholinotrifluorosulfurane, it may be stored in the absence of air moisture for several years. Diethylaminotrifluorosulfurane, frequently used to fluorinate various types of compounds, is a commercially available product.

The high reactivity and selectivity of DAST allows one to use them to substitute fluorine for the carbonyl oxygen in aldehydes and ketones, sulfur in thiocarbonyl compounds, hydroxy group in alcohols, rather reactive chlorine in compounds of various types, and to transform the carboxyl group to the fluoroformyl one [26,27,31].

Aliphatic aldehydes and ketones react with DAST with evolution of heat. The yields of fluorination products exceed 55% [27,33–41].

$$AlkCHO + R_2NSF_3 \rightarrow AlkCHF_2 + R_2NSOF$$

$$Alk_2CO + R_2NSF_3 \rightarrow Alk_2CF_2 + R_2NSOF$$

If a molecule has several functional groups, the reaction with DAST results in their selective exchange for fluorine atoms. Thus in the fluorination of dispiro 3.1.3.1 decan-5,10-dione by an equimolar amount of DAST in CCl_4 at 25 °C for 30 h, only one keto group is substituted. Treatment of the resulting difluoro-derivative with the fluorinating agent in the same conditions results in the fluorination of another carbonyl group [37].

In the reaction of DAST with oxo carboxylic acids under mild conditions, the carbonyl oxygen is substituted by fluorines. In the case of ethyl 3-oxocarboxyla-tes, the fluorination proceeds at 20 °C during 50 h, the yield of the difluoride being 75% [39].

$$PhCH_2COCH_2COOEt \xrightarrow{\text{DAST}} PhCH_2CF_2CH_2COOEt$$

However in the case of the COOH group, in the reaction of 4-oxo carboxylic acids, there occurs fluorination and subsequent cyclization leading to fluorolac-tones with a high yield [40].

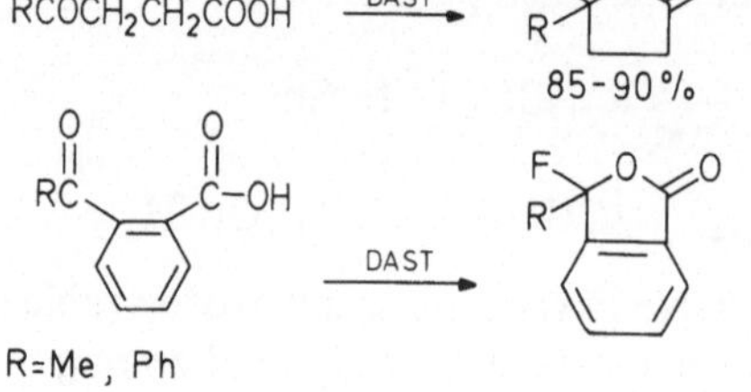

R=Me, Ph

As a rule, the reactions of aldehydes and ketones with SF_4 are conducted in the presence of Lewis acids. Many aldehydes and ketones are unstable in acids, and DAST are therefore indispensable reagents for fluorination of such compounds.

It has been shown in the case of fluorination of pivalic aldehyde that the yield of the reaction products essentially depends on the solvent polarity. Thus in nonpolar solvents such as $CFCl_3$ or pentane, the main product is gem-difluoride, the yields of the skeletal rearrangement products being no more than 13%. On passing to polar solvents, the yield of the difluoride decreases, and in diglyme it does not exceed 30% [27].

$$Me_3CCHO + Et_2NSF_3 \rightarrow Me_3CCHF_2 + CH_2{=}CMeCMeHF + FCMe_2CHFCH_3$$

Pentane	88%	2%	10%
Diglyme	30%	32%	38%

On the basis of these data, a scheme for fluorination of aldehydes with DAST has been suggested. The reaction is supposed to start with the electrophilic attack of the sulfurane sulfur atom at the carbonyl oxygen to form alkoxydialkylaminosulfurane as an intermediate, whose dissociation to the ion pair and transformations of the latter, depending on conditions, lead to the end products [27].

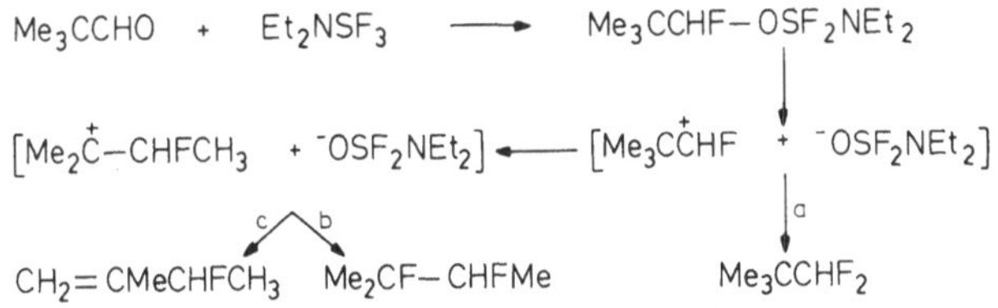

$$Me_3CCHO \;+\; Et_2NSF_3 \longrightarrow Me_3CCHF{-}OSF_2NEt_2$$

$$[Me_2\overset{+}{C}{-}CHFCH_3 \;+\; {}^-OSF_2NEt_2] \longleftarrow [Me_3C\overset{+}{C}HF \;+\; {}^-OSF_2NEt_2]$$

$$CH_2{=}CMeCHFCH_3 \quad Me_2CF{-}CHFMe \qquad Me_3CCHF_2$$

a - non-polar solvent ; - polar non -basic solvent
c - polar basic solvent

Aromatic aldehydes react with DAST less vigorously; as a rule, heating of the reaction mixture to 50 to 60 °C is required. The yields of benzal fluorides are 65–75% [26,27,42,43].

$$ArCHO + Alk_2NSF_3 \rightarrow ArCHF_2 + Alk_2NSOF$$

The carbonyl group in fatty aromatic ketones is substituted by fluorine in much more severe conditions than in aromatic aldehydes. Thus the reaction of DAST with acetophenone proceeds with heating of a mixture of the reagents at 85 °C for 20 h [27].

$$C_6H_5COCH_3 + Et_2NSF_3 \rightarrow C_6H_5CF_2CH_3$$

A similar reaction of DAST with bromoacetophenone requires still more prolonged heating. It proceeds in benzene at 65 °C for 48 h [39].

$$C_6H_5COCH_2Br + Et_2NSF_3 \rightarrow C_6H_5CF_2CH_2Br$$

It was impossible to substitute the carbonyl group by fluorine in benzophenone. Thus mixing of morpholinotrifluorosulfurane with benzophenone does not result in a reaction even upon heating to 110 °C.

Acids of different types react with DAST, forming the respective acyl fluorides. The reaction is highly exothermal, the yields of acyl fluorides being 60–90% [26,44].

$$AcOH + Alk_2NSF_3 \rightarrow AcF + Alk_2NSOF + HF$$

$$Ac = AlkCO, ArCO, R_2PO, CF_3SO_2$$

The oxygen atom of the fluoroformyl group may not be substituted by fluorine even upon heating to 110 °C, and at higher temperatures DAST start to decompose.

However there is an example of substitution of the carbonyl group by fluorine by the reaction with DAST. Heating of an excess of DAST with benzoic acid at 80 °C for 20 h in diglyme gave benzotrifluoride with a 50% yield [42].

DAST may be used for the synthesis of acyl fluorides from the respective chlorides and to substitute fluorine for the reactive chlorine atoms in other compounds [31].

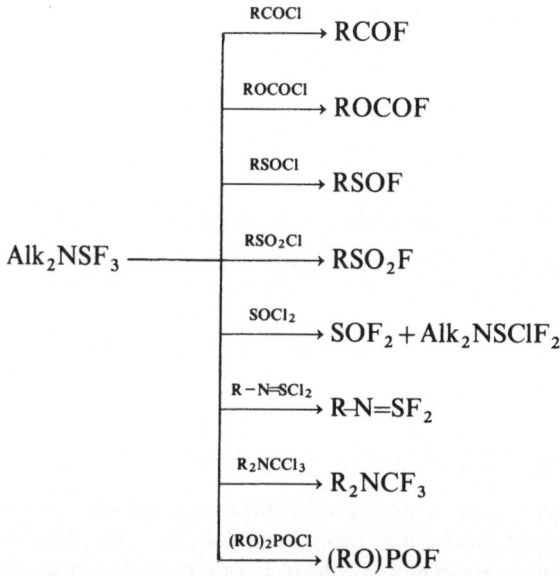

The yields of fluorination products in these reactions are 70–90%. Depending on the reactivity of the chloro-derivative and the method of isolation of the reaction products, the reactions are carried out by mixing the reagents in a solvent or without it.

The interaction of DAST with alcohols has been studied at length, and is a general method of substitution of the hydroxy group by fluorine. DAST in the fluorinations of alcohols has some advantages before other fluorinating agents used to substitute hydroxyl by fluorine, such as SF_4, the SeF_4–pyridine complex, α-fluoroalkylamines, HF, and HF-amine complexes [45–48].

This method is used to transform primary, secondary, and tertiary alcohols in mild conditions to the respective alkyl fluorides in high yields [27]. In the synthesis of low-boiling fluorides, a suitable solvent is diglyme, of high-boiling fluorides—pentane and dichloromethane.

The substitution of alcohol hydroxyl by fluorine under the action of various fluorinating agents is known to be accompanied by side reactions, such as rearrangements of various types and dehydration. Similar reactions are also characteristic for DAST. For instance, the reaction of i-BuOH with DAST leads to i-BuF and t-BuF in the ratio of 2:1, whereas its reaction with SeF$_4$-pyridine complex gives only the rearrangement product—$tert$-butyl fluoride [27].

$$(CH_3)_2CHCH_2OH + Et_2NSF_3 \rightarrow (CH_3)_2CHCH_2F + (CH_3)_3CF$$
$$\;\; 49\% \qquad\qquad 21\%$$

The reaction of DAST with borneol and iso-borneol capable of forming stable carbocations, leads to the rearranged fluorides with 72 and 74% yields respectively [27].

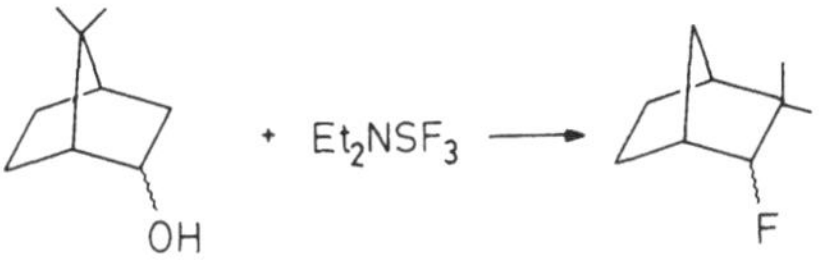

The dehydration of the hydroxy-derivatives in the reactions with DAST occurs to a less extent than in fluorinations with other agents. Thus the reaction of DAST with cyclooctanol leads to cyclooctyl fluoride and cyclooctene in the ratio of 70:30. The reaction of cyclooctanol with the Yarovenko reagent (Et$_2$NCF$_2$CHClF) leads exclusively to the dehydration product—cyclooctene [27,47].

Isomeric 2-buten-1-ol and 3-buten-2-ol react with DAST to form a mixture of 3-fluoro-1-butene and 1-fluoro-2-butene. The formation of the same products from isomeric alcohols may be attributed to the ability of the latter to form the same carbocation, indicating the S$_N$1 mechanism of the reaction [27].

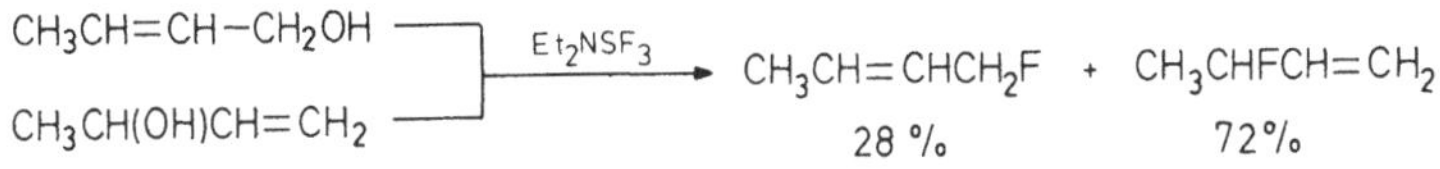

When one and the same molecule contains the hydroxyl and halogen, the latter remains intact in the reaction with DAST and does not produce any marked effect on the conditions of hydroxyl substitution [27,49–51].

$$Br(CH_2)_nCH_2OH + Et_2NSF_3 \rightarrow Br(CH_2)_nCH_2F$$

n = 1, 5, 7, 9

$$BrCH_2CH_2CHOHCH_2Br + Et_2NSF_3 \rightarrow BrCH_2CH_2CHFCH_2Br$$
$$76\%$$

When the starting molecule has both the alcoholic and phenolic hydroxyls, the latter remains intact in the reaction with DAST. Thus, *erythro*-3,4-bis(4-hydroxyphenyl)hexan-1-ol reacts with DAST at 20 °C forming *erythro*-1-fluoro-3,4-bis(4-hydroxyphenyl)hexane with a high yield [52].

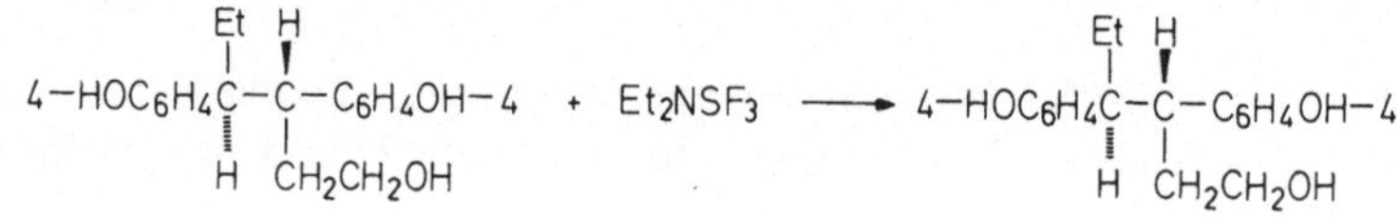

The hydroxyl in fatty aromatic alcohols is substituted as easily as in ordinary aliphatic alcohols [27,53,54].

$$C_6H_5CH_2OH + Et_2NSF_3 \rightarrow C_6H_5CH_2F$$

$$C_6H_5CH_2CH_2OH + Et_2NSF_2 \rightarrow C_6H_5CH_2CH_2F$$

If a molecule of alcohol being fluorinated contains a sufficiently reactive carbonyl group, it also undergoes fluorination; these reactions are often accompanied by HF elimination and formation of unsaturated compounds [55,56]. For example, in the fluorination of 2-hydroxycyclohexanone, the main reaction product is difluorocyclohexene [55].

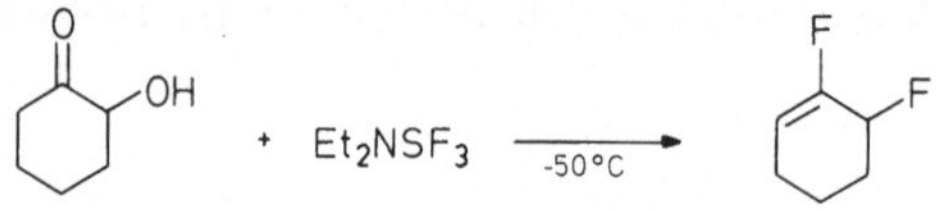

The fluorination of 7,12-dihydroxy-7(12)-methyl-7,12-dihydrobenz[a]anthracenes is often accompanied by dehydrohalogenation specific to anthracene derivatives [57].

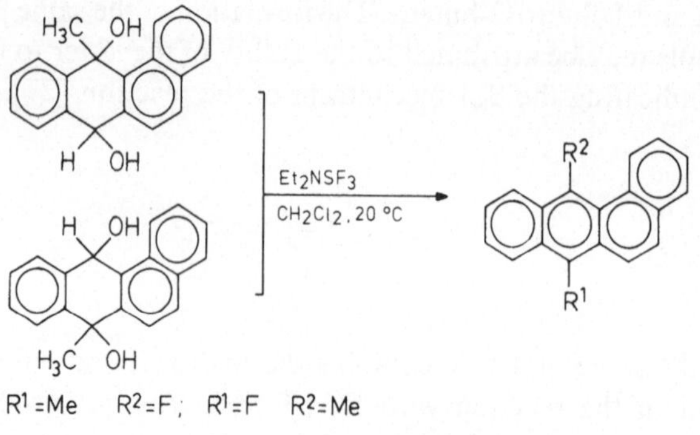

$R^1 = Me$ $R^2 = F$; $R^1 = F$ $R^2 = Me$

The ester and amide groups are stable against DAST. In the hydroxy-compounds containing these groups, only hydroxyl reacts. For example, in the reaction of ethyl lactate with DAST, fluorine is substituted for hydroxyl [27,58].

$$CH_3CHOHCOOEt + Et_2NSF_3 \rightarrow CH_3CHFCOOEt$$
$$78\%$$

In a similar way, with the phosphoryl group remaining intact, DAST react with α-hydroxyphosphonates. The reactions proceed in mild conditions, giving the respective α-fluoroalkyl phosphonates with good yields [59,60].

$$R'C_6H_4CR(OH)PO(OEt)_2 + Et_2NSF_3 \xrightarrow{CH_2Cl_2} R'C_6H_4CRFPO(OEt)_2$$

R=R'=H; R=H, R'=3-Cl, 4-Cl, 4-Me; R=Me, R'=H, 4-Cl

In a similar way DAST react with α-hydroxyphosphinates [60].

$$PhCH(OH)POPh_2 + Et_2NSF_3 \rightarrow PhCHFPOPh_2$$

However α-trimethylsiloxyalkyl phosphonates do not react with DAST, and α-fluoroalkyl phosphonates may not be obtained in this way [60].

In contrast to this, trimethylsiloxyalkanes both of linear and iso-structure react in very mild conditions with DAST, forming fluoroalkanes in high yields. The reactions proceed with complete reversal of the configuration [61].

$$CH_3(CH_2)_5CH(OSiMe_3)CH_3 \xrightarrow{DAST} CH_3(CH_2)_5CHFCH_3$$
$$89\%$$

2-Octanol and 2-octyl tosylate are fluorinated in a similar way [61]. However, in the fluorination of bis(trimethylsiloxy)alkanes having Me_3SiO groups at the primary and tertiary carbon atoms, fluorine is substituted only for the siloxy group at the primary carbon atom.

$$CH_3C(CH_3)(CH_2)_{11}OSiMe_3 \xrightarrow{DAST} CH_3C(CH_3)(CH_2)_{11}F$$
$$|\qquad\qquad\qquad\qquad\qquad\qquad\qquad |$$
$$OSiMe_3 \qquad\qquad\qquad\qquad\qquad OSiMe_3$$
$$50\%$$

In some cases DAST shows a different reactivity towards the primary and secondary hydroxy groups. For example, the interaction of D-(threo)-1-(4-nitrophenyl)-2-phthalimido-1,3-propanediol with DAST at 20 °C leads to D-(threo)-1-(4-nitrophenyl)-2-phthalimido-3-fluoropropan-1-ol [62].

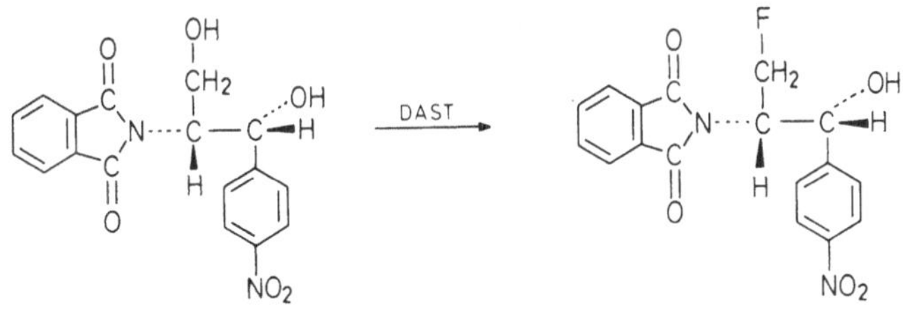

Mild and in many cases selective substitution of the hydroxy group by fluorine using DAST proved to be a convenient method for the synthesis of various physiologically active compounds [43,56,63–117]. DAST was used to synthesize a great number of 5α-androstanes. The fluorination proceeds selectively at room temperature, and only the hydroxy group is substituted by the fluorine atom, and the carbonyl group remains intact [65,66]. In the selective fluorination of the carbonyl group, the hydroxyl must be protected and then at approx. 80 °C the carbonyl group is smoothly transformed to the difluoromethylene one.

The route of the reaction of DAST with steroidal alcohols largely depends on the structure of the steroid [67]. If a molecule has the structure of 5-en-3-ol, the fluorination products are the respective 5-en-3-fluorosteroids. These reactions are supposed to form the intermediate carbocations stabilised by conjugation with the double bond. Substitution of the OH group by fluorine proceeds with preservation of configuration [67].

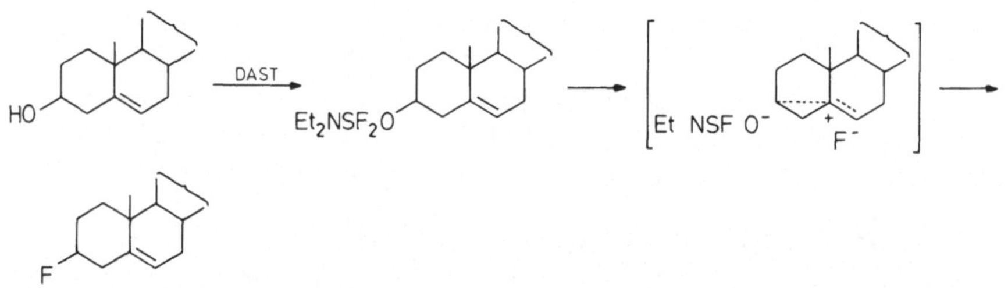

In the absence of the stabilising effect of double bond, the intermediate product is transformed by the S_N2 mechanism with reversal of configuration or via the carbocation, leading to rearrangements.

The reaction of DAST with cholestanol leads to 2-cholestene and 3α-fluorocholestene, and 3β-fluorocholestane is not formed [67].

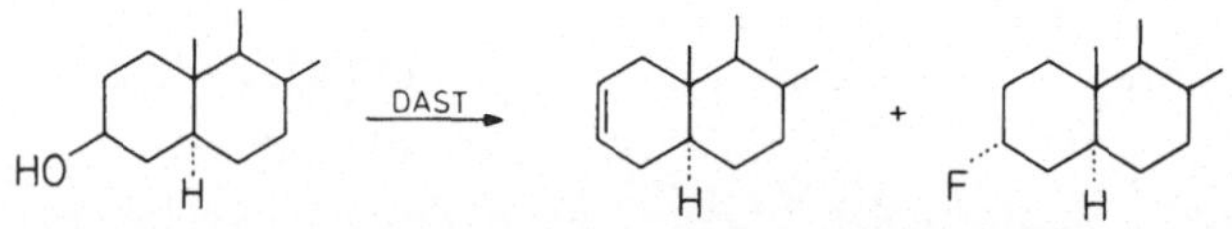

If the reaction involves the *trans-diaxial* elimination, no fluorination products are formed. The main product in the reaction of 5α-pregnan-3α,17α-diol-20-one with DAST is the unsaturated compound formed as a result of HF elimination [67].

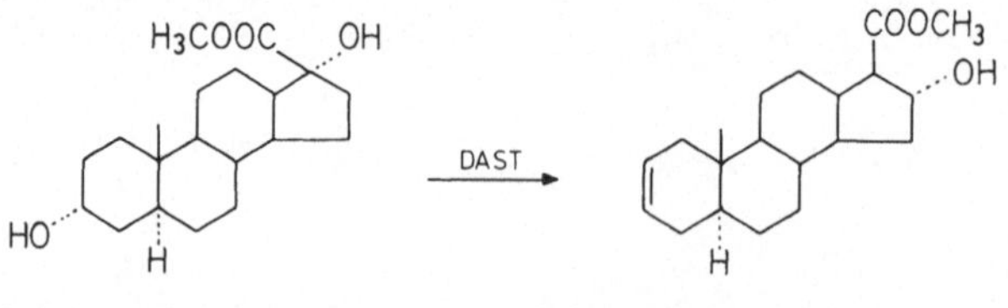

It is also impossible to carry out the fluorination of ergosterol and 7-dehydrositosterol. The reactions of these compounds with morpholinotrifluorosulfurane proceed with transformation of the position of double bonds and do not involve the substitution of the OH group in the 3-position by fluorine, giving the unsaturated compound [68–70].

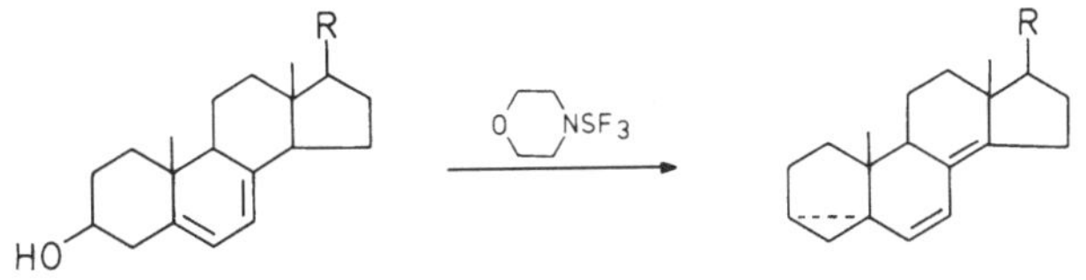

The reaction is assumed to proceed via the intermediate formation of cation A, which is subsequently rearranged to the isomeric cation B stabilised by the proton elimination at C_{14} to form the end product.

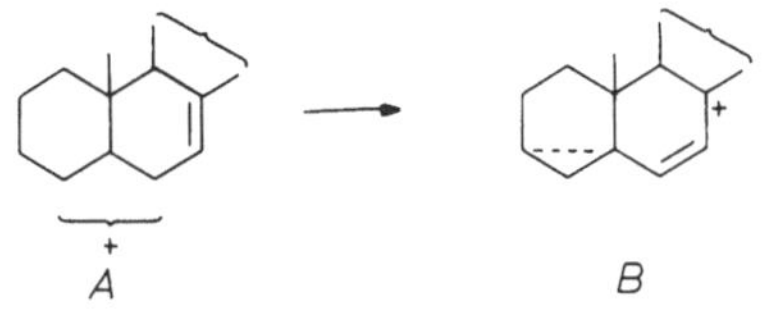

If the 5,7-double bonds in ergosterol and 7-dehydrositosterol are protected, hydroxyl in these compounds may be easily substituted by fluorine.

The reactions of these compounds with morpholinotrifluorosulfurane give the difluoro-derivatives with approx. 50% yield, which after removal of protection from double bonds give 3β-fluoroergosterol and 3β-fluoro-7-dehydrositosterol. Thermal isomerisation of the latter gives 3β-fluorovitamins D_2 and D_5 [68,70].

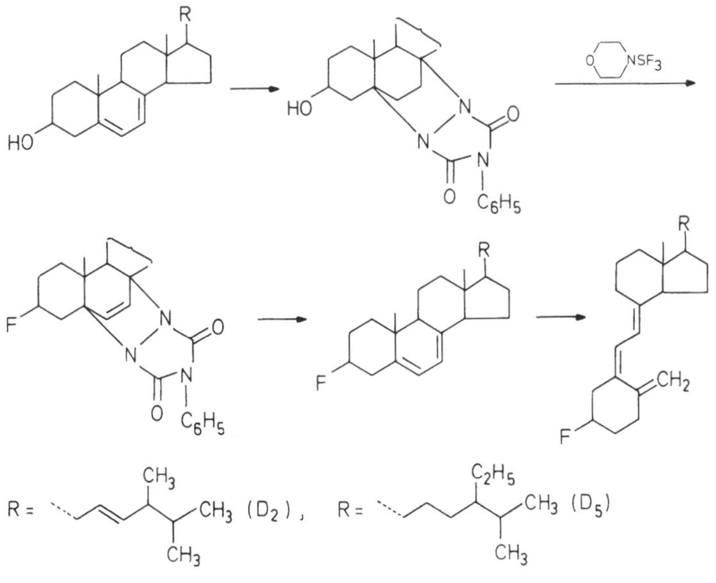

DAST are used in the synthesis of fluorine-containing carbohydrates [79–96], e.g.:

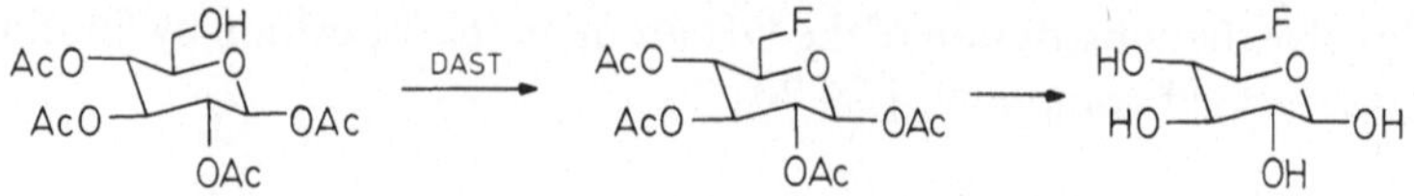

Recently, methods for the selective fluorination of various derivatives of α- and β-D-gluco- and mannopyranosides using DAST have been suggested. The fluorination of methyl-α-D-glucopyranoside with an excess of DAST without a solvent has been shown to proceed with substitution of two hydroxy groups to form methyl-4,6-didesoxy-4,6-difluoro-α-D-glucopyranoside. If the reaction is conducted in dichloromethane, selective monofluorination occurs to give methyl-6-fluoro-α-D-glucopyranoside with a good yield [87,88].

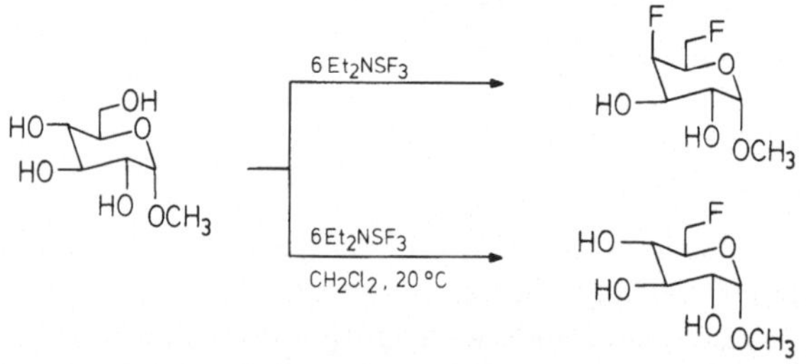

By contrast with α-glucosides, β-glucosides in dichloromethane and under the similar conditions as for α-glucosides, form the 3,6-didesoxy-3,6-difluoro-derivatives, but if the reaction time is reduced to 15 min, only the monofluoro-derivatives of β-glucosides may be obtained [88].

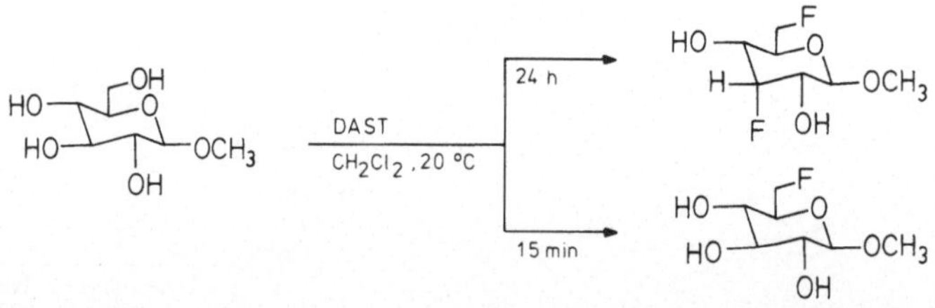

DAST find increasingly wide use in the synthesis of fluorine-containing prostaglandines and prostacyclines. Under the action of morpholinotrifluoro-sulfurane, the hydroxy group in prostaglandines A$_2$ and B$_2$ is substituted in mild conditions by fluorine. The ester and ketone groups remain intact [97–103].

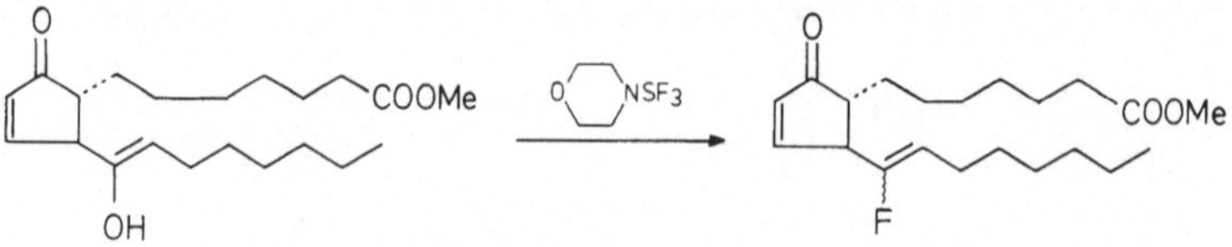

It is known that the steric position of fluorine often produces a critical effect on the type of biological activity. The attempted stereospecific substitution of the hydroxyl in prostaglandines by fluorine in the reaction of dimethylamino- and morpholinotrifluorosulfuranes resulted in the non-stereospecific formation of a mixture of α- and β-epimers (2 : 1) of C_{15}-fluorides. However treatment of methyl ester of prostaglandine A_2 with dimorpholinodifluorosulfurane leads to only one C_{15} epimer [101].

In the fluorination reactions, bis(dialkylamino)difluorosulfuranes are, as a rule, less active than DAST. For example, the reaction of 4-chlorobenzaldehyde with DAST starts at 60 °C and evolves much heat, whereas the similar reaction with dimorpholinodifluorosulfurane proceeds only under boiling in benzene for 1.5 h [29]. Bis(dialkylamino)difluorosulfuranes however may be effectively used to substitute the hydroxy group in highly reactive alcohols. The use of these reagents often allows to reduce side reactions—dehydration and isomerisation, as in the case of fluorination of 2-buten-1-ol and geraniol [27,104].

7.5 Preparations

1. A general procedure for the synthesis of DAST [26]
To a solution of SF_4 (0.12 mol) in 150 ml of anhydrous ether at -78 °C was added dropwise, with stirring, a solution of 0.1 mol of N,N-dialkyl-N-trimethyl-silylamine in 50 ml of ether. The reaction mixture was then gradually heated to 20 °C, an excess of SF_4 and ether was evaporated at reduced pressure, and the residue was fractionated. The yields of DAST were 60–70%. Diethylaminotri-fluorosulfurane: b.p. 43 to 44 °C (12 mm). Morpholinotrifluorosulfurane: b.p. 41 to 42 °C (0.5 mm).

2. Fluorination with aryltrifluorosulfuranes [13]
Phenyltrifluorosulfurane (16.6 g, 0.1 mol) and benzaldehyde (10.6 g, 0.1 mol) were mixed in a 50 ml flask linked to a 45 mm column. Upon mixing of the reagents, an exothermal reaction starts. The reaction mixture was heated on an oil bath to 100 °C, then the pressure in the flask was reduced and benzal fluoride was distilled off (10.2 g, 80%) (b.p. 68 °C (80 mm)).

3. Fluorination of aldehydes and ketones [26]
DAST (0.1 mol) was mixed without solvent at 0 °C with 0.1 mol of carbonyl compound. The mixture was carefully heated to the start of the exothermal reaction, then heated at 60 °C for 15 min, whereupon it was dissolved in 20 ml of dichloromethane or CCl_4. The solution was poured into ice-water, stirred for 10 min, and the organic layer was separated. The mixture was dried and the solvent distilled off. The fluorination products were purified by distillation or recrystallisation. Yields are 55–75%.

4. Fluorination of alcohols [27]

To a solution of DAST in an inert solvent (CH_2Cl_2, pentane, diglyme, etc.) was added with cooling to -50 to $-78\,°C$ an equimolar amount of alcohol. The reaction mixture was then heated to room temperature. The exothermal reaction usually proceeds at a low temperature, but in some cases it occurs upon heating. The low-boiling fluorination products were distilled off at reduced pressure. The reaction mixture containing high-boiling fluorides was treated with water, the organic layer was separated and dried, and the solvent was distilled off. The fluorination products were purified by distillation, recrystallisation or column chromatography. Yields of the fluoro-derivatives in these reactions were 60–90%.

5. Fluorination of acids [26]

To a solution or suspension of 0.01 mol of acid in 30 ml of dry ether was added dropwise, with stirring and cooling with ice water, a solution of 0.01 mol of DAST in 10 ml of ether. The mixture was stirred at $20\,°C$ for 15 min. The acyl fluorides were isolated by filtration or fractional distillation. Yields were 60–90%.

6. Fluorination of acyl chlorides [31]

DAST was added dropwise, with stirring and cooling (in the case of exothermal reaction), to the equimolar amount of acyl chloride. The reaction mixture was stirred at $20\,°C$ for 15 min, and then at $60\,°C$ till the gaseous products ceased to evolve (about 30 min).

7.6 References

1. Tyczkowski EA, Bigelow LA (1953) J. Amer. Chem. Soc. 75:3523
2. Chambelain DL, Kharasch N (1955) J. Amer. Chem. Soc. 77:1041
3. Ratcliffe CT, Shreeve JM (1968) J. Amer. Chem. Soc. 90:5403
4. Sauer DT, Shreeve JM (1971) J. Fluor. Chem. 1:1
5. Abe T, Shreeve JM (1973/1974) J. Fluor. Chem. 3:17
6. Haran G, Sharp DWA (1973/1974) J. Fluor. Chem. 3:423
7. Sprenger GH, Cowley AH (1976) J. Fluor. Chem. 7:333
8. Middleton WJ, Howard EG, Sharkey WH (1965) J. Org. Chem. 30:1375
9. Denney DB, Denney DZ, Hsu YF (1973) J. Amer. Chem. Soc. 95:4064
10. US Pat 3456024 (1969); (1969) Chem. Abs. 71:70078
11. Rosenberg RM, Muetterties EL (1962) Inorg. Chem. 1:756
12. Sheppard WA (1960) J. Amer. Chem. Soc. 82:4751
13. Sheppard WA (1962) J. Amer. Chem. Soc. 84:3058
14. Sheppard WA, Foster SS (1973) J. Fluor. Chem. 2:53
15. Herkes FE, Simons HE (1975) J. Org. Chem. 40:420
16. Hasek WR, Smith WC, Engelhardt VA (1960) J. Amer. Chem. Soc. 82:543
17. Markovski LN, Bobkova LS, Pashinnik VE, Iksanova SV (1981) Zh. Org. Khim. 17:486
18. Markovski LN, Bobkova LS, Jaremenko VV, Pashinnik VE (1983) Zh. Org. Khim. 19:1632

19. Pashinnik VE, Tovstenko VI, Bobkova LS, Markovski LN (1985) Zh. Org. Khim. 21:2072
20. Demitras GG, Kent RA, MacDiarmid AG (1964) Chem. Ind. 38:1712
21. Demitras GG, MacDiarmid AG (1967) Inorg. Chem. 6:1903
22. Halasz SP, Glemser O (1970) Chem. Ber. 103:594
23. Halasz SP, Glemser O (1971) Chem. Ber. 104:1247
24. Gibson JA, Ibbott DG, Janzen AF (1973) Can. J. Chem. 51:3203
25. Brown DH, Crosbie KD, Darragh JI, Ross DS, Sharp DWA: J. Chem. Soc. (A) 1970:914
26. Markovski LN, Pashinnik VE, Kirsanov AV: Synthesis 1973:787
27. Middleton WJ (1975) J. Org. Chem. 40:574
28. Middleton WJ, Bingham EM (1977) Org. Synth. 57:50
29. Markovski LN, Pashinnik VE, Kirsanova NA (1975) Zh. Org. Khim. 11:74
30. Markovski LN, Pashinnik VE, Kirsanova NA (1976) Zh. Org. Khim. 12:965
31. Markovski LN, Pashinnik VE: Synthesis 1975:801
32. Braun C, Dell W, Sasse HE, Ziegler ML (1979) Z. anorg. allg. Chem. 450:139
33. Adcock W, Gupta BD, Khor-Thong-Chak (1976) Aust. J. Chem. 29:2571
34. US Pat 4416822 (1983)
35. Swiss Pat 616433 (1980)
36. Siegemund G (1979) Lieb. Ann. Chem. 9:1280
37. Sharts CM, McKee ME, Steed RF, Shellhamer DF, Greeley AC, Green RC, Sprague LG (1979) J. Fluor. Chem. 14:351
38. Mursakulov IG, Samoshin VV, Binnatov RV, Pashinnik VE, Povolotzkii NI, Zefirov NS (1983) Zh. Org. Khim. 19:1336
39. Buss CW, Coe PL, Tatlow JC (1986) J. Fluor. Chem. 34:83
40. Patrick TB, Poon YE (1984) Tetrahedron Lett. 25:1019
41. May JA, Sartorelli AC (1979) J. Med. Chem. 22:971
42. US Pat 3914265 (1975); (1976) Chem. Abs. 84:42635
43. US Pat 3950329 (1976); (1976) Chem. Abs. 85:78142
44. Radchenko OA, Il'chenko AJ, Yagupolskii LM (1980) Zh. Org. Khim. 16:863
45. US Pat 2980740 (1961); (1961) Chem. Abs. 55:23342
46. Olah GA, Nojima M, Kerekes J (1974) J. Amer. Chem. Soc. 96:925
47. Jarovenko NN, Raksha MA (1959) Zh. Obshch. Khim. 29:2159
48. Olah GA, Nojima M, Kerekes J: Synthesis 1973:786
49. Cavalho JF, Prestwich GD (1984) J. Org. Chem. 49:1251
50. Olah GA, Singh BP, Liang G (1984) J. Org. Chem. 49:2922
51. Saito K, Digenis GA, Hawi AA, Chaney J (1987) J. Fluor. Chem. 35:663
52. Goswami R, Harsy SG, Heiman DF, Katzenellenbogen JA (1980) J. Med. Chem. 23:1002
53. Middleton WJ, Ringham EM (1977) Org. Synth. 57:72
54. Ishikawa N, Kitazume T, Takaoka A (1979) J. Synth. Org. Chem. Jap. 37:606
55. US Pat 4112815 (1980); (1980) Chem. Abs. 93:239789
56. Cross BE, Simpsom JC: J. Chem. Res. 1980:118
57. Newman MS, Khanna JM (1979) J. Org. Chem. 44:866
58. Esfahani M, Cavanaugh SR, Preffer P (1981) Biochem. Biophys. Res. Commun. 101:306
59. Blackburn GM, Parratt MJ: J. Chem. Soc. Chem. Commun. 1983:886
60. Blackburn GM, Kent DE: J. Chem. Soc. Perkin Trans. I 1986:913

61. Asai T, Yasuda A, Matsumura Y, Kato M, Uchida K (1986) Repts. Res. Lab. Asachi Glass Co. 36:49
62. US Pat 4235892 (1980); (1981) Chem. Abs. 94:139433
63. Cross BE, Erasmuson A, Filipone P: J. Chem. Soc. Perkin Trans. I 1981:1293
64. Cross BE, Erasmuson A: J. Chem. Soc. Chem. Commun. 1987:1013
65. Bird TGC, Felsky G, Fredericks PM, Jones ERH, Meakin GD: J. Chem. Res. 1979:388
66. Bird TGC, Fredericks PM: J. Chem. Soc. Chem. Commun. 1979:65
67. Rosen S, Faust J, Ben-Yakov H (1979) Tetrahedron Lett. 20:1823
68. Jakhimovich RI, Fursaeva NF, Pashinnik VE: Khim. prirod. soed. 1980:580
69. Jakhimovich RI, Fursaeva NF, Baum VK, Valinietze MU, Apyhovskaya LI, Nekrasova NB, Kovalev VE (1981) Khim. Farm. Zh. 15:75
70. Jakhimovich RI, Fursaeva NF, Pashinnik VE: Khim. prirod. soed. 1985:102
71. Robays MV, Busson R, Vanderhaeghe H: J. Chem. Soc. Perkin Trans. I 1986:251
72. Mann J, Pietrzak B: J. Chem. Soc. Perkin Trans. I 1987:385
73. Biollaz M, Kalvoda J (1977) Helv. Chim. Acta 60:2703
74. Prestwich GD, Shieh HM, Gayen AK (1983) Steroids 41:79
75. Ger. Offen. 2632550 (1977); (1977) Chem. Abs. 87:23616
76. Yang SS, Dorn CP, Jones H (1977) Tetrahedron Lett. 27:2315
77. Napoli JI, Fivizzani MA, Schnoes HK, Deluca HF (1979) Biochem. 18:1614
78. Kobayashi Y, Taguchi T (1985) J. Synth. Org. Chem. Jap. 43:1073
79. Card PJ (1985) J. Carbohyd. Chem. 4:451
80. Sharma M, Korytnyk W: Tetrahedron Lett. 1977:573
81. Sharma M, Korytnyk W (1980) Carbohyd. Res. 83:163
82. Sufrin JR, Bernacki RS, Morin MS, Korytnyk W (1980) J. Med. Chem. 23:143
83. US Pat 4284764 (1981); (1982) Chem. Abs. 96:7032
84. Somawardhana CW (1981) Carbohyd. Res. 94:14
85. Tewson TJ, Welch MJ (1978) J. Org. Chem. 43:1090
86. Klemm GH, Kaufman RJ, Sighu RS (1982) Tetrahedron Lett. 23:2927
87. Card PJ (1983) J. Org. Chem. 48:393
88. Card PJ, Reddy GS (1983) J. Org. Chem. 48:4734
89. Albert R, Dax K, Katzenbeisser U, Sterk H, Stutz AE (1985) J. Carbohyd. Chem. 4:521
90. Druekhammer DG, Wong Chi Huey (1985) J. Org. Chem. 50:5912
91. Kovac P, Yah HJC (1986) J. Carbohyd. Chem. 5:497
92. Kovac P (1986) Carbohyd. Res. 153:168
93. Faghih R, Escribano FC, Castillon S, Garcia J, Lukacs G, Olesker A, Trang Ton That (1986) J. Org. Chem. 51:4558
94. Kovac P, Yeh HJC, Jung GL, Glademans CPJ (1986) J. Carbohyd. Chem. 5:497
95. Street IP, Withers SG (1986) Can. J. Chem. 64:1400
96. Binder TP, Robyt JE (1986) Carbohyd. Res. 147:149
97. Bezuglov VV, Bergel'son LD (1980) Dokl. Akad. Nauk SSSR 250:468
98. Shevchenko VP, Myasoedov NF, Bezuglov VV, Bergel'son LD (1981) Bioorg. Khim. 7:448
99. Bezuglov VV, Bergel'son LD (1979) Bioorg. Khim. 5:1531
100. Bezuglov VV, Serkov IV, Gafurov RG, Llerena EM, Pashinnik VE, Markovski LN, Bergel'son LD (1984) Dokl. Akad. Nauk SSSR 279:378
101. Pashinnik VE, Markovski LN, Bezuglov VV, Serkov IV, Gafurov RG, Bergel'son LD (1984) in: Tezisy 2nd Vses. Soveshch., Ufa, USSR

102. Bezuglov VV (1986) in: Tezisy Vses. Symp. Tallin, USSR
103. Serkov IV, Gafurov RG, Pashinnik VE, Tovstenko VI, Markovskii LN, Bezuglov VV, Bergel'son LD (1986) in: Tezisy Vses. Symp. Tallin, USSR
104. Poulter CD, Wittins PL, Plummer TL (1981) J. Org. Chem. 46:1532
105. Asato AE, Liu RSH (1986) Tetrahedron Lett. 27:3337
106. US Pat 4188345 (1980)
107. US Pat 4226787 (1980)
108. US Pat 4229358 (1980)
109. US Pat 4263214 (1981)
110. US Pat 4229357 (1980)
111. US Pat 4230627 (1980)
112. ·US Pat 4224230 (1980)
113. Gani D, Hitchcock PB, Young DW: J. Chem. Soc. Chem. Commun. 1983:898
114. Tsushima T, Sato T, Tsuji T (1980) Tetrahedron Lett. 21:3591
115. An Seung-Ho, Bobek M (1986) Tetrahedron Lett. 27:3219
116. Biggadike K, Borthwick AD, Evans D, Exall AM, Kirk BE, Roberts SM, Stephenson L, Young P, Slawin AM, Williams DJ: J. Chem. Soc. Chem. Commun. 1987: 251
117. US Pat 4397783 (1983); (1983) Chem. Abs. 99:175473

Subject Index